T0210891

Mohua Banerjee
Shankara Narayanan Krishna (Eds.)

Logic and Its Applications

6th Indian Conference, ICLA 2015
Mumbai, India, January 8-10, 2015
Proceedings

Volume Editors

Mohua Banerjee
Indian Institute of Technology Kanpur
Department of Mathematics and Statistics
Kanpur, India
E-mail: mohua@iitk.ac.in

Shankara Narayanan Krishna
Indian Institute of Technology Bombay
Department of Computer Science and Engineering
Mumbai, India
E-mail: krishnas@cse.iitb.ac.in

ISSN 0302-9743 e-ISSN 1611-3349
ISBN 978-3-662-45823-5 e-ISBN 978-3-662-45824-2
DOI 10.1007/978-3-662-45824-2
Springer Heidelberg New York Dordrecht London

Library of Congress Control Number: 2014956240

LNCS Sublibrary: SL 1 – Theoretical Computer Science and General Issues

Typesetting: Camera-ready by author, data conversion by Scientific Publishing Services, Chennai, India

Printed on acid-free paper

Springer is part of Springer Science+Business Media (www.springer.com)

Preface

The sixth edition of the Indian Conference on Logic and Its Applications (ICLA) was held during January 8–10, 2015, at the Indian Institute of Technology (IIT) Bombay. This volume contains papers presented at the conference.

ICLA is a biennial conference organized under the aegis of the Association for Logic in India. The aim of ICLA is to bring together researchers working on pure and applied formal logic. Areas covered include relationships between logic and other branches of knowledge, history of logic, and systems of logic in the Indian tradition, especially in relation to modern logical studies.

We thank all who submitted papers to ICLA 2015. The contributions in this volume are on varied themes, including proof theory, set theory, model-checking, reasoning in the presence of uncertainty, and Indian systems. Each submission to ICLA 2015 had at least three reviews by Program Committee (PC) members or external experts. Several rounds of discussion (that included further reviews) by the PC members followed, after which final decisions of acceptance were made. We are immensely grateful to all the PC members for their efforts and support. We thank all the external reviewers for their invaluable help. ICLA 2015 also included three invited talks; we specially thank Steve Awodey, Michael Dunn, and Emmanuel Filiot for kindly accepting our invitation, and writing for the volume.

We used the EasyChair system to the hilt: from the submission stage, to the preparation of the proceedings. It was of great help.

Thanks are due to the Department of Computer Science, IIT Bombay, the Organizing Committee members in particular, and all the volunteers, for making this edition of ICLA possible. Our special thanks go to Ganesh Narwane for ensuring that several organizational issues that threatened to spiral out of control, were eventually tamed in time. We would also like to thank Kamal Lodaya, for helpful suggestions throughout.

We are grateful to the Editorial Board of Springer for agreeing to publish this volume in the LNCS series.

Mohua Banerjee
Shankara Narayanan Krishna

Conference Organization

Program Chairs

Mohua Banerjee	Indian Institute of Technology Kanpur
Shankara Narayanan Krishna	Indian Institute of Technology Bombay

Program Committee

S. Arun-Kumar	Indian Institute of Technology Delhi
Rupa Bandyopadhyay	Jadavpur University, India
Nathalie Bertrand	Inria Rennes Bretagne-Atlantique, France
Mihir K. Chakraborty	Jadavpur University, India
Ivo Düntsch	Brock University, Canada
Sujata Ghosh	Indian Statistical Institute, Chennai, India
John Horty	University of Maryland, USA
Juliette Kennedy	University of Helsinki, Finland
Benedikt Löwe	University of Hamburg, Germany and ILLC Amsterdam, The Netherlands
Paritosh Pandya	Tata Institute of Fundamental Research, Mumbai, India
R. Ramanujam	The Institute of Mathematical Sciences, Chennai, India
S. P. Suresh	Chennai Mathematical Institute, India
Zach Weber	University of Otago, New Zealand
Gregory Wheeler	LMU Munich, Germany

Local Organization

Organizing Chair

Ashutosh Trivedi	Indian Institute of Technology Bombay

Organizing Committee

Bharat Adsul	Indian Institute of Technology Bombay
Devendra Bhave	Indian Institute of Technology Bombay
Khushraj Madnani	Indian Institute of Technology Bombay
Ganesh Narwane	Homi Bhabha National Institute Mumbai
Abhisekh Sankaran	Indian Institute of Technology Bombay
Akshay S.	Indian Institute of Technology Bombay
Krishna S.	Indian Institute of Technology Bombay
Shetal Shah	Indian Institute of Technology Bombay
Chandrakant Talekar	Indian Institute of Technology Bombay

Additional Reviewers

Bharat Adsul
Jesse Alama
Can Baskent
Amita Chatterjee
Francesc Esteva
Lluis Godo
Wesley Holliday
Piyush Kurur
Antti Kuusisto
Kamal Lodaya
Bastien Maubert
Graham Priest
Saptarshi Sarkar
Sundar Sarukkai
François Schwarzentruber
Prabal Sen
Andrzej Szałas
Ashutosh Trivedi
Hans van Ditmarsch
Ayineedi Venkateswarlu
Yì N. Wáng
Anna Zamansky

Table of Contents

Homotopy Type Theory*

Steve Awodey

Carnegie Mellon University
awodey@cmu.edu

Abstract. Homotopy Type Theory is a new, homotopical interpretation of constructive type theory. It forms the basis of the recently proposed Univalent Foundations of Mathematics program. Combined with a computational proof assistant, and including a new foundational axiom – the Univalence Axiom – this program has the potential to shift the theoretical foundations of mathematics and computer science, and to affect the practice of working scientists. This talk will survey the field and report on some of the recent developments.

Overview

- *Homotopy Type Theory* is a recently discovered connection between Logic and Topology.
- It is based on an interpretation of intensional Martin-Löf type theory into homotopy theory.
- *Univalent Foundations* is an ambitious new program of foundations of mathematics based on HoTT.
- New constructions based on homotopical intuitions are added as Higher Inductive Types, providing classical spaces, (higher) quotients, truncations, etc.
- The new Univalence Axiom is also added. It implies that isomorphic structures are equal, in a certain sense.
- And a new "synthetic" style of axiomatics is used, simplifying and shortening many proofs.
- A large amount of classical mathematics has been developed in this new system: basic homotopy theory, higher category theory, real analysis, commutative algebra, cumulative hierarchy of set theory, ...
- Proofs are formalized and verified in computerized proof assistants (e.g. Coq).
- There is a comprehensive book containing the informal development.

Type Theory

Martin-Löf constructive type theory consists of:

- **Types:** $X, Y, \dots, A \times B,\ A \to B, \dots$

* Partially supported by the U.S. Air Force Office of Sponsored Research.

M. Banerjee and S.N. Krishna (eds.): ICLA 2015, LNCS 8923, pp. 1–10, 2015.

- **Terms:** $x : A,\ b : B,\ \langle a, b\rangle,\ \lambda x.b(x), \ldots$
- **Dependent Types:** $x : A \vdash B(x)$
  - $x : A, y : B(x) \vdash C(x, y)$
  - $\sum_{x:A} B(x)$
  - $\prod_{x:A} B(x)$
- **Equations** $s = t : A$

It was originally intended as a foundation for constructive mathematics, but is now used also in the theory of programming languages and as the basis of computational proof assistants.

Propositions as Types

The system has a dual interpretation:

- once as **mathematical** objects: types are "sets" and their terms are "elements", which are being constructed,
- once as **logical** objects: types are "propositions" and their terms are "proofs", which are being derived.

This is known as the **Curry-Howard correspondence**:

0	1	$A+B$	$A \times B$	$A \to B$	$\sum_{x:A} B(x)$	$\prod_{x:A} B(x)$
$\bot$	$\top$	$A \vee B$	$A \wedge B$	$A \Rightarrow B$	$\exists_{x:A} B(x)$	$\forall_{x:A} B(x)$

Gives the system its **constructive character**.

Identity Types

It's natural to add a primitive relation of **identity** between any terms of the same type:

$$x, y : A \vdash \mathtt{Id}_A(x, y)$$

Logically this is the proposition "x is identical to y".

But what is it **mathematically**?

The **introduction** rule says that $a : A$ is always identical to itself:

$$\mathtt{r}(a) : \mathtt{Id}_A(a, a)$$

The **elimination** rule is a form of Lawvere's law:

$$\frac{c : \mathtt{Id}_A(a, b) \qquad x : A \ \vdash \ d(x) : R(x, x, \mathtt{r}(x))}{\mathtt{J}_d(a, b, c) : R(a, b, c)}$$

Schematically:

$$\text{“ } a = b \ \& \ R(x, x) \ \Rightarrow \ R(a, b) \text{ ”}$$

The Homotopy Interpretation (Awodey-Warren)

Suppose we have terms of ascending identity types:

$$\begin{aligned} a,\ b &: A \\ p,\ q &: \mathtt{Id}_A(a,b) \\ \alpha,\ \beta &: \mathtt{Id}_{\mathtt{Id}_A(a,b)}(p,q) \\ \ldots &: \mathtt{Id}_{\mathtt{Id}_{\mathtt{Id}_{\ldots}}}(\ldots) \end{aligned}$$

Consider the following interpretation:

$$\begin{aligned} \text{Types} &\rightsquigarrow \text{Spaces} \\ \text{Terms} &\rightsquigarrow \text{Maps} \\ a : A &\rightsquigarrow \text{Points } a : 1 \to A \\ p : \mathtt{Id}_A(a,b) &\rightsquigarrow \text{Paths } p : a \Rightarrow b \\ \alpha : \mathtt{Id}_{\mathtt{Id}_A(a,b)}(p,q) &\rightsquigarrow \text{Homotopies } \alpha : p \Rrightarrow q \\ &\vdots \end{aligned}$$

This extends the familiar **topological interpretation** of the *simply-typed* λ-calculus:

$$\begin{aligned} \text{types} &\rightsquigarrow \text{spaces} \\ \text{terms} &\rightsquigarrow \text{continuous functions} \end{aligned}$$

to *dependently typed* λ-calculus with Id-types, via the **basic idea**:

$$p : \mathtt{Id}_X(a,b) \rightsquigarrow p \text{ is a path from point } a \text{ to point } b \text{ in } X$$

This forces *dependent types to be fibrations*, Id-*types to be path spaces*, and *homotopic maps to be identical.*

The Fundamental Groupoid of a Type (Hofmann-Streicher)

Like path spaces in topology, identity types endow each type in the system with the structure of a (higher-) groupoid:

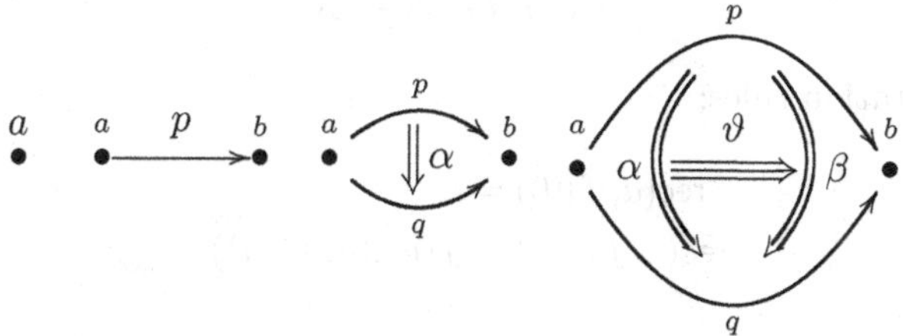

The laws of identity are the **groupoid operations**:

$$\begin{array}{rll} r : \mathtt{Id}(a,a) & \text{reflexivity} & a \to a \\ s : \mathtt{Id}(a,b) \to \mathtt{Id}(b,a) & \text{symmetry} & a \leftrightarrows b \\ t : \mathtt{Id}(a,b) \times \mathtt{Id}(b,c) \to \mathtt{Id}(a,c) & \text{transitivity} & a \to b \to c \end{array}$$

The **groupoid equations** only hold "up to homotopy", i.e. up to a higher identity term.

Homotopy n-types (Voevodsky)

The universe of all types is stratified by "homotopical dimension": the level at which the fundamental groupoid becomes trivial.

A type X is called:

contractible iff $\sum_{x:X} \prod_{y:X} \mathtt{Id}_X(x,y)$

A type X is a:

proposition iff $\prod_{x,y:X}$ $\mathsf{Contr}(\mathtt{Id}_X(x,y))$,
set iff $\prod_{x,y:X}$ $\mathsf{Prop}(\mathtt{Id}_X(x,y))$,
1-type iff $\prod_{x,y:X}$ $\mathsf{Set}(\mathtt{Id}_X(x,y))$,
(n+1)-type iff $\prod_{x,y:X}$ $\mathsf{nType}(\mathtt{Id}_X(x,y))$.

This gives a new view of the mathematical universe.

Higher Inductive Types (Lumsdaine-Shulman)

The natural numbers $\mathbb{N}$ are implemented as an (ordinary) inductive type:

$$\mathbb{N} := \left\{ \begin{array}{l} 0 : \mathbb{N} \\ s : \mathbb{N} \to \mathbb{N} \end{array} \right.$$

The **recursion property** is captured by an elimination rule:

$$\frac{a : X \qquad f : X \to X}{\mathsf{rec}(a,f) : \mathbb{N} \to X}$$

with computation rules:

$$\begin{aligned} \mathsf{rec}(a,f)(0) &= a \\ \mathsf{rec}(a,f)(sn) &= f(\mathsf{rec}(a,f)(n)) \end{aligned}$$

In other words, $(\mathbb{N}, 0, s)$ is the **free** structure of this type:

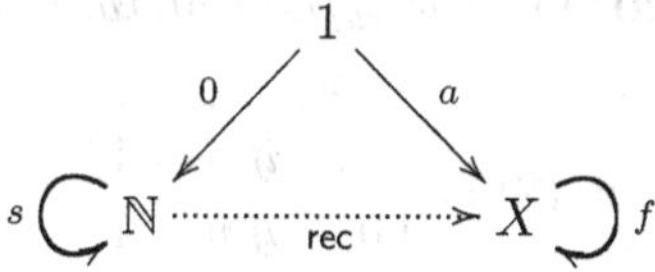

The map $\mathsf{rec}(a, f) : \mathbb{N} \to X$ is unique.

Theorem 1. *$\mathbb{N}$ is a set (i.e. a 0-type).*

Higher Inductive Types: The Circle S^1

The homotopical circle $\mathbb{S} = S^1$ can be given as an inductive type involving a "higher-dimensional" generator:

$$\mathbb{S} := \begin{cases} \mathsf{base} : \mathbb{S} \\ \mathsf{loop} : \mathsf{base} \rightsquigarrow \mathsf{base} \end{cases}$$

where we write "$\mathsf{base} \rightsquigarrow \mathsf{base}$" for "$\mathtt{Id}_{\mathbb{S}}(\mathsf{base}, \mathsf{base})$".

$$\mathbb{S} := \begin{cases} \mathsf{base} : \mathbb{S} \\ \mathsf{loop} : \mathsf{base} \rightsquigarrow \mathsf{base} \end{cases}$$

The recursion property of $\mathbb{S}$ is given by its elimination rule:

$$\frac{a : X \qquad p : a \rightsquigarrow a}{\mathsf{rec}(a, p) : \mathbb{S} \to X}$$

with computation rules:

$$\mathsf{rec}(a, p)(\mathsf{base}) = a$$
$$\mathsf{rec}(a, p)(\mathsf{loop}) = p$$

In other words, $(\mathbb{S}, \mathsf{base}, \mathsf{loop})$ is the **free** structure of this type:

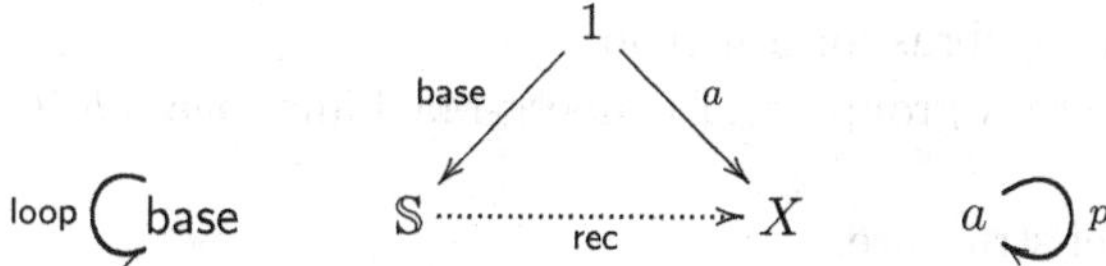

The map $\mathsf{rec}(a, p) : \mathbb{S} \to X$ is unique up to homotopy.

Here is a sanity check:

Theorem 2 (Shulman 2011). *The type-theoretic circle $\mathbb{S}$ has the correct homotopy groups:*

$$\pi_n(\mathbb{S}) = \begin{cases} \mathbb{Z}, & \text{if } n = 1, \\ 0, & \text{if } n \neq 1. \end{cases}$$

The proof combines classical homotopy theory with methods from constructive type theory, and uses Voevodsky's Univalence Axiom. It has been formalized in Coq.

Corollary 1. *There are* 1*-types. (This also uses univalence).*

Higher Inductive Types: The Interval *I*

The unit interval $\mathbb{I} = [0, 1]$ is also an inductive type, on the data:

$$\mathbb{I} := \begin{cases} 0, 1 : \mathbb{I} \\ p : 0 \rightsquigarrow 1 \end{cases}$$

again writing $0 \rightsquigarrow 1$ for the type $\mathtt{Id}_{\mathbb{I}}(0, 1)$.

Slogan:

In topology, we start with the **interval** and use it to define the notion of a **path.**
In HoTT, we start with the notion of a **path**, and use it to define the **interval**.

Higher Inductive Types: Conclusion

Many basic spaces and constructions can be introduced as HITs:

- higher spheres S^n, cylinders, tori, cell complexes, ...,
- suspensions ΣA,
- homotopy pullbacks, pushouts, etc.,
- truncations, such as connected components $\pi_0(A)$ and "bracket" types $[A]$,
- quotients by equivalence relations and general quotients,
- free algebras, algebras for a monad,
- (higher) homotopy groups π_n, Eilenberg-MacLane spaces $K(G, n)$, Postnikov systems,
- Quillen model structure,
- real numbers,
- cumulative hierarchy of sets.

Univalence

Voevodsky has proposed a new foundational axiom to be added to HoTT: the **Univalence Axiom.**

- It captures the informal mathematical practice of **identifying isomorphic objects.**
- It is very useful from a practical point of view, especially when combined with HITs.
- It is formally **incompatible** with the set-theoretic model of type theory, but provably **consistent** with homotopy type theory.
- Its status as a constructive principle is the focus of much current research.

Isomorphism and Equivalence

In type theory, the usual notion of *isomorphism* $A \cong B$ is definable:

$A \cong B \Leftrightarrow$ there are $f : A \to B$ and $g : B \to A$
such that $gfx = x$ and $fgy = y$.

Formally, there is the type of isomorphisms:

$$\mathrm{Iso}(A, B) := \sum_{f:A \to B} \sum_{g:B \to A} \Big(\prod_{x:A} \mathtt{Id}_A(gfx, x) \times \prod_{y:B} \mathtt{Id}_B(fgy, y)\Big)$$

Thus $A \cong B$ iff this type is inhabited by a closed term, which is then just an isomorphism between A and B.

- There is also a more refined notion of *equivalence* of types,

$$A \simeq B$$

  which adds a further "coherence" condition relating the *proofs* of $gfx = x$ and $fgy = y$.
- Under the homotopy interpretation, this is the type of *homotopy equivalences.*
- This subsumes *categorical equivalence* (for 1-types), *isomorphism of sets* (for 0-types), and *logical equivalence* (for (-1)-types).

Invariance

One can show that all *definable properties* $P(X)$ of types X *respect type equivalence*:

$$A \simeq B \text{ and } P(A) \text{ implies } P(B)$$

In this sense, *all properties are invariant.*

Moreover, therefore, *equivalent types* $A \simeq B$ *are indiscernable*:

$$P(A) \Rightarrow P(B), \text{for all } P$$

How is this related to **identity of types** A and B?

Univalence

To reason about **identity of types**, we need a *type universe* $\mathcal{U}$, with an identity type,

$$\mathtt{Id}_{\mathcal{U}}(A, B).$$

Since identity implies equivalence there is a comparison map:

$$\mathtt{Id}_{\mathcal{U}}(A, B) \to (A \simeq B).$$

The *Univalence Axiom* asserts that this map is an equivalence:

$$\mathtt{Id}_{\mathcal{U}}(A, B) \ \simeq \ (A \simeq B) \tag{UA}$$

So UA can be stated: *"Identity is equivalent to equivalence."*

The Univalence Axiom: Remarks

- Since UA is an equivalence, there is a map coming back:

  $$\mathtt{Id}_{\mathcal{U}}(A, B) \longleftarrow (A \simeq B)$$

  So **equivalent objects are identical.**
  (isomorphic sets, groups, etc., can be identified.)
- In the system with a universes $\mathcal{U}$, the UA is equivalent to the **invariance property**:

  $$A \simeq B \text{ and } P(A) \text{ implies } P(B)$$

  for all "properties" $P(X)$, i.e. type families $P : \mathcal{U} \longrightarrow \mathcal{U}$.
- UA implies that $\mathcal{U}$, in particular, is not a set (0-type).
- The computational character of UA is still an open question.

The Univalence Axiom: How it Works

To compute the fundamental group of the circle $\mathbb{S}$, we shall construct the universal cover: This will be a dependent type over $\mathbb{S}$, i.e. a type family

$$\mathsf{cov} : \mathbb{S} \longrightarrow \mathcal{U}.$$

To define a type family

$$\mathsf{cov} : \mathbb{S} \longrightarrow \mathcal{U},$$

by the recursion property of the circle, we just need the following data:

- a point $A : \mathcal{U}$
- a loop $p : A \leadsto A$

For the point A we take the integers $\mathbb{Z}$.

By Univalence, to give a loop $p : \mathbb{Z} \leadsto \mathbb{Z}$ in $\mathcal{U}$, it suffices to give an equivalence $\mathbb{Z} \simeq \mathbb{Z}$.

Since $\mathbb{Z}$ is a set, equivalences are just isomorphisms, so we can take the successor function $\mathsf{succ} : \mathbb{Z} \cong \mathbb{Z}$.

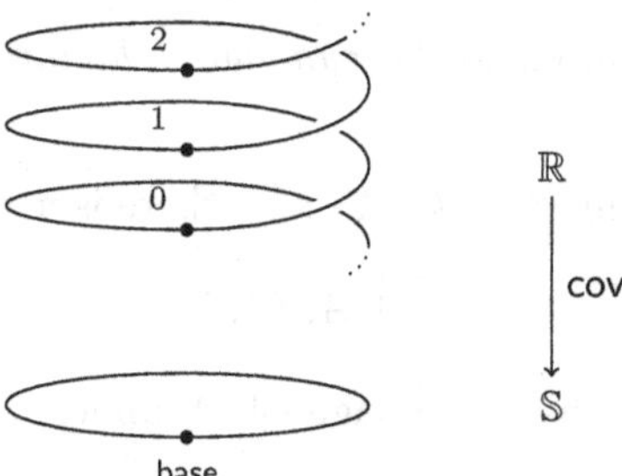

Definition 1 (Universal Cover of $\mathbb{S}^1$). *The dependent type* $\mathsf{cov} : \mathbb{S} \longrightarrow \mathcal{U}$ *is given by circle-recursion, with*

$$\mathsf{cov}(\mathsf{base}) := \mathbb{Z}$$
$$\mathsf{cov}(\mathsf{loop}) := \mathsf{ua}(\mathsf{succ}).$$

As in classical homotopy theory, we use the universal cover to define the "winding number" of any path $p : \mathsf{base} \rightsquigarrow \mathsf{base}$ by $\mathsf{wind}(p) = p_*(0)$. This gives a map

$$\mathsf{wind} : \Omega(\mathbb{S}) \longrightarrow \mathbb{Z},$$

which is inverse to the map $\mathbb{Z} \longrightarrow \Omega(\mathbb{S})$ given by

$$z \mapsto \mathsf{loop}^z.$$

Formalization of Mathematics

- The idea of logical foundations of math has great conceptual and philosophical interest, but in the past this was too lengthy and cumbersome to be of any use.
- Explicit formalization of math is finally *feasible*, because computers can now take over what was once too tedious or complicated to be done by hand.
- Future historians of mathematics will wonder how Frege and Russell could have invented formal logical foundations *before* there were any computers to run them on!
- Formalization can provide a *practical tool* for working mathematicians and scientists: increased certainty and precision, supports collaborative work, cumulativity of results, searchable library of code, ... I think that mathematics will eventually be fully formalized.
- UF uses a "synthetic" method involving high-level axiomatics and structural descriptions; allows shorter, more abstract proofs; closer to mathematical practice than the "analytic" method of ZFC. Use of UA is very powerful.

Final Example: The Cumulative Hierarchy

Given a universe $\mathcal{U}$, we can make the *cumulative hierarchy* V of sets in $\mathcal{U}$ as a HIT:

- for any small A and any map $f : A \to V$, there is a "set":

  $$\mathsf{set}(A, f) : V$$

  We think of $\mathsf{set}(A, f)$ as the image of A under f, i.e. the classical set $\{f(a) \mid a \in A\}$
- For all $A, B : \mathcal{U}$, $f : A \to V$ and $g : B \to V$ such that

  $$\big(\forall a : A\, \exists b : B\ f(a) = g(b)\big) \wedge \big(\forall b : B\, \exists a : A\ f(a) = g(b)\big)$$

  we put in a path in V from $\mathsf{set}(A, f)$ to $\mathsf{set}(B, g)$.
- The 0-truncation constructor: for all $x, y : V$ and $p, q : x = y$, we have $p = q$.

Membership $x \in y$ is then defined for elements of V by:

$$(x \in \mathsf{set}(A, f)) \ :=\ (\exists a : A.\ x = f(a)).$$

One can show that the resulting structure $(V, \in)$ satisfies most of the axioms of Aczel's constructive set theory CZF.

Assuming AC for sets (0-types), one gets a model of ZFC set theory.

The proofs make essential use of UA.

References and Further Information

More Information:

`www.HomotopyTypeTheory.org`

The Book:

Homotopy Type Theory: Univalent Foundations of Mathematics
The Univalent Foundations Program,
Institute for Advanced Study, Princeton, 2013

The Relevance of Relevance to Relevance Logic*

J. Michael Dunn

School of Informatics and Computing, and Department of Philosophy,
Indiana University – Bloomington

Abstract. I explore the question of whether the concept of relevance is relevant to the study of what Anderson and Belnap call "relevance logic." The answer should be "Of course!" But there are some twists and turns, as is shown by the fact that it has taken over 50 years to get here. Despite protests by R. K. Meyer that the concept of relevance is not part of what he calls "relevant logic," I suggest and defend interpreting the Routley–Meyer ternary accessibility relation using information states a, b, c, so $Rabc$ means "in the context a, b is relevant to c." Motivations are provided from Sperber and Wilson's work in linguistics on relevance.

1 Introduction

As the title suggests I will be looking at the question of whether the concept of relevance is relevant to the study of what Anderson and Belnap have called "relevance logic." The reader might think that a short abstract of my paper, if not the paper itself, would be "Of course!" After all Anderson and Belnap titled their magnum opus *Entailment: The Logic of Relevance and Necessity*, and this book (vol. I, p. xxii) opens with the claim that Wilhelm Ackermann's system of *strenge Implikation* "give us for the first time a mathematically satisfactory way of grasping the elusive notion of relevance of antecedent to consequent in "if ... then —" propositions; such is the topic of this book." Ackermann's system with a few important modifications became Anderson and Belnap's system E of entailment, which Anderson and Belnap promote as their system that captures both relevance and necessity. It is at one and the same time both what they term a relevance logic, and also a modal logic. They also present the system R of relevant implication, which was intended to be E stripped of modality. The system R has taken on a life of its own and in many ways has become the focus of relevance logic, and of course, one can always add necessity. Maksimova (1973) showed that one cannot get E back from R, defining entailment as a necessary relevant implication by adding what would seem to be the appropriate S4 type necessity operator. And when one sees how this breaks down, this can

* This is part of a larger joint project with Katalin Bimbó, supported by an Insight Grant from the Social Sciences and Humanities Research Council of Canada: "The Third Place is the Charm: The Emergence, the Development and the Future of the Ternary Relational Semantics for Relevance and Some Other Non-classical Logics." I thank Kata for reading the manuscript and for her corrections/suggestions.

M. Banerjee and S.N. Krishna (eds.): ICLA 2015, LNCS 8923, pp. 11–29, 2015.

be another ground for preferring the system R and just adding one's favorite necessity operator(s).

But prominent researchers, even within the Anderson and Belnap community, have challenged the appropriateness of the name "relevance logic." Robert K. Meyer in fact preferred to talk of "relevant logic." Meyer (1978) contains an opening section titled "Do relevant logics capture relevance."[1] He then says (p. 3): "The answer to the question with which this section opens is 'No'. The relevant logics do not capture relevance. They do not *begin* to capture relevance.... Despite the sub-title of *Entailment*, there is no 'Logic of relevance'." Meyer (1985) has a more nuanced discussion of the issues, and we discuss these below.

2 More of Meyer on Relevance

Setting aside the personal (consistency with his own early usage) and the grammatical reason ("relevant" is the adjectival form), Meyer has a variety of more substantive reasons, none of which he thinks in themselves make the logics E and R less interesting (though he does have other concerns with E in particular, hence his title "Farewell to Entailment"). He says (p. 607): "I argue rather that capturing relevance does not have much to do with the nature or purposes of Relevant Logics." Meyer's more substantive reasons include the elusiveness of the notion of "relevance," and the thought that (p. 610) "Relevance is not an *ingredient* of a theory of logical entailment. Insofar as logical sense can be assigned to the notion, it is a *consequence* of holding such a theory."

Meyer goes on to examine Anderson and Belnap's argument that *Entailment* gives a " 'mathematically satisfactory way' of isolating relevance as a component of good argument," saying that there are the "two main strings of its bow." The first string is what Anderson and Belnap have labeled the "Variable Sharing Property," that is if $A \to B$ is a theorem, then there is some sentential variable p that occurs in both A and B – thereby showing some connection of meaning, or *relevance.* Both the logics E and R have this property. The second string is harder to describe and depends on, as Meyer puts it, "tracking" the assumptions in a natural-deduction argument to make sure that if $A \to B$ is proven by assuming the antecedent A and showing the consequent B, that A was actually used in the derivation. This is supposed to show that A is *relevant* to B.

Both of these criteria have subtleties in their statement, and this might be enough to undermine them as a criterion for "relevance." To give one example, $[{\sim}p \wedge (p \vee q)] \to q$ has the Variable Sharing Property. Also the argument ${\sim}p \wedge (p \vee q) \therefore q$ is classically valid (it is the notorious Disjunctive Syllogism, which Anderson and Belnap rejected, at least as a relevant implication). And yet the

[1] Greg Restall (2006) explains why he calls the subject "relevant logic," although "nothing of substance hangs on this issue: Americans call our topic 'relevance logic,' and people of Commonwealth countries (primarily Australia and Scotland) call it 'relevant logic.' The split comes down to a disagreement between Nuel Belnap and Robert Meyer. Meyer brought his favored terminology 'relevant' with him to Australia, where it has stuck."

implication $[\sim p \wedge (p \vee q)] \to q$ is not provable in E or R. So the Variable Sharing Property, even together with validity, do not suffice for a "relevant implication."

The Tracking Criterion relies on some way of keeping track. Anderson and Belnap present a variation of Fitch's natural deduction system for classical logic, but introduced subscripts to keep track of how assumptions are used. The idea is that these subscripts are passed down from step to step and combined into a set of subscripts when a two-premise rule is applied. Thus if one infers B from A and $A \to B$ where A has a set of subscripts α and $A \to B$ has a set of subscripts β, then B is inferred with subscripts $\alpha \cup \beta$. The idea then is that if one assumes A with subscript i, and one derives B with subscripts β, then in order to derive $A \to B$ we must have $i \in \beta$, and then we can derive $A \to B$ with subscripts $\beta - \{i\}$. This works beautifully with the pure implicational fragment of R (and with a slight modification of E), and even stays nice when rules for negation are added.

But things get complicated when we add conjunction, and really weird when disjunction is added as well. The problem is that if we assume p with subscript 1 and then assume q with subscript 2, then one might think it would be natural to infer $p \wedge q$ with subscripts $\{1, 2\}$. But then one could derive $q \to (p \wedge q)$ with subscript $\{1\}$, and then prove $p \to (q \to (p \wedge q))$ as a theorem. From this one can prove $p \to (q \to p)$ (exercise). This is the notorious Positive Paradox of Implication, and if one plugs any theorem B in for p and any sentence A for q, then by modus ponens one can prove $A \to B$, where A and B may well not share a sentential variable.

Anderson and Belnap avoid this problem by not allowing the inference

$$\begin{array}{c} A_1 \\ B_2 \\ \hline A \wedge B_{1,2} \end{array}$$

but instead only allow $\wedge$-introduction when the premises have the same (set of) subscripts.

This fixes the problem with conjunction (though the uninitiated might find it ad hoc), but it creates another problem when we try to add disjunction. The way that disjunction makes things even worse is that Anderson and Belnap want to have as a theorem:

$$[A \wedge (B \vee C)] \to [(A \wedge B) \vee (A \wedge C)] \qquad \textit{Distribution.}$$

We will not bother to state the introduction and elimination rules for disjunction, but simply state that although they seem very natural, they give Anderson and Belnap no better way to prove Distribution that to simply postulate it. For more details please consult Dunn (1986), where you will also find a way to solve the problem.

One thing Meyer has on his side is that the Anderson-Belnap "relevance logics" do not have some relevance operation ρ in their vocabulary, so that one might define "A relevantly implies B" as say "Relevantly, A materially implies B": $\rho(A \supset B)$. Of course, the same kind of thing is true of formulations of modal logic that use strict (necessary) implication as a primitive and do not bother to have a connective $\Box$ for necessity. But at least for the standard normal modal logics, strict implication can be defined as $\Box(A \supset B)$. This kind of move is not open for relevance logics.

3 Hilbert-Style Formulations for Relevance Logics

Anderson and Belnap have a basic formal language that contains the unary connective $\sim$ of De Morgan negation, and the binary connectives of conjunction $\wedge$, disjunction $\vee$, and (relevant) implication $\rightarrow$.

The following sets of axioms (cf. Anderson and Belnap (1975) or Dunn (1986)) can be seen as forming various fragments of the relevance logic R: 1–4 the implicational fragment $R_{\rightarrow}$, 1–4 with 10–12 the implication-negation fragment $R^{\sim}_{\rightarrow}$, and 1–9 the positive fragment R+.

Implication:

$$A \rightarrow A \quad \text{Self-Implication} \quad (1)$$

$$(A \rightarrow B) \rightarrow [(C \rightarrow A) \rightarrow (C \rightarrow B)] \quad \text{Prefixing} \quad (2)$$

$$[A \rightarrow (A \rightarrow B)] \rightarrow (A \rightarrow B) \quad \text{Contraction} \quad (3)$$

$$[A \rightarrow (B \rightarrow C)] \rightarrow [B \rightarrow (A \rightarrow C)] \quad \text{Permutation} \quad (4)$$

Conjunction-Disjunction:

$$A \wedge B \rightarrow A, \quad A \wedge B \rightarrow B \quad \text{Conjunction Elimination} \quad (5)$$

$$[(A \rightarrow B) \wedge (A \rightarrow C)] \rightarrow (A \rightarrow B \wedge C) \quad \text{Conjunction Introduction} \quad (6)$$

$$A \rightarrow A \vee B, \quad B \rightarrow A \vee B \quad \text{Disjunction Introduction} \quad (7)$$

$$[(A \rightarrow C) \wedge (B \rightarrow C)] \rightarrow (A \vee B \rightarrow C) \quad \text{Disjunction Elimination} \quad (8)$$

$$A \wedge (B \vee C) \rightarrow (A \wedge B) \vee C \quad \text{Distribution} \quad (9)$$

Negation:

$$(A \rightarrow {\sim}A) \rightarrow {\sim}A \quad \text{Reductio} \quad (10)$$

$$(A \rightarrow {\sim}B) \rightarrow (B \rightarrow {\sim}A) \quad \text{Contraposition} \quad (11)$$

$${\sim}{\sim}A \rightarrow {\sim}A \quad \text{Double Negation} \quad (12)$$

A Hilbert-style axiom system is often taken to have only one rule of inference: $A, A \rightarrow B \therefore B$ (modus ponens), and this is the only rule for $R_{\rightarrow}$ and $R^{\sim}_{\rightarrow}$. However for R+ and R itself Anderson and Belnap have an additional rule of inference: $A, B \therefore A \wedge B$ (adjunction).

The system E of Entailment can be obtained by replacing axiom 4 (Permutation) with Restricted Permutation: where $\vec{B}$ abbreviates $B_1 \rightarrow B_2$,

$$[A \rightarrow (\vec{B} \rightarrow C)] \rightarrow [\vec{B} \rightarrow (A \rightarrow C)] \quad \text{Restricted-Permutation.}$$

There are a number of interesting axiom sets for R and E, as well as for many lesser known relevance logics. The reader who want to learn more "axiom chopping" is referred to Dunn (1986), as well as of course, Anderson and Belnap (1975). It is worth pointing out that a kind of necessity can be defined in E as $\Box A = (A \to A) \to A$, and it has roughly the structure of the Lewis modal system S4. This gives us a way to obtain an axiom set for R from that of E by simply adding

$$A \to \Box A \quad \text{Demodaliser}$$

It is interesting that one obtains the classical propositional paradox (with $A \to B$ provably equivalent to the material implication $A \supset B$) if one adds to either R or E

$$A \to (B \to A) \quad \text{Positive Paradox.}$$

We introduced the axioms for E because of E's historical importance, and because Meyer rails against it, but in the sequel we will focus on the system R.

4 The Semantics of Relevance Logic

Any of us, not just Anderson and Belnap, can lay down a set of axioms, or maybe even create a natural deduction system with nice rules. But what do they mean? From the beginnings of relevance logic there has been much controversy about its semantics. First there was complaint that it had none, and then there was complaint that it had one, particularly the so-called "Routley–Meyer semantics" for relevance logic – which used the novelty of a ternary accessibility relation. We shall speak of the Routley–Meyer semantics even though a number of other logicians produced similar semantics at about the same time.

Copeland, and van Benthem in his review of Copeland, raised questions about whether this is a semantics in name only, or merely just a formal device. Similar issues had of course already been raised in connection with the binary accessibility relation in the so-called "Kripke semantics" for modal logic. But there have actually been various interpretations made of both the binary and ternary accessibility relations. The best recent place to read about interpretations of the ternary accessibility relation is Beall et al. (2012).

We first must explain a little about relevance logic and the Routley–Meyer semantics in particular. Routley and Meyer published "Semantics of Entailment I, II, III" in the years 1972 and 1973.[2] Routley and Meyer use a frame $(K, R, *, 0)$. K is a set, $0 \in K$, $R \subseteq K^3$, and $*$ is a unary operation on K. Routley and Meyer call the members of K "set ups," and put various constraints on a frame, but we shall not explore these in detail now. We do though note that they defined

[2] And "Semantics of Entailment IV" written in 1972 but published as an appendix in Routley, Meyer, Plumwood, and Brady (1982). As with the "Kripke semantics," there were a lot of "competitors" in the early 1970's with essentially the same, or very similar ideas, including (in alphabetical order) Fine, Gabbay, Maksimova, and Urquhart.

a binary relation $a \leq b$ as $R0ab$, and gave R properties that assure that $\leq$ is a quasi-order (reflexive and transitive).[3]

They take a valuation v to be a function that assigns to each pair (p, a) (p an atomic sentence, $x \in K$) a member of $\{T, F\}$. They then inductively define a function I that assigns to each pair (A, x) (A an arbitrary formula) a member of $\{T, F\}$. We shall get to this inductive definition, but we shall write $x \models A$ rather than $I(A, x) = T$.

But there is an important restriction. Routley and Meyer require the Hereditary Condition on atomic sentences: if $a \leq b$ and $a \models p$, then $b \models p$. It can then be shown by induction that this extends as well to compound formulas. The Hereditary Condition is needed to show that $0 \models A \to A$.

This is important because validity on a frame is defined by the condition that a sentence holds at 0 for all valuations.[4] Routley and Meyer show that a sentence A is a theorem of R iff A is valid on all frames satisfying the following conditions:

p1. $R0aa$
p2. $Raaa$
p3. $\exists x(Rabx \text{ and } Rxcd) \Rightarrow \exists x(Racx \text{ and } Rxbd)$. (They nicely write this as $R^2abcd \Rightarrow R^2acbd$.)
p4. $R^2 0bcd \Rightarrow Rbcd$[5]
p5. $Rabc \Rightarrow Rac^*b^*$
p6. $a^{**} = a$

The Routley–Meyer valuation clauses can now be stated as follows:

(vp) $x \models p$ iff $x \in V(p)$ (Atomic)
(v$\sim$) $x \models {\sim}A$ iff *not* $x^* \models A$ (Negation)
(v$\wedge$) $x \models A \wedge B$ iff $x \models A$ and $x \models B$ (Conjunction)
(v$\vee$) $x \models A \vee B$ iff $x \models A$ or $x \models B$ (Disjunction)
(v$\to$) $x \models A \to B$ iff $\forall a, b$, if $Rxbc$ and $a \models A$ then $b \models B$ (Relevant implication)

In the present paper we shall be examining an interpretation of relevant implication in terms of, guess what? Relevance! Strangely while this is the most naive or natural interpretation, it seems not to have been explored or even mentioned until now. The whole idea of a *relevant* implication $A \to B$ is that there is supposed to be some sort of relevance between the truth of the antecedent A and the consequent B. What could be more natural than to interpret $Rabc$ as that in the context of the information a, the information b is relevant to the

[3] They actually use the notation $<$ but because the relation turns out to be reflexive it has become standard to use $\leq$.

[4] It is easy to miss this important point. Similar points hold for Urquhart's and Fine's semantics, and I will share with you that when Urquhart first explained his semantics to me back in 1970, I thought I had found a mistake since it seemed that $A \to A$ was invalid.

[5] There is a typo when this axiom is listed in Routley and Meyer (1973). They have an "a" instead of a "0." And of course spelled out it means $R0bx$ and $Rxcd$ imply $Rbcd$.

information c? That is a relevant implication $A \to B$ is true in the context x just when if we pick any other pieces of information y and z such that y is relevant to z in the context x, then if y determines A, then z determines B.

5 What is Relevance?

The concept of relevance occurs in a number of contexts. The most relevant for us (sorry for the pun) are commonsense reasoning, probability and statistics, information retrieval (particularly, search of unstructured data bases such as the WWW to take an extreme example), library science, the law, cognitive science, epistemology, and linguistics and philosophy of language (particularly, pragmatics). We cannot possibly run through these all (just google the word "relevance" – I believe Google has a good relevancy algorithm) so I want to jump to what I believe is the most relevant work for our purposes here, which relates to the pragmatics of language.

Pragmatics is one of the three dimensions of natural language, as introduced by Charles Morris (1946), the other two being syntax and semantics. Put quickly, syntax has to do with grammar, semantics has to do with meaning, and pragmatics has to do with use. Paul Grice in his famous John Locke Lectures at Oxford (1967), printed in Grice (1989), introduced the idea of "conversational maxims," saying (p. 27) "Under the category of Relation I place a single maxim, namely, 'Be relevant.'" Grice goes on to say "Though the maxim itself is terse, its formulation conceals a number of problems that exercise me a good deal: questions about what different kinds and focuses of relevance there may be, how these shift in the course of a talk exchange, how to allow for the fact that subjects of conversation are legitimately changed, and so on." Grice in fact does offer anything, at least directly, to answer such questions, and in fact says. "I find the treatment of such questions exceedingly difficult, and I hope to revert to them in later work." To the best of my knowledge Grice never published such later work. But the linguists Deirdre Wilson and Dan Sperber (1986) turned to the task of clarifying the concept of relevance, and they have carried on with it for over a quarter of a century – see Wilson and Sperber (2012).

There is clearly much to be said about their Herculean task and the Herculean response to it. But I content myself here with this quotation from Wilson and Sperber (2004).

> Intuitively, an input (a sight, a sound, an utterance, a memory) is relevant to an individual when it connects with background information he has available to yield conclusions that matter to him: say, by answering a question he had in mind, improving his knowledge on a certain topic, settling a doubt, confirming a suspicion, or correcting a mistaken impression. In relevance-theoretic terms, an input is relevant to an individual when its processing in a context of available assumptions yields a positive cognitive effect. A positive cognitive effect is a worthwhile difference to the individual's representation of the world – a true conclusion, for example.

> False conclusions are not worth having. They are cognitive effects, but not positive ones (Sperber & Wilson 1995: §3.1–2).The most important type of cognitive effect achieved by processing an input in a context is a contextual implication, a conclusion deducible from the input and the context together, but from neither input nor context alone. For example, on seeing my train arriving, I might look at my watch, access my knowledge of the train timetable, and derive the contextual implication that my train is late (which may itself achieve relevance by combining with further contextual assumptions to yield further implications).

It is interesting that Sperber and Wilson seem to barely mention Anderson and Belnap's work on relevance logic,[6] and conversely Anderson and Belnap never cite Sperber and Wilson's work.[7] Gabbay and Woods (2003) is the only work I know of that makes a connection between them, present company excluded. I should mention prominently, and that is why this is not just a footnote, that Gabbay and Woods' book deserves careful study, and is an interesting alternative in many ways to the work of Sperber and Wilson and certainly the only work comparable in extent and detail to theirs. I admit to coming across it only very recently, and I have not had time to give it the careful study that it deserves.

The whole idea of a *relevant* implication $A \to B$ is that there is supposed to be some sort of relevance between the truth of the antecedent A and the consequent B. What could be more natural then that to interpret $Rabc$ as that in the context of the information a, the information b is relevant to the information c. If one adopts Fine's hybrid approach to the ternary relation, this might be symbolized as $a \bullet b \leq c$. In the words of Sperber and Wilson, "an input is relevant to an individual when its processing in a context of available assumptions yields a positive cognitive effect." "The most important type of cognitive effect achieved by processing an input in a context is a contextual implication, a conclusion deducible from the input and the context together, but from neither input nor context alone."

6 Absolute Versus Contextual Relevance

Let us write $a \leq b$ to mean that the state of information a is relevant all by itself to the state of information b. We shall call this *absolute* relevance, and will get to *relative* relevance, where relevance depends on a context, in just a few

[6] The only exception I know is Wilson and Sperber (1986), where in motivating their own work to give an explicit account of relevance, they give a series of quotations intended to demonstrate "considerable scepticism over whether any such account is in principle possible." They quote from p. xxi of Anderson and Belnap (1975): "The difficulty of treating relevance with the same degree of mathematical sophistication and exactness characteristic of extensional logic has led many influential philosopher-logicians to believe that it was *impossible* to find a satisfactory treatment of the topic." This is ironic in that Anderson were clearly doing the same for their own work.

[7] They are not even in the extremely comprehensive bibliography by Robert G. Wolfe in *Entailment,* vol. II.

paragraphs. What are the formal properties of absolute relevance? Our choice of notation suggests that it should be reflexive, transitive, and not generally symmetric. It seems entirely plausible that an information state a should be relevant to itself, and that if information state a is relevant to information state b, and b is relevant to information state c, then a is relevant to c. Readers of detective stories would surely agree on this.

Let us now review the Routley–Meyer conditions on R with relevance in mind. For the sake of clarity, we give equivalents of the Routley–Meyer conditions as they are in Dunn (1986).

R1. $R0aa$ Reflexivity
R2. $Raaa$ Total Reflexivity
R3. $Rabc \Rightarrow Rbac$ Commutativity
R4. $\exists x(Rabx$ and $Rxcd) \Rightarrow \exists y(Rayd$ and $Rbcy)$ (which can be written $R^2(ab)cd \Rightarrow R^2a(bc)d$) Associativity
R5. $a' \leq a$ and $Rabc \Rightarrow Ra'bc$ Monotony (in first position)

Note that because of Commutativity, we also have:

R5′ $b' \leq b$ and $Rabc \Rightarrow Rab'c$ Monotony (in second position)[8]

Let us first see if $\leq$ is reflexive and transitive, so it matches these properties of absolute relevance. These are well-known properties of the Routley–Meyer semantics, but we shall derive them anyway because of their importance here ((I almost said "relevance") and also strangely hard to find in the literature). Reflexivity is immediate, since by the definition of $\leq$, it is just R1. For transitivity, assume (1) $a \leq b$, i.e., $R0ab$, and (2) $b \leq c$, i.e., $R0bc$. We must show $a \leq c$, i.e., $R0ac$. This follows immediately using R5′.

Let us now discuss the Routley–Meyer conditions one by one.

R1: The first question to ask is what is the "0"? Let us call "0" the "null context." I assume as a degenerate case that even in a lack of context, an information state is relevant to itself. In Fine's approach, this amounts to saying that the null information state 0, when combined with any information state a, is included in the information state a.

R2: This states that a is relevant to itself when taken as its own context. This amounts to saying that the information state a when combined with itself gives no more information than just a. This is a bit questionable as we shall see.

R3: This states that which of the information states a and b is taken as context, and which as input, does not matter. To take Sperber and Wilson's example of the train, it doesn't matter if I see the train arriving and look at my watch, or vice versa.

R4: b is relevant to x (in context a) and c is relevant to d (in context x). Then we must find some input y such that y is relevant to d in context a and c

[8] Also if we had fusion $\circ$ as primitive (instead of defining it as $\sim(B \to \sim A)$) we would have to add R6″: $c \leq c'$ and $Rabc \Rightarrow Rabc'$.

is relevant to y in context b. It is easiest to consider this in the Fine framework. We have $a \bullet b \leq x$ and $x \bullet c \leq d$. We need a y such that $a \bullet y \leq d$ and $b \bullet c \leq y$. The obvious answer is that $x = a \bullet b$ and that $(a \bullet b) \bullet c \leq a \bullet (b \bullet c)$. This last means that it does not matter given three pieces of information in a given order, which two we combine first – as long as we preserve their order.

R5: Let us suppose that a' is absolutely relevant to a and that in the context a, b is relevant to c. It then seems that, in the context a', b should be relevant to c. Note that canonically the information states $a, b, c, \ldots$ are certain sets of sentences ("prime theories" – but we will not go into the details here – see e.g., Dunn (1986)) and that $a \leq b$ is just $a \subseteq b$, i.e., the theory a is included in the theory b. Another way to look at this is that the information in a is included in the information in state b. Honesty compels me to raise the question as to whether this conforms to intuitions we might have about a being absolutely relevant to b. Is a subtheory relevant to a theory? I believe the answer is 'yes'. Consider the example of the role of Peano arithmetic in the more general theory of say the positive and negative integers. But what about the other way around? The intuitions here are admittedly slippery, but while it seems to me that the theory of the natural numbers is "integral" (absolutely relevant) to the theory of the integers, this is no way holds the other way around.

Put quickly, for sets generally (not just for theories), if $b \subseteq c$, then if there is any change in b, say in an extreme case that its members were to cease to exist, then there is of course a change in c. This is a case of Absolute Relevance.

Contextual Relevance will be the way that we interpret the Routley–Meyer accessibility relation R, but we need to be sensitive about how we define it. Union (i.e., $a \cup b$) is perhaps the first idea that springs to mind, and if we interpret $Rabc$ as $a \cup b \subseteq c$ then this interpretation does satisfy all of the requirements R1–R5.[9] But there are some definite peculiarities in understanding Contextual Relevance using union. In order for $a \cup b \subseteq c$, we must have $b \subseteq c$, and so strangely Contextual Relevance implies Absolute Relevance, but not vice versa (since maybe $a \nsubseteq c$). Note also that the context $a \subseteq c$, and so the context a is also absolutely relevant to c. These are downright strange relationships.

But there are various natural ways to combine information states, and union is only one of these. Let us denote a combination of the information states a and b as $a \bullet b$, and then general idea is that we can interpret the Routley–Meyer ternary accessibility relation $Rabc$ as something like $a \bullet b \leq c$, and there is no reason to think that any of $a, b, a \cup b$ are included in c.

[9] Food for thought. Propositions viewed as sets of information states (or possible worlds) are disjunctive in character. Each member of a proposition p can be viewed as a way that p might be true. If proposition p is included in proposition q, then p entails q. Information states viewed as sets, say of sentences as with Urquhart, are conjunctive in character. So if $a \subseteq b$, then b contains more information that a, and viewing a and b as theories (or as sets of axioms for a theory), we can speak of b "entailing" a.

7 Ways of Combining Information

Here are three natural ways of combining information $a \bullet b$:

1. Data Combining Interpretation: the piece of information a when combined with b equals (or is included) in c. (Urquhart, Fine).
2. Program Applied to Data Interpretation: view information state a as "input" (static) and view the information state b as a "program" (dynamic). Information state c is a potential result of running the program on that input. (Dunn).
3. Program Combining Interpretation: view a and b both as programs, and view the result of composing these two programs as equal to (or included in) c. (Dunn).

The kind of interpretation in 1 comes from Urquhart (1972) who talks of combining pieces of information. Fine talks of theories instead of pieces of information. Interpretations 2 and 3 can be found in various forms in Dunn (2001a, 2001b, 2001c, 2003). Mares (1997) contains another "informational" interpretation of the ternary accessibility relation.

Imagine pieces of information as piles of paper on your desk. Interpretation 1 has to do with viewing the pieces of paper as containing data and combining them together into a single pile, and of course, this can be done in different ways. The simplest being to just treat them as sets and not care about the order in which they are placed, or whether there are duplicates. Another way might be to regard them as multisets, and disregard the order while carefully noting the number of duplicates. Maybe, the order could matter too as with sequences. And maybe, the way the are grouped into files, say with file folders could matter. We will not explore all of these here, but list them to provoke thoughts. For interpretation 2, think of the pieces of paper in the first pile (a) as a kind of program containing instructions about what to do with sentences on pieces of paper, and the idea is just to apply those instructions to the sentences in the other pile (b). For interpretation 3, the idea is to treat the sentences in both piles as instructions, and to compose the instructions from the first pile (a) with those in pile (b) so as to get new instructions.

Re 1, Urquhart (1972) took the simplest mode of combining pieces of information. He took pieces of information to be sets of sentence and took the standard operation of set union to be the way of combining them. He did this independently about the same time as Routley and Meyer came up with their ternary relation, and in fact avoided the need for it by defining $x \models A \to B$ iff for every a, if $a \models A$ then $x \cup a \models B$. The ternary relation is implicit and can be defined as $Rxab$ iff $x \cup a = b$. Urquhart's way of doing things is often referred to as the "operational semantics" for relevance logic and is contrasted with the Routley–Meyer "relational semantics." There is only one thing wrong with the operational semantics, and that is that it is not complete for any of the well-known relevance logics, say R, and in fact it is not nicely axiomatizable at all, as Fine showed. Fine produced his own semantics for the system R and other relevant logics, sometimes called the "relational-operational" semantics, or as I like to think of it "best of both." We will not go into the details of the Fine semantics, but one way that Fine presents it is to compare it to the Routley–Meyer semantics so there is a binary operation $\circ$ and a partial-order $\leq$ so that he can define $x \models A \to B$ iff for every a, b, if $x \circ a \leq b$ and $a \models A$ then $b \models B$. If you let your eyes go out of focus a bit, you will see, as Fine suggests, that this is essentially the Routley–Meyer definition with $x \circ a \leq b$ in place of $Rxab$. Fine sometimes writes $a \leq_x b$, which would serve perfectly for the notion of contextual relevance.

8 Comparing the Fine and Urquhart Semantics for Relevance Logic

Let us consider Fine's and Urquhart's two different ways of defining the truth of a relevant implication using $\bullet$ to combine *information states* (we shall use this term as an abstraction to cover Urquhart's *pieces of information* and Fine's *theories*):

(F) $x \models A \to B$ iff for $\forall a, b$, if $x \bullet a \leq b$ and $a \models A$, then $b \models B$ (Fine),
(U) $x \models A \to B$ iff $\forall a$, if $a \models A$, then $x \bullet a \models B$ (Urquhart).

(U) kind of hides the contextual relevance, whereas (F) sticks it in your face. But in fact they are equivalent. Indeed, (U) is a special case of (F), if we instantiate b to $x \bullet a$. But (U) can be seen conversely to imply (F). It suffices to show that the right hand side of (U) implies the right hand side of (F). So let us assume the right hand side of (U): $\forall a, b$, if $a \models A$ then $x \bullet a \models B$. We will show the right-hand side of (F): $\forall a$, b,if $x \bullet a \leq b$ and $a \models A$ then $b \models B$. So for this purpose further assume $x \bullet a \leq b$ and $a \models A$. From $a \models A$ and the right hand side of (U), we derive $x \bullet a \models B$. But from this and $x \bullet a \leq b$, by using the Hereditary Condition, we can show that $b \models B$, as needed. Of course, we have to show the Hereditary Condition, but this is routine.

There are clearly ways of combining pieces of information that do not have all of the properties of $\cup$. What are the properties of $\cup$? It is well-known that from an equational perspective these are:

$a \circ a = a$ (Idempotence)
$a \circ b = b \circ a$ (Commutation)
$a \circ (b \circ c) = (a \circ b) \circ c$ (Association).

These three properties characterize a semi-lattice, and any semi-lattice is isomorphic to a collection of sets closed under $\cup$.

Let's start with idempotence: $a \circ a = a$, and consider the inequality half: $a \circ a \leq a$ (*Square decreasing*). This is just the Routley–Meyer condition $Raaa$, but that does not make it sacrosanct. If the pieces of paper I am combining are dollar bills, there is more information in my hand ("I have two dollars") when I show two of them than when I show just one of them ("I have one dollar"). Girard

(1987) uses similar examples to motivate his Linear Logic since the contraction axiom depends on this property and linear logic lacks that. Another example might be if we think of a as the code for a program and we apply that very same program to itself as input. There is no reason to believe it will turn out that code as output.

Another important property of $\cup$ is $a \circ b = b \circ a$ (Commutation). The computer program example destroys that too.

The third property of $\cup$ is $a \circ (b \circ c) = (a \circ b) \circ c$ (Association). This is a little harder to dismiss. Even if I have three bills in two different pockets (say a in the left, and b and c in the right), and then rearrange them (both a, b in the left, and c in the right) I still have the same documentation as to my wealth. But again one can devise a programming counter-example (though interestingly not when one interprets a, b, c all as the programs that they code and views $\bullet$ as composition of programs).

Let us reexamine the three ways of combining information that we listed above. Here is a simplified way to think about them.

For interpretation 1 (Data Combining Interpretation), imagine that $a = \{p\}, b = \{q\}$. Then $a \bullet b = \{p, q\}$.

For interpretation 2 (Program Applied to Data Interpretation), imagine that $a = \{p\}, b = \{p \rightarrow q\}$. Then $a \bullet b = \{q\}$.

For interpretation 3 (Program Combining Interpretation), imagine that $a = \{p \rightarrow q\}, b = \{q \rightarrow r\}$. Then $a \bullet b = \{p \rightarrow r\}$.

The problem put quickly is that I may be in different kinds of mental states as I acquire new cognitive input. If I am in a merely receptive state and I acquire two pieces of information, say p and q, I may merely "file them away" into $a \bullet b = \{p, q\}$. This may be true even if I acquire p and $p \rightarrow q$ and my mind is not very active and/or they are deeply buried with other pieces of information. On the other hand, if my mind is very active (and confident of its powers) I might conclude $a \bullet b = \{q\}$ and at the same time discard the premises p and $p \rightarrow q$ that led to the information q. And of course a similar story may be told about interpretation 3.

9 Negation

This is a bit of digression, but it is justified by the fact that treatment of negation, not just implication, is critical in relevance logic. In classical logic, from the premise $A \wedge {\sim}A$ any other sentence B can be derived as a conclusion, because in classical logic $(A \wedge {\sim}A) \supset B$ is a theorem. In general $(A \wedge {\sim}A) \rightarrow B$ is not a theorem in the systems R, E, and other relevance logics, because of the Variable Sharing Property. This has at least as much to do with negation as it does with implication, and indeed even in the FDE (first-degree entailment) fragment of E and R this is not provable.

There are at least three treatments of what is standardly called "De Morgan" negation in relevance logic, and there is also another negation entirely, classical, or "Boolean" negation which Meyer and Routley showed could be conservatively added to the system R. Put quickly:

("Routley Star") $x \models {\sim}A$ iff not $x^* \models A$

(Four-valued) $x \models_1 {\sim}A$ iff $x \models_0 A$
$x \models_0 {\sim}A$ iff $x \models_1 A$

(Perp) $x \models {\sim}A$ iff $\forall a \in A, a \perp x$.

Depending on subtleties, the first two were both in my dissertation Dunn (1966). I discussed various representations of De Morgan lattices (the algebraic counterpart to first-degree entailments) and showed them all equivalent. The "Routley Star" was in this context anticipated by Białynicki and Rasiowa in their representation of De Morgan lattices (they called them quasi-Boolean algebras). The four-valued semantics was implicit in another representation of De Morgan lattices using "topics" but was not made explicit until Dunn (1969) and not published until Dunn (1976). Two important qualifications though – these did not address the issue of nested implications. We will discuss this some more in the next section. The Perp treatment of negation arose also in the representation of De Morgan lattices, but much later. See Dunn (1993). We shall not discuss it here. For a discussion of these in a more general setting see Dunn (1999).

The Routley Star should really be called the Routleys' Star, since it was introduced in Routley and Routley (1972) as a semantics for first degree entailments. They required that it satisfy:

$a^{**} = a$ (Period two)

Routley and Meyer go further and combine it with their ternary accessibility relation to provide a semantics for R, E, etc. and require that * satisfy in addition to Period Two the following:

If $Rabc$ then Rac^*b^*. (Antilogism)

x^* can be understood as the sentences not denied by x, which helps us informally understand the validity of

$(A \to B) \to ({\sim}B \to {\sim}A)$ (Contraposition),

which is formally determined by Antilogism. An informal understanding of Antilogism in terms of contextual relevance goes something like this. Antilogism says that if b is relevant to c in the context a, then c^* is relevant to b^* in the same context a, that is, the information not denied by c is relevant to the information not denied by b. Go figure!

The "Routley-Star" has come under a lot of criticism both from those within and outside of the relevance logic community, and was more of a focus of Copeland's (1979) critical review than the ternary accessibility relation.

The 4-valued approach (Dunn (1976)) assigns each sentence a subset of the set of truth values $\{1, 0\}$ instead of just a single one of the truth values $1, 0$.[10] There are clearly 4 such subsets, and hence 4 values: $\{\{1\}, \{0\}, \{\}, \{1, 0\}\}$.

Belnap (1977a, 1977b) labeled these T, F, N, B, for True, False, Neither, Both. We can understand these as subsets of $\{1, 0\}$, which is the basic approach of Dunn (1976). We then can do "double-entry bookkeeping" with $x \models_1 A$ meaning that the information state a is assigning the sentence A at least the value 1, and $x \models_0 A$ meaning that a is assigning A at least the value 0. We start with a valuation v that assigns to each atomic sentence p some subset of $\{1, 0\}$. From this we can define $V_1 = \{x : 1 \in v(p)\}$, $V_0 = \{x : 0 \in v(p)\}$. Then:

(vp) $x \models_1 p$ iff $x \in V_1(p)$ (Atomic)
$x \models_0 p$ iff $x \in V_0(p)$

Clauses for $\sim$, $\wedge$, $\vee$ then are as follows:

(v$\sim$) $x \models_1 {\sim}A$ iff $x \models_0 A$ (Negation)
$x \models_0 {\sim}A$ iff $x \models_1 A$
(v$\wedge$) $x \models_1 A \wedge B$ iff $x \models_1 A$ and $x \models_1 B$ (Conjunction)
$x \models_0 A \wedge B$ iff $x \models_0 A$ or $x \models_0 B$
(v$\vee$) $x \models_1 A \vee B$ iff $x \models_1 A$ or $x \models_1 B$ (Disjunction)
$x \models_0 A \vee B$ iff $x \models_0 A$ and $x \models_0 B$.

This in fact gives a complete semantics for First Degree Entailments (FDE), those formulas of R and E that do not contain nested implications, i.e., formulas of the form $A \to B$ where A and B do not contain $\to$.

Plus, the sharp-eyed reader will have noticed, we need to have two clauses for relevant implications as well. This gets complicated.

Of course we could just continue and write down:

(v$\to$) $x \models_1 A \to B$ iff $\forall a, b$, if $Rxab$ and $a \models_1 A$ then $b \models_1 B$
$x \models_0 A \to B$ iff $\exists a, b$, $Rxab$ and $a \models_1 A$ then $b \models_0 B$.

This might be fine, except it does not seem to give a completeness theorem for R. Fortunately, Mares (2004) has found a way to get a variant of it to work. But it needs the complication of adding a "neighborhood semantics," much like the neighborhood semantics of various non-normal modal logics. We leave it as an open problem whether sense can be made of this addition in terms of relevance.

[10] Though an alternative was suggested of viewing a valuation as a relation of a sentence to $1, 0$ instead of a function taking just one of these.

10 What is an Information State?

This is also a bit of a digression (no pun intended). Routley–Meyer, Urquhart, and Fine, all have in their semantics a set of *elements* corresponding to Kripke's possible worlds in his semantics for modal logic (often called "possible worlds semantics"). But these elements cannot be taken to be possible worlds, for they are often incomplete and/or inconsistent. Thus, for some element a and sentence A we may have neither $a \models A$ nor $a \models {\sim}A$, and for another sentence we may have both $a \models A$ and $a \models {\sim}A$.

Routley–Meyer coined the name "set ups" for their elements. Urquhart used the term "pieces of information," and Fine used the term "theories." We have tended to use the term "information state."[11] This is clearly an abstraction, but a concrete way to think of it is as the state of a storage system in a computer. This can be visualized for an antique computer as the setting of switches (either on or off), or for a hard drive as changes in direction of the magnetic field from place to place. No matter the physical method of storage, an information state may be thought of as a finite sequence of bits, either 1 or 0. Propositions can be thought of as sets of information states, and each proposition P is true of false in a given information state a depending on whether $a \in P$ or $a \notin P$, and in principle, this goes back to Shannon. It is related to Carnap and his state descriptions – see for example Dunn (2001a). In this setting the Routley–Meyer valuation clause

(v$\vee$) $x \models A \vee B$ iff $x \models A$ or $x \models B$

makes perfect sense.

But for a more ordinary conception of an information state the left-to-right direction is problematic. I remember that I left my keys upstairs on the dresser, or in the basement on the workbench, but I don't remember which. Or suppose I am about to throw a coin. I have the information that it will turn up heads or tails, but I do not have the information as to which. Or suppose the dog knows that his master took either the left fork of the path or the right fork, but does not know which (he hasn't yet sniffed). Or in playing the game *Clue*, figure out that that the murder was committed by Miss Scarlet with a knife in the study, or by Professor Plum with a candlestick in the dining room, etc.

Urquhart, calling the elements of his semantics "pieces of information," runs head on into this problem. Fine, calling his elements "theories" would seem to also, for clearly theories can contain disjunctions without containing either disjunct (look at the theories above about my keys, etc.). But Fine has a way out. He has special kinds of theories. Besides the set T of ordinary theories he has the subset S "of all theories that contain a disjunct of any contained

[11] But back in the old days of Dunn (1976), I used the term "situation." And Mares (2004) with his "situated semantics" continues to use the term "situation," partly to build an alliance with the Barwise–Perry Situation Theory. Situations would seem to provide plausible terminology for "situated relevance": situation b is relevant to situation c in the context of situation a.

disjunction, that answer every either-or question they raise." In the section he wrote in *Entailment* II, he suggests (among other names) that these may be called "prime," and this is the name that we use. It derives from the notion of a prime filter in lattice theory, and Routley–Meyer used the same notion in constructing the canonical models used in proving their semantics complete.

This suggests that we generalize the notion of an information state so that it is not just a finite sequence of 0's and 1's, but that we also let the notation N (for neither) sometimes occur to indicate that the information state is not complete. There are also circumstances where we might have conflicting information. In a computer network, sites are sometime cloned for various purposes, good or bad (e.g., safety backup, facilitating access, phishing). It is entirely possible that the cloning introduces errors so that one site has 1 in a certain position in its current state, and a supposedly duplicate site shows a 0 at the same time in the same position. For that we might use the notation B (for both). Such a 4-valued information state was described by Dunn (2008), and clearly is an abstraction from the so-called Belnap–Dunn 4-valued logic.

11 (Tentative) Conclusion

I call this concluding section "tentative," because I hope that this paper has opened all kinds of doors to further research. You all are invited to walk through them. All of the variations on the contextual relevance relation, handled in the right way, give rise to different logics, some well-known, others new.[12] I am tempted to follow the example of Fermat here and say: "I have discovered a truly marvellous proof of this, which this margin – I mean paper – is too narrow to contain." But the truth is I have barely begun to peak through many of the doorways.

This is an embarrassment of riches. We started out (at least according to Meyer), with no logic of relevance, and now we have too many logics of relevance. Will the real logic of relevance please stand up?

At any rate, we have given an interpretation of the Routley–Meyer accessibility relation *Rabc* in terms of contextual relevance that allows us to regard the logic R of relevant implication as also a logic of relevance. Of course, we must admit that there are various other uses of relevance outside of logic, but that point is not relevant.

Suppose I were to claim that classical logic is the logic of the two truth-values truth and falsity. It seems to me that this is correct even though truth and falsity have many applications outside of logic.

References

Anderson, A., Belnap, N.: Entailment: The Logic of Relevance and Necessity, vol. I. Princeton University Press (1975)

Anderson, A., Belnap, N., Dunn, J.M.: Entailment: The Logic of Relevance and Necessity, vol. II. Princeton University Press (1992)

[12] A good place to start is to consider the list in Routley and Meyer (1973) of various properties of the ternary relation with the corresponding logical axioms they validate.

Beall, J.C., Brady, R., Dunn, J.M., Hazen, A.P., Mares, E., Meyer, R.K., Priest, G., Restall, G., Ripley, R., Slaney, J., Sylvan (formerly Routley, R.): On the Ternary Relation and Conditionality. The Journal of Philosophical Logic 41, 595–612 (2012)

Belnap, N.D.: A Useful Four-valued Logic. In: Dunn, J.M., Epstein, G. (eds.) Modern Uses of Multiple-Valued Logic, pp. 8–37. D. Reidel, Dordrecht (1977a)

Belnap, N.D.: How a Computer Should Think. In: Ryle, G. (ed.) Contemporary Aspects of Philosophy, pp. 30–55. Oriel Press Ltd., Stocksfield (1977b)

van Benthem, J.: Review of "On when a semantics is not a semantics" by B. J. Copeland. The Journal of Symbolic Logic 49, 994–995 (1984)

Białynicki, A., Rasiowa, H.: On the Representation of Quasi-Boolean Algebras. Bulletin de l'Académie Polonaize des Sciences 5, 259–261

Copeland, B.J.: On When a Semantics Is Not a Semantics: Some Reasons for Disliking the Routley–Meyer Semantics for Relevance Logic. Journal of Philosophical Logic 8, 399–413 (1979)

Dunn, J.M.: The Algebra of Intensional Logics, Ph.D. Dissertation, University of Pittsburgh, University Microfilms (1966)

Dunn, J.M.: Natural Language versus Formal Language. In: Joint Symposium of the Association for Symbolic Logic and the American Philosophical Association (December 1969), On-line with historical introduction at `http://www.philosophy.indiana.edu/people/papers/natvsformal.pdf`

Dunn, J.M.: Intuitive Semantics for First Degree Entailments and Coupled Trees. Philosophical Studies 29, 149–168 (1976)

Dunn, J.M.: Relevance Logic and Entailment. In: Gabbay, D., Guenthner, F. (eds.) Handbook of Philosophical Logic, vol. 3, pp. 117–224. D. Reidel, Dordrecht (1986); Updated as Dunn, J.M., Restall, G.: Relevance Logic. In: Gabbay, D., Guenthner, F. (eds.) Handbook of Philosophical Logic, 2nd edn., vol. 6, pp. 1–128. Kluwer Academic Publishers (2002)

Dunn, J.M.: Star and Perp: Two Treatments of Negation. In: Tomberlin, J. (ed.) Philosophical Perspectives. Language and Logic, vol. 7, pp. 331–357 (1993)

Dunn, J.M.: A Comparative Study of Various Semantical Treatments of Negation: A History of Formal Negation. In: Gabbay, D., Wansing, H. (eds.) What is Negation, pp. 23–61. Kluwer Academic Publishers (1999)

Dunn, J.M.: The Concept of Information and the Development of Modern Logic. In: Stelzner, W., Stöckler, M. (eds.) Zwischen traditioneller und moderner Logik: Nichtklassische Ansatze (Non-classical Approaches in the Transition from Traditional to Modern Logic), pp. 423–447. Mentis-Verlag, Paderborn (2001a)

Dunn, J.M.: Ternary Relational Semantics and Beyond: Programs as Data and Programs as Instructions. Logical Studies 7, 1–20 (2001b), `http://www.logic.ru/LogStud/`

Dunn, J.M.: Representation of Relation Algebras Using Routley–Meyer Frames. In: Anderson, C.A., Zelëny, M. (eds.) Logic, Meaning and Computation: Essays in Memory of Alonzo Church. Draft in Indiana Univ. Logic Group Preprint Series, IULG-93-28, pp. 77–108. Kluwer, Dordrecht (2001c)

Dunn, J.M.: Ternary Semantics for Dynamic Logic, Hoare Logic, Action Logic, etc. In: Smirnov's Readings, vol. 4, pp. 70–71. Institute of Logic, Russian Academy of Sciences (2003)

Dunn, J.M.: Information in Computer Science. In: van Benthem, J., Adriaans, P. (eds.) Handbook of the Philosophy of Information, pp. 581–608. Elsievier, Amsterdam (2008)

Dunn, J.M., Meyer, R.K.: Combinatory Logic and Structurally Free Logic. Journal of the Interest Group in Pure and Applied Logic 5, 505–537 (1997)

Fine, K.: Models for Entailment. Journal of Philosophical Logic 3, 347–372 (1974)
Gabbay, D.M., Woods, J.: Agenda Relevance: A Study in Formal Pragmatics in their A Practical Logic of Cognitive Systems, vol. 1. Elsevier, Amsterdam (2003)
Girard, J.-Y.: Linear Logic. Theoretical Computer Science 50, 1–102 (1987)
Grice, P.: Logic and Conversation, pp. 3–143. Harvard University, William James Lectures (1967) (In Grice (1989))
Grice, P.: Studies in the Way of Words. Harvard University Press, Cambridge (1989)
Maksimova, L.: A Semantics for the System E of Entailment. Bulletin of the Section of Logic, Polish Academy of Sciences, Institute of Philosophy and Sociology 2, 18–21 (1973)
Mares, E.D.: Relevant Logic and the Theory of Information. Synthese 109, 345–360 (1996)
Mares, E.D.: 'Four-valued' Semantics for the Relevant Logic R. Journal of Philosophical Logic 33, 327–341 (2004)
Mares, E.D.: Relevant Logic: A Philosophical Interpretation. Cambridge University Press (2004)
Meyer, R.K.: Why I Am Not a Relevantist, Research paper, no. 1, Australian National University Logic Group, Research School of Social Sciences, Canberra (1978)
Meyer, R.K.: A Farewell to Entailment. In: Dorn, G., Weingartner, P. (eds.) Foundations of Logic and Linguistics: Problems and Solutions, pp. 577–636. Plenum Press, New York (1985)
Morris, C.W.: Foundations of the Theory of Signs. In: Neurath, O. (ed.) International Encyclopedia of Unified Science, vol. 1(2). University of Chicago Press, Chicago (1938)
Restall, G.: Relevant and Substructural Logics. In: Gabbay, D.M., Woods, J. (eds.) Handbook of the History of Logic. Logic and Modalities in the Twentieth Century, vol. 7, pp. 289–398. Elsevier, Amsterdam (2006)
Routley, R., Meyer, R.K.: The Semantics of Entailment, II, III. Journal of Philosophical Logic 1, 53–73, 192–208 (1972)
Routley, R., Meyer, R.K.: The Semantics of Entailment. In: Leblanc, H. (ed.) Truth, Syntax and Modality, Proceedings of the Temple University Conference on Alternative Semantics. North-Holland, Amsterdam (1973)
Routley, R., Meyer, R.K., Plumwood, V., Brady, R.: Relevant Logics and their Rivals. In: Part I, The Basic Philosophical and Semantic Theory. Ridgeview Publishing Company, Atascadero (1983)
Routley, R., Routley, V.: Semantics of First Degree Entailment. Noûs 6, 335–359 (1972)
Sperber, D., Wilson, D.: Postface to the second edition of Relevance: Communication and Cognition. Blackwell, Oxford (1995)
Urquhart, A.: Semantics for Relevant Logics. The Journal of Symbolic Logic 37, 159–169 (1972)
Wilson, D., Sperber, D.: On Defining Relevance. In: Grandy, R., Warner, R. (eds.) Philosophical Grounds of Rationality: Intentions, Categories, Ends, pp. 143–158. Oxford University Press (1986)
Wilson, D., Sperber, D.: Relevance Theory. In: Horn, L.R., Ward, G. (eds.) The Handbook of Pragmatics, pp. 607–632. Blackwell, Oxford (2004)
Wilson, D., Sperber, D.: Meaning and Relevance. Cambridge University Press (2012)

Logic-Automata Connections for Transformations

Emmanuel Filiot*

Université Libre de Bruxelles, Brussels, Belgium

Abstract. Pioneered by Büchi, Elgot and Trakhtenbrot, connections between automata and logics that define languages of words and trees are now well-established. During the last decade, some of these powerful connections have been extended to binary relations (transformations) of words and trees. This paper is a survey of known automata-logic connections for transformations.

1 Introduction

The connections between mathematical logics and computational models have a long research history, which goes back to the foundations of theoretical computer science and the seminal works of Church and Turing [12,43]. In particular, Turing has shown how to express the behaviour of a universal machine in first-order logic, and then proved that first-order logic is undecidable, as a consequence of the undecidability of the halting problem. The Curry-Howard isomorphism is another important example of connection that shows correspondences between the formulas of a logic and the types of a computational model, and between proofs and programs [17,29].

Further connections between mathematical logic and automata theory have been discovered in the 60s by Büchi [10], Elgot [20] and Trakhtenbrot [42], who have shown that the class of finite word languages definable in monadic second-order logic corresponds, in an effective way, to the class of languages definable by finite state automata, and thus to regular languages. While logical formalisms have a high-level descriptive power, automata are easier to analyse algorithmically. For instance, checking whether the language defined by a finite state automaton is empty can be decided in linear-time. Therefore, as an application of Büchi-Elgot-Trakhtenbrot's theorem, monadic second-order logic (interpreted on finite words) has decidable satisfiability problem. Since this seminal result, many other similar connections have been shown, most notably for regular languages of infinite words and trees [11,35,36] and first-order definable languages of words [38]. More details can be found in the following survey: [41], [44] and [18].

A language of finite words over an alphabet Σ is a mapping from the set of words Σ^* to $\{0,1\}$. A *transformation* of finite words is a binary relation R on Σ^*, and therefore it generalises the concept of languages. It is *functional* if R

* FNRS Research Associate (*Chercheur Qualifié*).

M. Banerjee and S.N. Krishna (eds.): ICLA 2015, LNCS 8923, pp. 30–57, 2015.

is a function. Although transformations are as fundamental as languages, much less is known on the relation between automata and logic for transformations. Nevertheless, some important results have been obtained in the last decade. In this paper, we survey some of them.

A transformation R of finite words over an alphabet Σ can be seen as a language, for instance the language $\{u\#v \mid (u,v) \in R\}$, where $\# \notin \Sigma$. However, the formalisms from language theory, such as automata, are not well-suited to describe transformations defined on this encoding and therefore, proper extensions have been introduced to define transformations. Automata have been for instance extended to *automata with outputs*, usually called *transducers*. Perhaps the most studied transducer model is that of *finite state transducers* [31,37]. Finite state transducers (FST) extend finite state automata with an output mechanism. Whenever an FST reads an input symbol, it moves to the next symbol, updates its internal state, and write a partial output word. The final output word is the concatenation, taken in order, of all the partial output words produced while processing the whole input word.

The expressiveness of FST is limited and other, more powerful, state-based models have been introduced and studied, such as two-way transducers and more recently, streaming string transducers [2]. On the logic side, monadic-second order logic has been extended in a natural way to *MSO-transducers* by B. Courcelle, to define transformations of logical structures [14,15]. The predicates of the output structure are defined by MSO formulas interpreted over a fixed number of copies of the input structure. The first automata-logic connection, or one should say transducer-logic connection, has been shown in [21] by J. Engelfriet and H.J. Hoogeboom. They have extended Büchi-Elgot-Trakhtenbrot's theorem to functional transformations by showing that any transformation definable by a deterministic two-way finite state transducer is definable by an MSO-transducer (interpreted over finite words), and conversely. Moreover, this correspondence is effective, i.e., an MSO-transducer can be effectively constructed from a deterministic two-way finite state transducer and conversely. An important consequence of this result is the decidability of equivalence of MSO-transducers, since the equivalence problem for deterministic two-way transducers is decidable [16].

Since then, other transducer-logic connections have been established for finite word transformations and other structures such as infinite words and finite trees. Functional MSO-transformations of finite words have been shown to correspond to transformations definable by streaming string transducers [2], and this result has been extended to infinite words [3] and to non-functional MSO-transformations [21,5]. Engelfriet-Hoogeboom's theorem has been extended to finite trees [8,22,23]. First-order definable transformations of finite words have been considered in [26] and [32]. Some of these connections have been considered under a stronger semantics, the origin semantics, in [9].

This paper surveys some of these important results. All the transducer-logic connections presented in this paper are effective. The setting of functional transformations of finite words is presented in details, in contrast to the other results, which come nevertheless with the main bibliographic references. In Section 2, we present some preliminary notions. In Section 3, we define first-order and monadic second-order logics interpreted on finite words, and define MSO-transducers, for which we give several examples. In Section 4, we introduce the main state-based models of transformations used in this paper. In Section 5, we present the main transducer-logic connection for transformations of finite words. Finally in Section 6, we briefly survey some extensions of the finite word setting.

2 Word Transformations

We define the preliminary notions used all over this paper.

Words An *alphabet* Σ is a finite set of symbols, called letters. A *word* w over Σ is a finite sequence of letters $(\sigma_1, \dots, \sigma_n)$, denoted $w = \sigma_1 \dots \sigma_n$. The empty word (empty sequence) is denoted by ϵ. The *length* of a non-empty word $w = \sigma_1 \dots \sigma_n$ is defined by $|w| = n$, and $|\epsilon| = 0$. We denote by $\text{dom}(w) = \{1, \dots, |w|\} \subseteq \mathbb{N}^*$ the domain of w. In particular, $\text{dom}(\epsilon) = \varnothing$. For all $i \in \text{dom}(w)$, i is called a *position* of w and $w(i)$ denotes the i-th letter of w. The set of words over Σ is denoted by Σ^*, while the set of non-empty words over Σ is denoted by Σ^+.

Given two words $w_1 = \sigma_1 \dots \sigma_n$ and $w_2 = \beta_1 \dots \beta_m$, their *concatenation*, denoted $w_1.w_2$ (or simply $w_1 w_2$), is defined by $w_1.w_2 = \sigma_1 \dots \sigma_n \beta_1 \dots \beta_m$. In particular, $\epsilon w = w\epsilon = w$ for all words $w \in \Sigma^*$. For all $w \in \Sigma^*$ and $n \in \mathbb{N}$, we denote by w^n the concatenation of w, n times. In particular, $w^0 = \epsilon$, $w^1 = w$ and $w^2 = ww$.

Transformations A *transformation* R of finite words over an alphabet Σ is a binary relation over Σ^*, i.e. $R \subseteq \Sigma^* \times \Sigma^*$. For all words $u \in \Sigma^*$, we let $R(u)$ be the set of images of u by R, i.e. $R(u) = \{v \in \Sigma^* \mid (u, v) \in R\}$. The word u is usually called an *input word* while the words v such that $(u, v) \in R$ are called *output words.* We denote by $\text{dom}(R)$ the domain of R, and by $\text{range}(R)$ its range, i.e. $\text{dom}(R) = \{u \in \Sigma^* \mid R(u) \neq \varnothing\}$ and $\text{range}(R) = \{v \in \Sigma^* \mid \exists u \in \Sigma^*, v \in R(u)\}$.

A transformation R is *functional* if R is a function, i.e. for all words $u \in \Sigma^*$, the cardinality of $R(u)$ is smaller than or equal to 1, i.e. $|R(u)| \leq 1$. Functional transformations are rather denoted by $f, g, h \dots$. For a functional transformation f, we write $f(u) = v$ instead of $f(u) = \{v\}$, for all $(u, v) \in f$.

Example 1. Let $\Sigma = \{a, b\}$. The following examples of (functional) transformations of finite words over Σ are running examples in this paper.

- The transformation $f_{del} : \Sigma^* \to \Sigma^*$ deletes all letters a, i.e. for all input words $u = \sigma_1 \dots \sigma_n$, $f_{del}(u) = \sigma_{i_1} \dots \sigma_{i_k}$ such that $\{i_1 < \dots < i_k\} = \{i \in \text{dom}(w) \mid w(i) \neq a\}$. E.g. $f_{del}(abaabb) = bbb$.
- The transformation f_{double} doubles every input letter, i.e. for all $u = \sigma_1 \dots \sigma_n$, $f_{double}(u) = \sigma_1\sigma_1 \dots \sigma_n\sigma_n$, e.g. $f_{double}(abaa) = aabbaaaa$.

- The transformation f_{copy} copies input words twice, i.e. for all $u \in \Sigma^*$, $f_{copy}(u) = uu$.
- The transformation f_{rev} reverses input words, i.e. $f_{rev}(\sigma_1 \dots \sigma_n) = \sigma_n \dots \sigma_1$. E.g. $f_{rev}(abaa) = aaba$.
- The transformation $f_{1/2}$ is defined over a^* by, for all $n \geq 0$, $f_{1/2}(a^n) = a^{\lfloor n/2 \rfloor}$. E.g. $f_{1/2}(a^8) = a^4$ and $f_{1/2}(a^3) = a$.
- The transformation f_{exp} exponentiates the number of a symbols in a word of the form a^n, e.g. $f_{exp}(a^n) = a^{2^n}$, and $f_{exp}(w)$ is undefined if w contains at least one b.

3 Logical Transducers for Word Transformations

In this section, we introduce *logical transducers*, a logic-based formalism introduced by B. Courcelle [14] to define transformations of logical structures. We refer the reader to [15] for more details and results about logical transducers. Although logical transducers can generally define transformations of arbitrary logical structures, we specialise them to finite word transformations in this section. We first introduce the notion of word logical structures, and then the classical first-order and monadic second-order logics, interpreted over (logical structures of) words.

3.1 Words as Logical Structures

A word w over an alphabet Σ can be seen as a *logical structure*[1] $\tilde{w}$ over the signature $\mathcal{S}_\Sigma = \{(L_a)_{a\in\Sigma}, \preceq\}$, where $(L_a)_{a\in\Sigma}$ are monadic predicates that define the labels of the positions in w, and $\preceq$ is a binary predicate that defines the order on word positions. Formally, $\tilde{w} = (\text{dom}(w), (L_a^{\tilde{w}})_{a\in\Sigma}, \preceq^{\tilde{w}})$ is the logical structure whose domain is dom(w), and such that the predicates are interpreted as follows:

$$L_a^{\tilde{w}} = \{i \in \text{dom}(w) \mid w(i) = a\} \qquad \preceq^{\tilde{w}} = \{(i,j) \mid i,j \in \text{dom}(w) \wedge i \leq j\}$$

When it is clear from the context, we rather write w instead of $\tilde{w}$.

A structure on $\mathcal{S}_\Sigma$ is also called a $\mathcal{S}_\Sigma$-structure. We denote by $\mathcal{M}(\mathcal{S}_\Sigma)$ the set of $\mathcal{S}_\Sigma$-structures. Note that a $\mathcal{S}_\Sigma$-structure may not be isomorphic to any word. However for all $w \in \Sigma^*$, $\tilde{w} \in \mathcal{M}(\mathcal{S}_\Sigma)$. Given a structure in $M \in \mathcal{M}(\mathcal{S}_\Sigma)$, we denote by dom($M$) its domain.

3.2 First-Order and Monadic Second-Order Logics on Words

Given an alphabet Σ, *monadic second-order formulas* (*MSO formulas*) over the signature $\mathcal{S}_\Sigma$ are built over first-order variables $x, y \dots$ and second-order variables $X, Y \dots$. They are defined by the following grammar:

$$\phi ::= \exists X \cdot \phi \mid \exists x \cdot \phi \mid \phi \wedge \phi \mid \neg\phi \mid x \in X \mid L_a(x) \mid x \preceq y \mid (\phi)$$

[1] See for instance [19] or [40] for a definition of logical structures.

Universal quantifiers and other Boolean connectives are defined naturally: $\forall x \cdot \phi \equiv \neg\exists x \cdot \neg\phi$, $\forall X \cdot \phi \equiv \neg\exists X \cdot \neg\phi$, $\phi_1 \vee \phi_2 \equiv \neg(\phi_1 \wedge \phi_2)$ and $\phi_1 \rightarrow \phi_2 \equiv \neg\phi_1 \vee \phi_2$. We also define the formulas true and false: $\top \equiv \forall x \cdot (L_a(x) \vee \neg L_a(x))$ and $\bot \equiv \neg\top$. We do not define the semantics of MSO formulas, neither the standard notion of free and bound variables, but rather give examples and refer the reader to [19] or [40] for formal definitions.

Given an MSO formula ϕ, we write $\phi(x_1, \dots, x_n, X_1, \dots, X_m)$ to emphasise the fact that the free first-order variables of ϕ are exactly $x_1, \dots, x_n$, and its free second-order variables are $X_1, \dots, X_m$. Given a $\mathcal{S}_\Sigma$-structure M and an MSO sentence ϕ, we write $M \models \phi$ when M satisfies ϕ. Let $i_1, \dots, i_n \in \mathrm{dom}(M)$, $I_1, \dots, I_m \subseteq \mathrm{dom}(M)$. For a formula $\phi(x_1, \dots, x_n, X_1, \dots, X_m)$, we write $M \models \phi(i_1, \dots, i_n, I_1, \dots, I_m)$ to denote the fact that M together with the interpretation of x_j by i_j, $j = 1, \dots, n$ and X_j by I_j, $j = 1, \dots, m$, satisfy ϕ.

Given an MSO sentence ϕ, we write $[\![\phi]\!]$ the set of words that satisfy ϕ, i.e. $[\![\phi]\!] = \{w \in \Sigma^* \mid \tilde{w} \models \phi\}$. Given a language $L \subseteq \Sigma^*$, if there exists an MSO sentence ϕ such that $[\![\phi]\!] = L$, we say that L is MSO-definable, and that ϕ *defines* L.

First-order logic First-order (FO) formulas over $\mathcal{S}_\Sigma$ are MSO formulas in which no second-order variable occurs.

Example 2. Let $\Sigma = \{a, b\}$. The formula $\exists x \cdot \top$ defines the set of non-empty words. The formula $\exists x \cdot L_a(x)$ define the set of words over Σ that contain at least one position labelled a, i.e. the language $\Sigma^* a \Sigma^*$.

The formula $S(x, y) \equiv x \preceq y \wedge x \neq y \wedge \forall z \cdot (x \preceq z \preceq y \rightarrow (x = z \vee y = z)$ defines the successor relation.

The formula $\forall x \forall y \cdot (L_a(x) \wedge S(x, y) \rightarrow L_b(y))$ defines the set of words such that any occurrence of the letter a is followed by the letter b. The formulas

$$\mathrm{first}(x) \equiv \neg\exists y \cdot S(y, x) \qquad \text{and} \qquad \mathrm{last}(x) \equiv \neg\exists y \cdot S(x, y)$$

are such that for all $w \in \Sigma^+$ and $i \in \mathrm{dom}(w)$, $w \models \mathrm{first}(i)$ iff $i = 1$, and $w \models \mathrm{last}(i)$ iff $i = |w|$.

The language $a^* b^*$ is definable by the following formula:

$$\forall x \forall y \cdot (L_a(x) \wedge S(y, x) \rightarrow L_a(y))$$

More generally, it is known that the class of MSO-definable languages is the class of regular languages [41]. The MSO formula

$$\mathrm{part}(X_1, \dots, X_n) \equiv (\forall x \cdot \bigvee_{i=1}^{n} x \in X_i) \wedge \forall x \cdot \bigwedge_{i \neq j} (x \notin X_i \vee x \notin X_j)$$

holds true whenever $X_1, \dots, X_n$ defines a partition of the domain. Finally, one can define the set of words of even length in MSO, but one needs second-order variables X_o and X_e to capture, respectively, odd and even positions of the word, as defined by the formula

$$\begin{aligned}\phi_{o/e}(X_o, X_e) \equiv\ & \mathrm{part}(X_o, X_e) \wedge \forall x \cdot (\mathrm{first}(x) \rightarrow x \in X_o) \\ & \wedge \forall x \forall y \cdot S(x, y) \rightarrow (x \in X_o \rightarrow y \in X_e) \wedge (x \in X_e \rightarrow y \in X_o)\end{aligned}$$

Then, the set of words of even length is defined by the sentence

$$\exists X_o \exists X_e \cdot \phi_{o/e}(X_o, X_e) \wedge \forall x \cdot (\text{last}(x) \rightarrow x \in X_e).$$

3.3 Logical Transducers: Definition

Logical transducers define functional transformations from input to output word structures. The output structure is defined by taking a fixed number k of copies of the input structure domain. Some node of these copies can be filtered out by formulas with one free first-order variable. In particular, the nodes of the c-th copy are the input positions that satisfy some given formula $\phi^c_{pos}(x)$. The predicates L_a and $\preceq$ of the output structure are defined by formulas with respectively one and two free first-order variables, interpreted over the input structure. More precisely, position labelled a of the c-th copy are defined by a given formula $\phi^c_{L_a}(x)$, interpreted over the input word. If this formula holds true, it means that the c-th copy of x, if it exists, is labelled a in the output word. The order relation between two output positions is defined by formulas with two free variables interpreted over the input word. For instance, the order relation between positions of the c-th copy and the d-th copy (c and d can be equal) is defined by a formula $\phi_{\preceq}(x, y)^{c,d}$ interpreted over the input structure. If this formula holds true, it means the the c-th copy of x occurs before the d-th copy of y in the output word. Let us formally define logical transducers.

Definition 1. *Let Σ be an alphabet. A logical MSO-transducer (MSOT) on the signature $\mathcal{S}_\Sigma$ is a tuple*

$$T = (k, \phi_{dom}, (\phi^c_{pos}(x))_{1 \leq c \leq k}, (\phi^c_{L_a}(x))_{1 \leq c \leq k, a \in \Sigma}, (\phi^{c,d}_{\preceq}(x, y))_{1 \leq c,d \leq k})$$

where $k \in \mathbb{N}$ and the formulas ϕ_{dom}, ϕ^c_{pos}, $\phi^c_{L_a}$ and $\phi^{c,d}_{\preceq}$ for all $c, d \in \{1, \ldots, k\}$ and $a \in \Sigma$ are MSO formulas over $\mathcal{S}_\Sigma$.

Semantics A logical MSO-transducer T defines a function from $\mathcal{S}_\Sigma$-structures to $\mathcal{S}_\Sigma$-structures, denoted by $[\![T]\!] : \mathcal{M}(\mathcal{S}_\Sigma) \rightarrow \mathcal{M}(\mathcal{S}_\Sigma)$. The domain of $[\![T]\!]$ consists of all structures M such that $M \models \phi_{dom}$. Given a structure $M \in$ *dom*$([\![T]\!])$, the *output structure* N such that $(M, N) \in [\![T]\!]$ is defined by $N = (D^N, (L^N_a)_{a \in \Sigma}, \preceq^N)$ where:

- $D^N \subseteq \text{dom}(M) \times \{1, \ldots, k\}$ is defined by

$$D^N = \{(i, c) \mid i \in \text{dom}(M),\ c \in \{1, \ldots, k\},\ M \models \phi^c_{pos}(i)\}$$

  We rather denote by i^c the elements of D^M.
- for all $a \in \Sigma$, the interpretation L^N_a is defined by

$$L^N_a = \{i^c \mid i \in \text{dom}(M),\ c \in \{1, \ldots, k\},\ M \models \phi^c_{L_a}(i)\} \cap D^N$$

- the interpretation $\preceq^N$ is defined by

$$\preceq^N = \{(i^c, j^d) \mid i, j \in \text{dom}(M),\ c, d \in \{1, \ldots, k\},\ M \models \phi^{c,d}_{\preceq}(i, j)\} \cap (D^N \times D^N)$$

Remark 1. Note that the size of the output structure N is linearly bounded by the size of M, as it is at most $k.|dom(M)|$. We say that MSO-transducers define *linear-size increase* transformations.

Logical transducers as word-to-word transformers Note that in general, an MSO-transducer T over $\mathcal{S}_\Sigma$ may not define a word-to-word transformation, as the output structure of an input word structure may not be a word. We say that T is an *MSO-transducer of finite words over* Σ if for all words $w \in \Sigma^*$ such that $\tilde{w} \in \text{dom}(T)$, $[\![T]\!](\tilde{w})$ is a word, i.e., there exists $v \in \Sigma^*$ such that $[\![T]\!](\tilde{w})$ is isomorphic to $\tilde{v}$. This property is decidable:

Proposition 1. *It is decidable whether an MSO-transducer over $\mathcal{S}_\Sigma$ is an MSO-transducer of finite words over Σ.*

Proof. Let $T = (k, \phi_{dom}, (\phi^c_{pos}(x))_{1\leq c\leq k}, (\phi^c_{L_a}(x))_{1\leq c\leq k, a\in\Sigma}, (\phi^{c,d}_{\preceq}(x,y))_{1\leq c,d\leq k})$. We construct a formula is_word$_T$ which is satisfiable in Σ^* iff T is an MSO-transducer of finite words over Σ. The result follows since MSO over finite words is decidable, by Büchi-Elgot-Trakhtenbrot's Theorem.

Before giving the construction, let us introduce the following useful shortcuts. We write $\forall x^c \cdot \phi$ instead of $\forall x \cdot \bigwedge_{c=1}^k \phi$ and $\exists x^c \cdot \phi$ instead of $\exists x \cdot \bigvee_{c=1}^k \phi$. We also write $[\forall x^c] \cdot \phi$ instead of $\forall x^c \cdot (\phi^c_{pos}(x) \rightarrow \phi)$ to mean that x^c is quantified over output nodes that belong to the domain of the output structure. By $x^c = y^d$ we denote the formula $x = y$ if $c = d$, and $\bot$ if $c \neq d$. Therefore, by $x^c \neq y^d$ we denote the formula $x \neq y$ if $c = d$, and $\top$ if $c \neq d$.

It is also convenient to define the output successor relation. For all $c, d \in \{1, \dots, k\}$, we let

$$\phi^{c,d}_S(x,y) \equiv \phi^{c,d}_{\preceq}(x,y) \wedge x^c \neq y^d \wedge \forall z^e \cdot (\phi^{c,e}_{\preceq}(x,z) \wedge \phi^{e,d}_{\preceq}(z,y) \rightarrow z^e = x^c \vee z^e = y^d)$$

Finally, we can construct the expected formula:

$$\begin{array}{rll} \text{is\_word}_T \equiv \phi_{dom} \rightarrow & & \\ & (1) & [\forall x^c] \cdot \bigwedge_{a\neq b\in\Sigma} \neg\phi^i_{L_a}(x) \vee \neg\phi^i_{L_b}(x) \\ \wedge & (2) & [\forall x^c] \cdot \bigvee_{a\in\Sigma} \phi^c_{L_a}(x) \\ \wedge & (3) & [\forall x^c \forall y^d \forall z^e] \cdot (\phi^{c,d}_S(x,y) \wedge \phi^{c,e}_S(x,z) \rightarrow y^d = z^e) \\ \wedge & (4) & [\forall x^c \forall y^d \forall z^e] \cdot (\phi^{d,c}_S(y,x) \wedge \phi^{e,c}_S(z,x) \rightarrow y^d = z^e) \\ \wedge & (5) & [\forall x^c \forall y^d] \cdot (x^c \neq y^d) \rightarrow ([\exists z^e] \cdot \phi^{e,c}_S(z,x) \vee [\exists z^e] \cdot \phi^{e,d}_S(z,y)) \\ \wedge & (6) & [\exists x^c \forall y^d] \cdot \neg\phi^{d,c}_S(y,x) \end{array}$$

Subformula (1) ensures that each output node is labeled by at most one letter. Subformula (2) ensures that each output node is labeled by at least one letter. Subformula (3) ensures that the output successor relation is a function. Subformula (4) ensures that the inverse of the output successor relation is a function. Finally, Subformulas (5) and (6) ensures that there is exactly one output node without predecessor. □

In the rest of this section, by MSO-transducer and MSOT we always mean an MSO-transducer of finite words.

FO-transducers of finite words. An FO-transducer (FOT) T is defined as an MSO-transducer, except that each formula of T is an FO formula over $\mathcal{S}_\Sigma$.

Definability. We say that a transformation R of finite words is definable by a logical transducer T if $R = [\![T]\!]$. We say that R is MSOT-definable (resp. FOT-definable) if it is definable by an MSO-transducer (resp. FO-transducer) of finite words.

3.4 Logical Transducers: Examples

In this section, we give several examples of transformations that can be defined by MSO-transducers.

Example 3. We show that all transformations of Example 1 but f_{exp} are MSOT-definable. They are illustrated in Fig. 1. Only the successor relations are depicted. Input nodes filtered out by formulas $\phi^c_{pos}(x)$ are represented by fuzzy nodes.
• The transformation f_{del} on $\Sigma = \{a, b\}$ is definable by the transducer

$$T_{del}{=}(1, \phi_{dom} \equiv \top, \phi^1_{pos}(x) \equiv \neg L_a(x), (\phi^1_{L_\sigma}(x) \equiv L_\sigma(x))_{\sigma\in\Sigma}, \phi^{1,1}_{\preceq}(x,y) \equiv x \preceq y)$$

Given an input word $u \in \Sigma^*$, let $v \in \Sigma^*$ such that $\tilde{v} = [\![T_{del}]\!](\tilde{u})$. Then $dom(\tilde{v}) = \{i^1 \in \text{dom}(u) \mid u(i) = b\}$, as defined by ϕ^1_{pos}, and $\preceq^{\tilde{v}} = \{(i^1, j^1) \mid i^1, j^1 \in \text{dom}(\tilde{v}), i \leq j\}$.
• To define transformation f_{double}, one needs to take two copies of the input structure. It is defined by the transducer T_{double} with $k = 2$ and for $i \in \{1, 2\}$:

$$\begin{array}{llll}\phi_{dom} \equiv \top & \phi^i_{pos}(x) \equiv \top & \phi^i_{L_a}(x) = L_a(x) & \phi^i_{L_b}(x) = L_b(x) \\ \phi^{1,1}_{\preceq}(x,y) \equiv x \preceq y & \phi^{1,2}_{\preceq}(x,y) \equiv x \preceq y & \phi^{2,1}_{\preceq}(x,y) \equiv x \prec y & \phi^{2,2}_{\preceq}(x,y) \equiv x \preceq y\end{array}$$

Note that the output predicate $\preceq$ from copy 2 to copy 1 is only defined when x occurs strictly before y. It implies that an output node y^d is a successor of x^d iff one of the two following conditions hold: (i) $c = 1$ and $d = 2$ and $x = y$, or (ii) $c = 2$ and $d = 1$ and y is a successor of x in the input word. If one wants to restrict the domain of T_{double} to words in a^*, it suffices to define the domain formula by $\phi_{dom} \equiv \forall x \cdot L_a(x)$.
• Let us consider transformation f_{copy}. Again, one needs two copies of the input structure. It is similar to T_{double} except the way the output order is defined:

$$\begin{array}{llll}\phi_{dom} \equiv \top & \phi^i_{pos}(x) \equiv \top & \phi^i_{L_a}(x) = L_a(x) & \phi^i_{L_b}(x) = L_b(x) \\ \phi^{1,1}_{\preceq}(x,y) \equiv x \preceq y & \phi^{1,2}_{\preceq}(x,y) \equiv x \preceq y & \phi^{2,1}_{\preceq}(x,y) \equiv \bot & \phi^{2,2}_{\preceq}(x,y) \equiv x \preceq y\end{array}$$

Note that compared to T_{double}, only the definition of $\phi^{2,1}_{\preceq}$ differs. We indeed completely disallow a node from copy 2 to be smaller than a node from copy 1.

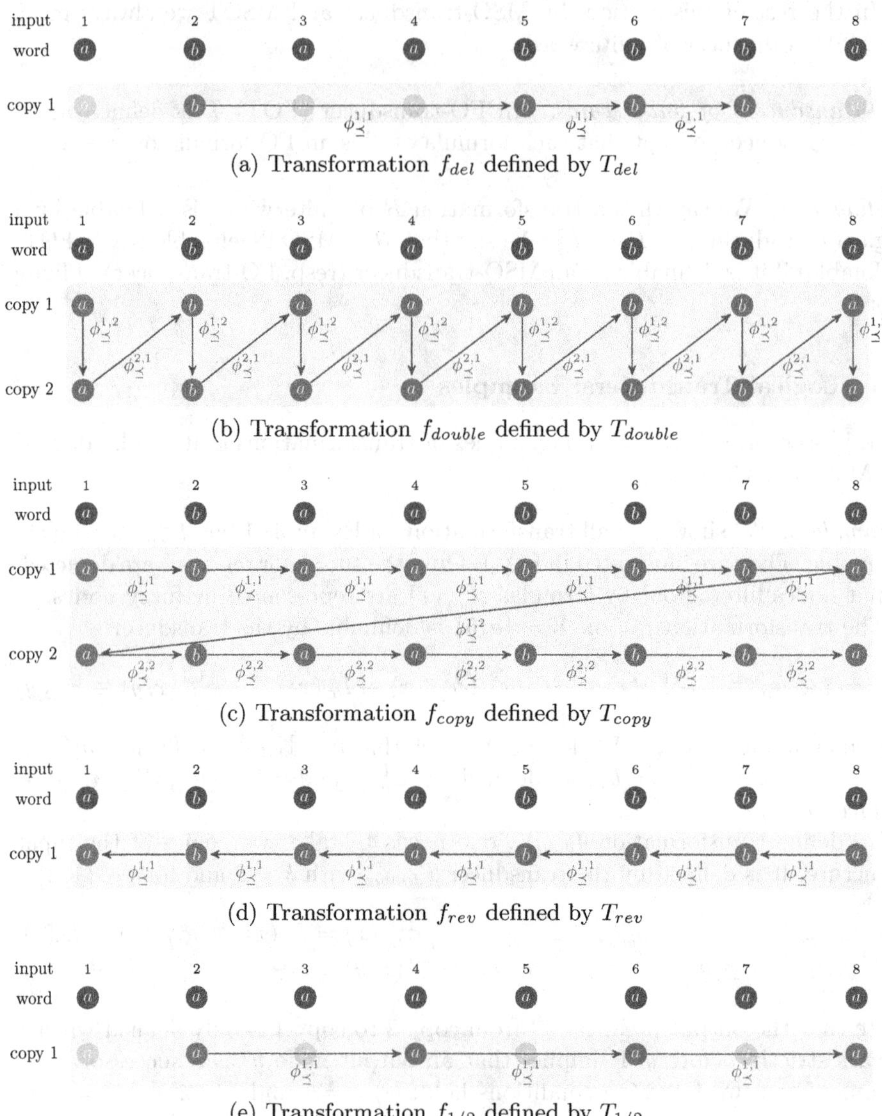

(a) Transformation f_{del} defined by T_{del}

(b) Transformation f_{double} defined by T_{double}

(c) Transformation f_{copy} defined by T_{copy}

(d) Transformation f_{rev} defined by T_{rev}

(e) Transformation $f_{1/2}$ defined by $T_{1/2}$

Fig. 1. Transformations of Example 1 defined by MSO-transducers

• The transformation f_{rev} is defined by the transducer T_{rev}: it suffices to take only one copy of the input structure and to inverse the order relation. Formally, T_{rev} is defined by $k = 1$ and:

$$\phi_{dom} \equiv \top \quad \phi^i_{pos}(x) \equiv \top \quad \phi^1_{L_a}(x) = L_a(x) \quad \phi^1_{L_b}(x) = L_b(x) \quad \phi^{1,1}_{\preceq}(x, y) \equiv y \preceq x$$

• To define with an MSO-transducer the transformation $f_{1/2} : a^n \mapsto a^{\lfloor n/2 \rfloor}$, one takes one copy of the input domain, sets the domain formula to $\phi_{dom} \equiv \forall x{\cdot}L_a(x)$, and filters out all odd positions of the input word, which is possible by the MSO formula $\phi^1_{pos}(x) \equiv \exists X_o \exists X_e \cdot \phi_{o/e}(X_o, X_e) \wedge x \in X_e$, where $\phi_{o/e}$ has been defined in Example 2. Finally, the order relation is just defined by $\phi^{1,1}_{\preceq}(x, y) \equiv x \preceq y$.
• The transformation $f_{exp} : a^n \mapsto a^{2^n}$ is not MSOT-definable, because it is not linear-size increase, while MSOT-definable transformations are, by Remark 1.

Remark 2. Let us mention an other, logic-based, transformation formalism, called *first-order translations*, that has been introduced by N. Immerman in [30], as a way to define reductions between problems. In first-order translations, the domain of the output structure is a set of k-tuples of elements of the input domain, for some k. It is defined by a first-order formula with k free variables. Predicates of arity n of the output structure are defined, similarly to Courcelle's logical transducers, by formulas with kn free variables, interpreted over the input structure. In contrast to logical transducers which are linear-size increase, first-order translations can map a structure to a polynomially larger output structure. First-order translations have been introduced as a logical way to define reductions between decision problems and nothing is known about their expressiveness as a formalism to define transformations. In this paper, we rather focus on (Courcelle) logical transducers, for which connections with state-based formalisms have been established. Nevertheless, let us mention the two papers [32] and [34], where the particular case of length-preserving FO-translations with $k = 1$ has been studied, as well as their connections with finite state transducers. See Section 5.2 for more details.

4 State-Based Models for Word Transformations

In this section, we introduce some of the main state-based models for defining (finite) word transformations for which connections with logics are known. These models are automata models extended with outputs, and are usually called *transducers*. We present three models: finite state transducers, two-way finite state transducers, and streaming string transducers.

4.1 Finite State Transducers

Finite state transducers (FST) extend finite state automata with partial output words on their transitions. Whenever an FST reads an input letter, it moves deterministically to the next state and appends a word to the output tape. Formally, an FST on an alphabet Σ is a tuple $T = (Q, q_0, F, \delta)$ such that Q is a finite set of states, $q_0 \in Q$ is the initial state, $F \subseteq Q$ is a set of accepting states, and $\delta : Q \times \Sigma \to Q \times \Sigma^*$ is the transition function.

A *run* of T is a sequence $r = p_0\sigma_1 p_1 \dots \sigma_n p_n \in (Q\Sigma)^*Q$ such that $p_0 = q_0$ and for all $i \in \{1, \dots, n\}$, there exists $v_i \in \Sigma^*$ such that $\delta(p_{i-1}, \sigma_i) = (p_i, v_i)$. Given $u \in \Sigma^*$, one says that r is a *run on* u if $u = \sigma_1 \dots \sigma_n$. The output of r,

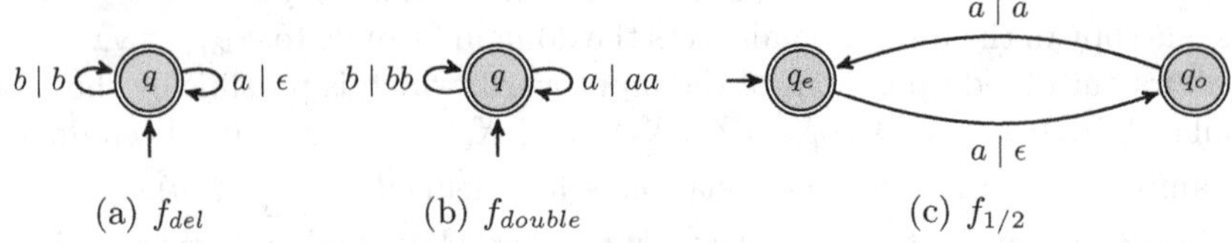

Fig. 2. Examples of finite state transducers

denoted by $O(r)$, is defined as the word $O(r) = v_1 \dots v_n$. The run r is *accepting* if $p_n \in F$.

An FST T realises a functional transformation $[\![T]\!] : \Sigma^* \rightarrow \Sigma^*$ defined by

$$[\![T]\!] = \{(u, v) \mid \text{there exists an accepting run } r \text{ of } T \text{ on } u \text{ such that } v = O(r)\}$$

Note that indeed, since T is deterministic, $[\![T]\!]$ is a function. The extension of FST with non-determinism allows one to define relations instead of functions.

A *non-deterministic finite state transducer* (NFT) over an alphabet Σ is a tuple $T = (Q, q_0, F, \Delta)$ where Q, q_0, F are defined as for FST, and $\Delta : Q \times \Sigma \times Q \rightarrow \Sigma^*$ is a (partial) function that defines the transitions[2]. A run of T is defined similarly as a run of an FST, as a sequence $r = p_0\sigma_1 p_1 \dots \sigma_n p_n$ such that $p_0 = q_0$ and for all $i \in \{1, \dots, n\}$, $\delta(p_{i-1}, \sigma_i, p_i)$ exists and is equal to some $v_i \in \Sigma^*$. The output $O(r)$ is defined by $O(r) = v_1 \dots v_n$. The other notions defined for FST carry over to NFT. Note that $[\![T]\!]$ may not be a function, since there can be several accepting runs on an input word. However, whether an NFT defines a function is decidable in PTime (see, for instance, [7]). NFT defining functions are known as *functional NFT*.

Example 4. Fig. 2 illustrates three FST that define the functions f_{del}, f_{double} and $f_{1/2}$ respectively. On these figures, the vertical arrow represents the initial state, the double circles the accepting states, and the arrows labelled $\sigma \mid v$ the transitions that read $\sigma \in \Sigma$ and produce $v \in \Sigma^*$. The other functions, f_{copy}, f_{rev} and f_{exp} are not definable by finite state transducers (even NFT). As we will see in Section 5.3, any NFT-definable functional transformation is definable by an MSO-transducer. We define in Section 5.3 a restriction on MSO-transducer that captures exactly NFT-definable functions.

4.2 Two-Way Finite State Transducers

Two-way finite state transducers (2FST) extend (one-way) finite state transducer with a bidirectional input head. Depending on the current state and letter, a 2FST updates its internal state and moves its input head either left or right. In order to detect the first and last positions of the input word, 2FST are assumed to run on words that are nested with begin and end markers $\vdash$, $\dashv$ respectively.

[2] These NFT are sometimes called *real-time* NFT, in contrast to a more general class of NFT that allow productive ϵ-transitions.

Formally, a *two-way finite state transducer (2FST)* over an alphabet Σ is a tuple $T = (Q, q_0, \delta, \delta_{halt})$ where Q is a finite set of states, $q_0 \in Q$ is the initial state, and δ is the transition relation[3], of type $\delta : Q \times (\Sigma \cup \{\vdash, \dashv\}) \to Q \times \{+1, -1\} \times \Sigma^*$, such that $\delta(q, \vdash) \in Q \times \{+1\} \times \Sigma^*$, and $\delta(q, \dashv) \in Q \times \{-1\} \times \Sigma^*$, for all $q \in Q$. Finally, δ_{halt} is the halting function, of type $\delta_{halt} : Q \times (\Sigma \cup \{\vdash, \dashv\}) \to \Sigma^*$. In order to ensure determinism, it is required that $\mathrm{dom}(\delta) \cap \mathrm{dom}(\delta_{halt}) = \varnothing$.

In order to see how a word $u \in \Sigma^*$ is evaluated by T, it is convenient to see the input as a tape containing $\vdash u \dashv$. Initially the head of T is on the first cell in state q_0 (the cell labelled $\vdash$). When T reads an input symbol, depending on the transitions in δ, its head moves to the left (-1) if the head was not in the first cell, or to the right ($+1$) if the head was not in the last cell, then it updates its state, and appends a partial output word to the final output. T stops as soon as it can apply the halting transition δ_{halt}, and produces a last partial output word.

A *configuration* of T is a pair $(q, i) \in Q \times \mathbb{N}$ where q is a state and i is a position on the input tape. A *run* r of T is a finite sequence of configurations. Let $u = \sigma_1 \dots \sigma_n \in \Sigma^*$, let $\sigma_0 = \vdash$ and let $\sigma_{n+1} = \dashv$. A run $r = (p_1, i_0) \dots (p_m, i_m)$ is *accepting on* u if (*i*) $p_1 = q_0$, $i_0 = 0$; (*ii*) $\delta_{halt}(p_m, \sigma_{i_m})$ is defined and equal to v_m for some $v_m \in \Sigma^*$; (*iii*) for all $k \in \{0, \dots, m-1\}$, $\delta(p_k, \sigma_{i_k})$ is defined and equal to $(p_{k+1}, i_{k+1} - i_k, v_k)$ for some $v_k \in \Sigma^*$. The *output* of r is defined by $O(r) = v_1 \dots v_m$. Like FST, the (functional) transformation defined by T, denoted by $[\![T]\!]$, is the set of pairs (u, v) such that there exists an accepting run r of T on u such that $O(r) = v$.

Example 5. Unlike FST, 2FST can define the functions f_{copy} and f_{rev}, as shown in Fig. 3. Therefore, there are strictly more expressive than FST. However, checking whether a 2FST is equivalent to some FST is decidable [25]. 2FST define linear-size increase transformations, because it can be proved that due to determinism, the number of times an input position can be visited is in $O(|Q|)$. Therefore, f_{exp} is not 2FST-definable.

4.3 Streaming String Transducers

Recently, an appealing transducer model for word transformations, whose expressiveness is exactly the same as 2FST, has been proposed in [2], as an extension of FST with registers, called *streaming string transducers (SST)*. Partial output words are stored in a fixed number of registers that can be concurrently updated and combined in different ways to define the output word. Moreover, SST are deterministic and, unlike 2FST, are one-way (left-to-right), making them easier to manipulate algorithmically. It is has been applied, for instance, to the automatic verification of some important classes of list-processing programs [1].

[3] In the literature, some definitions also include stay transition, i.e. transitions where the input head does not move. In the deterministic case, these transitions can however be removed without loss of expressiveness.

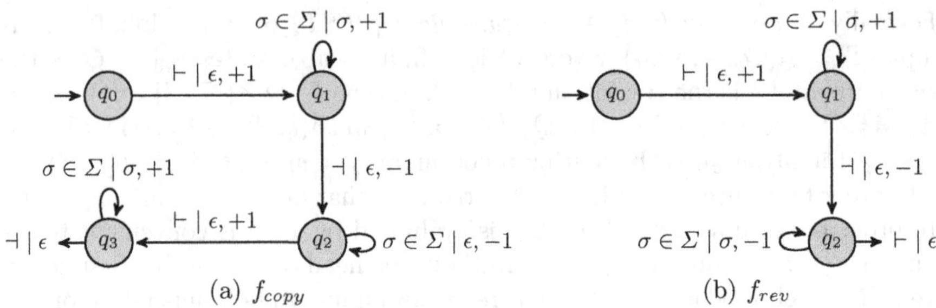

(a) f_{copy} (b) f_{rev}

Fig. 3. Examples of two-way finite state transducers

Let $\mathcal{X}$ be a finite set of registers denoted by the capital letters $U, V, W \dots$ and Σ be an alphabet. A substitution s is defined as a mapping $s : \mathcal{X} \to (\Sigma \cup \mathcal{X})^*$. A valuation is defined as a substitution $v : \mathcal{X} \to \Sigma^*$. Let $\mathcal{S}_{\mathcal{X},\Sigma}$ be the set of all substitutions. Any substitution s can be extended to $\hat{s} : (\Sigma \cup \mathcal{X})^* \to (\Sigma \cup \mathcal{X})^*$ in a straightforward manner. The composition $s_1 s_2$ of two substitutions s_1 and s_2 is defined as the standard function composition $\hat{s_1} \circ s_2$.

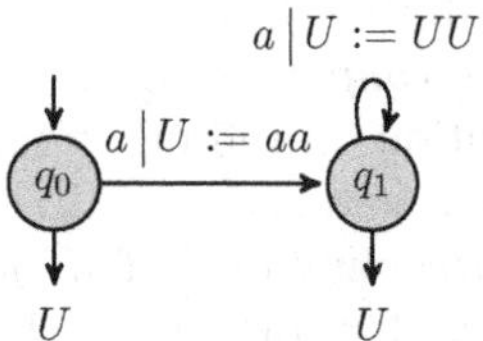

Fig. 4. SST for f_{exp}

A *streaming string transducer* (SST) over Σ is a tuple $T = (Q, q_0, \delta, \mathcal{X}, \rho, O)$ where Q is a finite set of states with initial state q_0; $\delta : Q \times \Sigma \to Q$ is a transition function; $\mathcal{X}$ is a finite set of registers; $\rho : \delta \to \mathcal{S}_{\mathcal{X},\Sigma}$ is a register update function; and $O : Q \rightharpoonup \mathcal{X}^*$ is a (partial) output function.

Like FST, a *run* r of an SST T is an alternating sequence of states and letters $r = p_0\sigma_1 p_1 \dots \sigma_n p_n$ such that $p_0 = q_0$ and for all $i \in \{0, \dots, n-1\}$, $\delta(p_i, \sigma_{i+1})$ is defined and equal to p_{i+1}. The run r is *accepting* if $p_n \in \mathrm{dom}(O)$. We let $|r| = n$ the length of r. In particular, a run of length 1 follows exactly one transition. The sequence $\langle s_{r,i} \rangle_{0 \le i \le |r|}$ of substitutions induced by r is defined inductively as: $s_{r,0}$ is the identity function over $\mathcal{X}$, and $s_{r,i} = s_{r,i-1}\rho(p_{i-1}, \sigma_i)$ for $1 \le i \le |r|$. We denote $s_{r,|r|}$ by s_r.

If the run r is accepting, we can extend the output function O to the run r by $O(r) = s_\epsilon s_r(O(p_n))$, where s_ϵ substitutes all registers by their initial value ϵ.

As for FST and 2FST, the functional transformation $[\![T]\!]$ defined by an SST T is the set of pairs (u, v) such that there exists an accepting run r of T on u such that $v = O(r)$.

Example 6. The definition of SST is best understood with some examples. Any FST can be encoded as an SST with a single register. As an example, consider the SST that defines the function $f_{1/2}$ in Fig. 5(a). It has only one register U. The register update substitutions are represented on the edges by the the assignment operator :=, while the output function is represented by the vertical arrows leading to an expression, here U. The function $f_{1/2}$ can also be defined with only one state, but using one additional register V, as depicted by

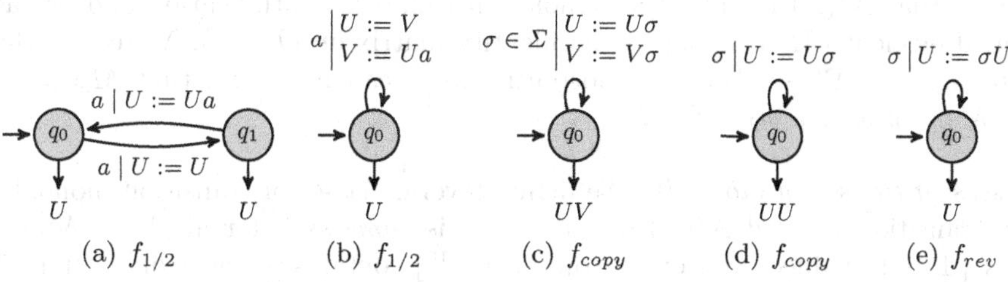

Fig. 5. Examples of streaming string transducers

Fig. 5(b). The function f_{copy} can be defined with one or two registers, as depicted by Fig. 5(c) and Fig. 5(d). Finally, f_{rev} is defined by the SST of Fig. 5(e). Unlike MSO-transformations, SST-definable transformations may not be linear-size increase, as shown by the SST of Fig.4 which defines the transformation f_{exp}. To capture MSO-transformations, various syntactic restrictions on SST register updates have been defined in several papers [2,3,4,6], which can be defined as restrictions of a uniform notion of *transition monoids* for SSTs [26], as presented in the next section.

4.4 Transition Monoids for Streaming String Transducers

The transition monoid of an SST is a set of matrices M_w, for all words $w \in \Sigma^*$, that represent the state and variable flow of the SST over w [26]. Let $T = (Q, q_0, \delta, \mathcal{X}, \rho, O)$ be an SST over an alphabet Σ, let two states $q_1, q_2 \in Q$, two registers $U_1, U_2 \in \mathcal{X}$, a word $w \in \Sigma^*$ and $n \in \mathbb{N} \cup \{\bot\}$. Intuitively, if $n \geq 0$, the pair (q_1, U_1) n-flows to (q_2, U_2) on reading w if there exists a run of T on w from q_1 to q_2, on which the sequence of register updates makes U_1 contributes n times to the content of U_2. The pair (q_1, U_1) $\bot$-flow to (q_2, U_2) if there is no run on w from q_1 to q_2. For example, for the SST of Fig. 5(a), (q_0, U) 1-flows to (q_1, U) on reading a, as well as a^k for all odd integers k. On Fig. 5(b), (q_0, U) 1-flows to (q_0, V) on reading a. On Fig. 4, (q_0, U) 2^k-flows to (q_0, U) on reading a^k for all $k \geq 0$.

Formally, (q_1, U_1) *n-flows to* (q_2, U_2) on w ($n \geq 0$), denoted $(q_1, U_1) \overset{w}{\underset{n}{\rightsquigarrow}} (q_2, U_2)$, if there exists a run r of T on w, from state q_1 to state q_2, such that U_1 occurs n times in $s_r(U_2)$, where s_r is the substitution defined by r. The pair (q_1, U_2) $\bot$-flows to (q_2, U_2) if there is no run of T on w, from state q_1 to state q_2. We denote by M_w the $(\mathbb{N} \cup \{\bot\})$-valued square matrices of dimension $|Q|.|\mathcal{X}|$ defined, for all $q_1, q_2 \in Q$ and all $U_1, U_2 \in \mathcal{X}$, and all $n \in \mathbb{N} \cup \{\bot\}$, by

$$M_w[q_1, U_1][q_2, U_2] = n \text{ iff } (q_1, U_1) \overset{w}{\underset{n}{\rightsquigarrow}} (q_2, U_2)$$

Definition 2 (SST Transition Monoid). *The transition monoid of an SST T is the set of matrices $\mathcal{M}(T) = \{M_w \mid w \in \Sigma^*\}$.*

Note that $\mathcal{M}(T)$ is indeed a monoid with matrix multiplication and M_ϵ as neutral element (M_ϵ is equal to the identity matrix on $Q \times \mathcal{X}$). Moreover, the mapping $w \in \Sigma^* \mapsto M_w$ is a morphism, as it can be shown that $M_{w_1 w_2} = M_{w_1} M_{w_2}$ for all $w_1, w_2 \in \Sigma^*$ [26].

Classes of transition monoids. We define several classes of transition monoids. The transition monoid $\mathcal{M}(T)$ of an SST T is *copyless* if for all $M \in \mathcal{M}(T)$, M is $\{\bot, 0, 1\}$-valued, and every row $M[q, U][.]$ contains at most one 1, for all $q \in Q$ and $U \in \mathcal{X}$. In other words, U can be copied in at most one other register (included itself). The monoid $\mathcal{M}(T)$ is *restricted copy* if all $M \in \mathcal{M}(T)$ is $\{\bot, 0, 1\}$-valued. In other words, a register can be copied more than once, but these copies must not be combined later on. Finally, we will also consider SST whose transition monoid is *finite*. The registers with a finite transition monoid can be copied, but not on loops. Note that any copyless transition monoid is restricted copy, and any restricted copy transition monoid is finite. It has been shown, as we will see, that the corresponding classes of SST are, however, of equal expressive power.

Several restrictions on register updates that have been defined in several papers, with ad-hoc definitions, are nicely captured by these simple classes of transition monoids. The copyless restriction of [2] corresponds to SST with copyless transition monoid. The restricted copy restriction of [6] corresponds to SST with restricted copy transition monoids. Finally, the bounded copy restriction of [3] corresponds to SST with finite transition monoid.

5 Automata-Logic Connections for Word Transformations

In this section, we present the main known automata-logic connections for word transformations.

5.1 MSO Transformations

The first automata-logic connection for word transformations has been discovered by J. Engelfriet and H.J. Hoogeboom [21]:

Theorem 1 (J. Engelfriet, H.J. Hoogeboom [21]). *Let $f : \Sigma^* \to \Sigma^*$. The function f is MSOT-definable iff it is 2FST-definable.*

The proof of MSOT⇒2FST is based on intermediate models of 2FST which can perform "MSO jumps" $\phi(x, y)$, where $\phi(x, y)$ is an MSO formula that defines a function from x positions to y positions. Intuitively, the machine can move from position x to position y providing $\phi(x, y)$ holds true. 2FST with MSO jumps are then converted, based on Büchi-Elgot-Trakhtenbrot's theorem, into 2FST with regular look-around. These 2FST can move only to positions which are in the 1-neighbourhood of the current position, but their move can be based on a regular property of the current prefix and suffix of the word. Finally, it is shown that 2FST with regular look-around are equivalent to 2FST.

The converse, 2FST⇒MSOT, is shown by first constructing an MSOT that takes as input a word structure, and output an edge-labeled graph (edges are labelled by words of bounded length) whose nodes are exactly the configurations (q, i) in which the 2FST is successively (where q is a state and i a position in the input word). This MSOT is then composed with an MSOT that transforms an edge-labeled graph into a node-labeled graph (nodes are here labelled with letters from Σ). The result follows as MSOT are closed under composition [15,14].

As we have seen in Fig. 4, SST can define functions which are exponential-size increase (f_{exp}), and therefore, not MSOT-definable. However, any MSOT-definable transformation is SST-definable, and, by weakening the expressiveness of SST, it is possible to capture exactly MSOT:

Theorem 2 (R. Alur, P. Černý [2], R. Alur, E. Filiot, A. Trivedi [3]). *Let $f : \Sigma^* \to \Sigma^*$. The following statements are equivalent:*

1. *f is MSOT-definable.*
2. *f is definable by an SST with copyless transition monoid.*
3. *f is definable by an SST with finite transition monoid.*

The equivalence between (1) and (2) was shown in [2]. The proof of (1) ⇒ (2) of [2] goes through the intermediate model of 2FST and relies on Theorem 1. More precisely, it is shown that any 2FST can be encoded as an SST, by extending to transducers the classical Sheperdson's construction that transforms a two-way finite automaton as a (one-way) finite automaton [39]. The resulting SST may not be necessarily copyless, and the main challenge is to show that it can be converted into a copyless SST. Conversely, it is shown how to directly encode an SST as an MSO-transducer. In [3], it has been shown that SST with finite transition monoid (called bounded copy) are equivalent to SST with copyless transition monoid. Although the proof of (1) ⇒ (2) in [2] relies on 2FST, a direct construction was also given in [6], in which the states of the SST are MSO-types of bounded quantifier rank.

5.2 FO Transformations

Recall that first-order transformations are transformations definable by FO-transducers, which are defined as MSO-transducers, except that only first-order formulas can be used. In automata theory for languages, first-order definable languages are captured by *aperiodic automata*, i.e. finite automata whose transition monoid is aperiodic [40,18]. A survey on first-order definable languages can be found in [18]. A monoid $(M, \cdot, e)$ (with operator $\cdot$ and neutral element e) is *aperiodic* if there exists $k \in \mathbb{N}$ such that for all $m \in M$, $m^k = m^{k+1}$. Recall that the language $(aa)^*$ is not FO-definable, while the language $(ab)^*$ is.

As we have seen, the functions f_{del}, f_{double}, f_{copy}, and f_{rev} are FOT-definable. However, the function $f_{1/2}$ is not FOT-definable [26], although its domain, a^*, is. So clearly, the FOT-definability of a function not only depends on the FO-definability of its domain. It can be seen on an example. In Fig. 5(a), the transition monoid of the underlying automaton of the SST defining $f_{1/2}$

(the automaton obtained by dropping the register updates) is not aperiodic. Therefore, in order to get FOT-definable functions, a first restriction would be to require that the underyling automaton of an SST is aperiodic. However, it is not sufficient, as shown by the SST of Fig. 5(b), whose underlying automaton has aperiodic transition monoid. However, the register flow, which alternates between U and V registers on reading a, is not aperiodic. If one requires that the transition monoid of an SST, which also speaks about the register flow, is aperiodic, then one gets exactly FOT-definable functions:

Theorem 3 (E. Filiot, K. Shankara Narayanan and A. Trivedi [26]). *Let $f : \Sigma^* \to \Sigma^*$. The function f is FOT-definable iff it is definable by an SST with aperiodic and restricted copy transition monoid.*

This theorem should not be understood as an effective characterization of FOT-definable functions. Indeed, it could be the case that an SST which defines an FOT-definable function has not an aperiodic transition monoid. As an example, consider an SST which alternates on reading the letter a between two states, both accepting, and realizes the identity function with a single register. Its transition monoid is not aperiodic, but the function it defines if FOT-definable. It is also the case for automata: a non-aperiodic automaton may define a first-order language. However for automata, FO-definable languages can be algebraically characterised, as show by M.P. Schützenberger : a language L is FO-definable iff its syntactic monoid is aperiodic [38]. This characterisation is effective if L is given as a finite automaton, and decidable in PSPace [18].

Finally, like for automata, deciding whether the transition monoid of an SST is aperiodic and restricted copy is PSPace-C [26].

FO-translations in one-free variable Let us mention another transducer-logic connection that has been shown for a less expressive class of functions, namely the FO-translations of [30] restricted to FO-formulas in one-free variable and length-preserving functions. Such a translation assumes a total order $<$ on Σ and is defined by a tuple T of FO-formulas in one-free variable x, say $T = (\phi_\sigma(x))_{\sigma\in\Sigma}$,

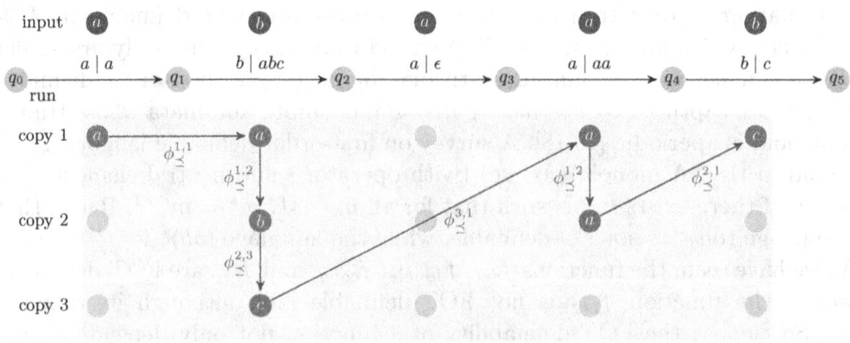

Fig. 6. Encoding of NFT by MSO-transducer

such that $\phi_{\sigma_m}(x) = \bigwedge_{\sigma<\sigma_m} \neg\phi_\sigma(x)$, where σ_m is the maximal element of Σ. Then, for all $w \in \Sigma^*$ of length n, $T(w) = \sigma_1 \dots \sigma_n$ such that for all $i \in \{1,\dots,n\}$, $w \models \phi_{\sigma_i}(i) \wedge \bigwedge_{\sigma<\sigma_i} \neg\phi_\sigma(i)$. Note that T defines a total and length-preserving function, which is actually definable by a one-copy (Courcelle) FO-transducer. It is shown in [32] that a function is definable by an FO-translation in one-free variable iff it is definable by an aperiodic NFT, where aperiodic NFT are the NFT whose underlying automata is aperiodic (i.e. they have aperiodic transition monoid). This result was later on, in [34], generalised to V-translations and NFT whose transition monoid is in V, where V is a pseudovariety of (finite) monoids (for instance, the pseudovariety of finite aperiodic monoids).

5.3 Order-Preserving MSO Transducers

An MSO-transducer T with k copies is *order-preserving* if for all words $w \in \mathrm{dom}(T)$, all positions $i,j \in \mathrm{dom}(w)$, and copies $c,d \in \{1,\dots,k\}$, if $w \models \phi_{\preceq}^{c,d}(i,j)$, then $i \preceq j$ holds true. Note that this can be syntactically ensured by requiring that formulas $\phi_{\preceq}^{c,d}(x,y)$ are of the form $x \preceq y \wedge \phi$. We show in Theorem 4 that order-preserving MSO-transducers characterise exactly the NFT-definable functions. This theorem can also be obtained as a consequence of a result shown in [9] for order-preserving transformations with origin semantics. However, we give here a direct proof, which illustrates the techniques that are usually used to derive automata-logic connections. Origin semantics is discussed later in Section 5.4.

Theorem 4. *A function $f : \Sigma^* \to \Sigma^*$ is NFT-definable iff it is definable by an order-preserving MSO-transducer.*

Proof. We first show the "only if" direction. Let $T = (Q, q_0, F, \Delta)$ be an NFT over Σ that defines a f. We construct an order-preserving MSO-transducer T' such that $[\![T]\!] = [\![T']\!] = f$. Since T defines a function, it is known that T is equivalent to an unambiguous NFT [31], i.e. an NFT such that there exists at most one accepting run on every input word. Therefore, we assume that T is unambiguous.

Let $K \in \mathbb{N}$ be the maximal length of the words occurring on the transitions of T, i.e. $K = \max\{|w| \mid \exists p,q \in Q \exists\sigma \in \Sigma,\ \Delta(p,\sigma,q) = w\}$. Clearly, for all words $u \in \mathrm{dom}(T)$, $|[\![T]\!](u)| \leq K.|u|$. In order to define $[\![T]\!]$ by an MSO-transducer, one needs to take K copies of the input structure. Let us intuitively explain how these copies will be used. Let $u = \sigma_1 \dots \sigma_n \in \mathrm{dom}(T)$ and $r = q_0\sigma_1q_1\dots\sigma_nq_n$ be **the** accepting run of T on u, and let $O(r) = v_1 \dots v_n$, where each word v_i is produced by the i-th transition of r. Every position $j \in \mathrm{dom}(v_i)$ will be encoded by the j-th copy of the input position i, i.e., by the node i^j. Of course, the j-th copy exists since $|v_i| \leq K$. For all j such that $|v_i| < j \leq K$, one needs to filter out all nodes i^j, by the formula $\phi_{pos}^j(x)$. The encoding is illustrated on Fig. 6.

The definition of T' relies on the existence of MSO formulas $\phi_t(x)$, for all tuples $t = (p,\sigma,v,q) \in Q \times \Sigma \times \Sigma^* \times Q$, such that for all words $u \in \Sigma^*$ and all positions $i \in \mathrm{dom}(u)$, $u \models \phi_t(i)$ iff $u \in \mathrm{dom}(T)$ and the run of T on u, say

$r = q_0\sigma_1 q_1 \dots \sigma_n q_n$, is such that $q_{i-1} = p$, $\sigma_i = \sigma$, $q_i = q$ and $\Delta(p,\sigma,q) = v$. We do not prove the existence of such formulas in this paper, it is obtained as a direct consequence of Büchi-Elgot-Trakhtenbrot's Theorem. Another direct consequence of this theorem is the existence of an MSO formula $\phi_{\mathrm{dom}(T)}$ which defines the domain of T (which is a regular language).

We define formally $T' = (k, \phi_{dom}, (\phi^c_{pos})_{1\leq c\leq k}, (\phi^c_{L_\sigma})_{1\leq c\leq k,\sigma\in\Sigma}, (\phi^{c,d}_{\preceq})_{1\leq c,d\leq k})$:

$$
\begin{array}{ll}
k & = K \\
\phi_{dom} & \equiv \phi_{\mathrm{dom}(T)} \\
\phi^c_{pos}(x) & \equiv \bigwedge\{\phi_t(x) \rightarrow |v| \geq c \mid t := (p,\sigma,v,q) \text{ s.t. } \Delta(p,\sigma,q) = v\} \\
\phi^c_{L_\sigma}(x) & \equiv \bigwedge\{\phi_t(x) \rightarrow (c \leq |v| \wedge v(c) = \sigma) \mid t := (p,\sigma,v,q) \text{ s.t. } \Delta(p,\sigma,q) = v\} \\
\phi^{c,d}_{\preceq}(x,y) & \equiv (x = y \wedge c \leq d) \vee x \prec y
\end{array}
$$

where the subformulas $|v| \geq c$, $c \leq |v| \wedge v(c) = \sigma$ and $c \leq d$ (which do not actually belong to MSO syntax) are either defined by $\top$ or $\bot$ depending on whether the Boolean expressions they represent hold true or not. Finally, note that T' is indeed order-preserving since clearly, if $u \models \phi^{c,d}_{\preceq}(i,j)$, then $i \leq j$.

Conversely, given an order-preserving MSO-transducer T, one shows how to construct an equivalent NFT T'. We start by a first observation. Let $u = \sigma_1 \dots \sigma_n \in \mathrm{dom}(T)$ and $v = [\![T]\!](u)$. Since T is order-preserving, there is no backward edges in the output word structures produced by T. Therefore the word v can be decomposed into $v_1 \dots v_n$ such that each word v_i is the subword of v induced by the non-filtered copies of the input position i, i.e. the set $N_i = \{i^1, \dots, i^k\} \cap \{i^j \mid u \models \phi^j_{pos}(i)\}$. It can be seen on Fig. 6: each vertical block of the output structure correspond to a partial output word, and the result output word of the transformation is obtained by concatenating, in order, these partial words. Let Σ^*_k be all the words over Σ of length at most k. For all $w \in \Sigma^*_k$, one can define an MSO formula $\Psi_w(x)$ such that $u \models \Psi_w(i)$ iff $u \in \mathrm{dom}(T)$ implies that $w = v_i$, where v_i is defined by the decomposition explained before. Note that the definition of order-preservation does not require that copies are ordered according to $\leq$ (on integers), i.e., whenever formula $\phi^{c,d}_{\preceq}(x,y)$ holds, it implies that $x \preceq y$, but not necessarily $c \leq d$, unlike the example of Fig. 6. In other words, it could be that the first position of v_i corresponds to node i^2 while the second position corresponds to node i^1 of the output structure. Therefore, in order to define $\Psi_w(x)$, one has to quantify, by a huge disjunction, over possible orders over these copies. Formally, for all tuples $t = (j_1, \dots, j_{|w|})$ such that $j_\ell \in \{1, \dots, k\}$ for all $\ell \in \{1, \dots, |w|\}$, and $j_{\ell_1} \neq j_{\ell_2}$ for all $\ell_1 \neq \ell_2 \in \{1, \dots, |w|\}$, we define an intermediate formula $\Psi^{(j_1,\dots,j_{|w|})}_w(x)$ which holds true whenever the x-th partial output word is w, and each position of w correspond to nodes x^{j_ℓ}. Then, $\Psi_w(x)$ is obtained as the disjunction of the formulas $\Psi^{(j_1,\dots,j_{|w|})}_w(x)$ for all tuples:

$$
\begin{array}{l}
\Psi^t_w(x) \equiv \bigwedge_{\ell=1}^{|w|} \phi^{j_\ell}_{pos}(x) \wedge \phi^{j_\ell}_{L_{w(\ell)}}(x) \wedge \bigwedge_{j\in\{1,\dots,k\}\setminus\{j_1,\dots,j_{|w|}\}} \neg\phi^j_{pos}(x) \bigwedge_{\ell=1}^{|w|-1} \phi^{j_\ell,j_{\ell+1}}_{\preceq}(x,x) \\
\Psi_w(x) \equiv \bigvee\{\Psi^t_w(x) \mid t = (j_1,\dots,j_{|w|}) \in \{1,\dots,k\}^{|w|}, j_{\ell_1} \neq j_{\ell_2}, \forall \ell_1 \neq \ell_2\}
\end{array}
$$

In other words, for all $u \in \mathrm{dom}(T)$, we have $u \models \Psi_w(i)$ iff w is the i-the subword of the output of u. This formula is of particular interest if we want to define an NFT, because it tells us exactly what partial output word must be produced while reading the i-th input letter.

The rest of the proof explains how one can construct an automaton A on the alphabet $\Gamma = \Sigma \times \Sigma_k^*$ which accepts all words $(\sigma_1, v_1) \dots (\sigma_n, v_n) \in \Gamma^*$ such that for all $i \in \{1, \dots, n\}$, v_i is the (unique) word such that $\sigma_1 \dots \sigma_n \models \Psi_{v_i}(i)$. The final NFT T' is obtained from this automaton A by replacing every transition $(q, (\sigma, v), p)$ of A by the transition $q \xrightarrow{\sigma|v} p$ of T'. Clearly, T' is equivalent to T.

In order to construct A, we again apply Büchi-Elgot-Trakhtenbrot's theorem, since the language of A is MSO-definable, by the sentence (on $\mathcal{S}_\Gamma$)

$$\phi_A \equiv [\phi_{dom}]_\Gamma \wedge \forall x \cdot \bigwedge_{(\sigma,v)\in\Gamma} L_{(\sigma,v)}(x) \rightarrow [\Psi_v(x)]_\Gamma$$

where $[\phi_{dom}]_\Gamma$ (resp. $[\Psi_v(x)]_\Gamma$) is obtained from ϕ_{dom} (resp. $\Psi_v(x)$), which is on the signature $\mathcal{S}_\Sigma$, by replacing every atom $L_\beta(y)$ by $\bigvee_{w \in \Sigma_k^*} L_{(\beta,w)}(y)$. Clearly, a word $s = (\sigma_1, v_1) \dots (\sigma_n, v_n)$ over $\Sigma \times \Sigma_k^*$ satisfies $[\phi_{dom}]_\Gamma$ iff $u = \sigma_1 \dots \sigma_n \models \phi_{dom}$. Similarly, $s \models [\Psi_v(i)]_\Gamma$ iff $u \models \Psi_v(i)$, and therefore the correctness follows. □

Again, the power of transducer-logic connections is illustrated by the following definability problem. It is decidable whether a deterministic two-way finite state transducer T is equivalent to a (one-way) functional finite state transducer [25]. As a consequence of this result, the connection between 2FST and MSOT (Theorem 1), and the previous theorem (Theorem 4), we obtain the following corollary:

Corollary 1. *Given an MSO-transducer T, it is decidable whether T is equivalent to some order-preserving MSO-transducer.*

Proof. It suffices to translate T into a deterministic two-way transducer T', by applying the (effective) encoding of Theorem 1, then to apply the procedure of [25] which decides whether T' is equivalent some NFT, and finally to apply Theorem 4. □

5.4 Transformations with Origin

One of the difficulties in the theory of transducers is to deal with the asynchronicity in the production of the output. For instance, two FST may not produce there output letters at the same time when reading the same input position, but still, they can be equivalent. This asynchronicity, which is implicitly present in the Post correspondence problem (PCP), can even, in some cases, lead to undecidable problems. For instance, the equivalence problem of non-deterministic FST is undecidable, and the undecidability proof is non surprisingly based on PCP [28]. The asynchronicity between output words have been captured by a notion of *output delay*, which have been used, for instance, to make elegant proofs of the decidability of equivalence of functional FST [7].

Recently introduced by M. Bojanczyk, a stronger semantics have been given to transformations, that takes into account the origin of the output letters in the input word. Several classical problems have been revisited and most of them become trivial with this new semantics [9]. Some problems, still open in the classical semantics, have also been solved in the origin semantics, such as getting a machine-independent characterisation of MSO-transformations, as well as an effective characterisation of first-order transformations. We present some of these new results in this section.

Definition 3. *Let Σ be an alphabet, and $u, v \in \Sigma^*$. An origin function for v in u is a mapping $o : \mathrm{dom}(v) \to \mathrm{dom}(u)$. A word transformation with origin over Σ is a set of pairs $(u, (v, o))$ such that $u, v \in \Sigma^*$ and o is an origin function for v in u.*

Intuitively, $o(i)$ is the position in the input word at which the position i of the output word has been produced. Origin semantics can be defined for the transducer models we have seen so far. For instance, the origin transformation (or o-transformation) defined by an FST T is the set of pairs, denoted by $[\![T]\!]_o$, of the form $(u, (v, o))$ such that $(u, v) \in [\![T]\!]$, and for all $i \in \mathrm{dom}(v)$, $o(i)$ the position of u where T has produced $v(i)$, i.e. where T has triggered a transition on reading $u(o(i))$ which has produced a partial word containing the letter $v(i)$.

With this semantics, the origin equivalence problem for FST, i.e. deciding whether two FST T_1, T_2 satisfy $[\![T_1]\!]_o = [\![T_2]\!]_o$, can be easily solved, because T_1 and T_2 can be seen respectively as two automata A_1, A_2 over $\Sigma \times \Sigma_k^*$, where k is the longest output word occurring on the transitions of T_1 and T_2. Indeed, $[\![T_1]\!]_o = [\![T_2]\!]_o$ iff the automata A_1 and A_2 are equivalent, i.e., define the same language. This trick, however, cannot be used for two-way FST, making origin semantics more interesting in this context.

It is also possible to give a natural origin semantics to transformations realised by MSO-transducers T. In that case, $(u, (v, o)) \in [\![T]\!]_o$ if $(u, v) \in [\![T]\!]_o$ and, if i^c is the k-th node of v (wrt to $\preceq$), where $i \in \mathrm{dom}(u)$ and c is a copy, then we let $o(k) = i$.

In the origin world, it is possible to characterise algebraically first-order definable transformations with origin, while this problem is open in the classical setting. Moreover, this characterisation is effective when the transformation is defined by, say, an MSO-transducer.

Let $u \in \Sigma^*$, $o : \mathrm{dom}(u) \to \mathbb{N}$, and $X \subseteq \mathbb{N}$. The abstraction $u/_X$ of u by X is the word over $\Sigma_\bot := \Sigma \cup \{\bot\}$ obtained by replacing in u each letter at position $i \in \mathrm{dom}(u)$ by $\bot$ if $o(i) \in X$, where $\bot$ is a fresh symbol not in Σ. For instance, if $u = aba$ with origin $o(1) = 3$, $o(2) = 2$ and $o(3) = 1$, then $u/_{\{2,3\}} = \bot\bot a$.

We let $\sim_\bot$ the equivalence relation on $\Sigma_\bot^*$ induced by the equation $\bot = \bot\bot$. E.g. $\bot a \bot b \sim_\bot \bot\bot a \bot\bot\bot b$, but $a \not\sim_\bot \bot a$.

Given an o-transformation f, one defines its reverse $rev(f)$. For all $u \in \Sigma^*$, if $(v, o) = f(u)$, then $rev(f) = (f_{rev}(v), rev(o))$, where $rev(o)(i) = o(|v| - i + 1)$.

Similarly to the syntactic equivalence relation for languages, one can define *left-* and *right-* equivalence relations for transformations with origin f, resp. denoted by $\mathcal{L}^f$ and $\mathcal{R}^f$.

Definition 4. *Let f be a transformation with origin on Σ. Let $v_1, v_2 \in \Sigma^*$. Then, $v_1 \mathcal{L}^f v_2$ if for all $w \in \Sigma^*$:*

1. $v_1 w \in \mathrm{dom}(f)$ *iff* $v_2 w \in \mathrm{dom}(f)$
2. *if* $v_1 w \in \mathrm{dom}(f)$, *then* $f(v_1 w)/_{\mathrm{dom}(v_1)} \sim_\perp f(v_2 w)/_{\mathrm{dom}(v_2)}$.

Symmetrically, $\mathcal{R}^f = \mathcal{L}^{rev(f)}$.

Example 7. For instance, consider the o-transformation g that maps any $u \in \Sigma^*$ to $(f_{rev}(u)u, o)$, where, naturally, for all $i \in \{1, \ldots, |u|\}$, $o(i)=|u|-i+1$ and $o(|u|+i)=i$. Then, there are only two equivalence classes for both equivalence relation: $\{\epsilon\}$ and Σ^+. Indeed, for all $w \in \Sigma^*$,

$$f(v_1 w)/_{\mathrm{dom}(v_1)} = (f_{rev}(w) f_{rev}(v_1) v_1 w)/_{\mathrm{dom}(v_1)} = f_{rev}(w) \perp^{2|v_1|} w.$$

Therefore $f(v_1 w)/_{\mathrm{dom}(v_1)} \sim_\perp f_{rev}(w)w$ if $v_1 = \epsilon$ and $f(v_1 w)/_{\mathrm{dom}(v_1)} \sim_\perp f_{rev}(w) \perp w$ otherwise. A similar arguments applies for $\mathcal{R}^f$.

It is then possible to characterise MSOT-definable o-transformations by a Myhill-Nerode like theorem, and to give an effective characterisation of FOT-definable o-transformations.

Theorem 5 (M. Bojanczyk [9][4]). *Let f be a transformation with origin over Σ. The following two statements hold true:*

1. *f is MSOT-definable iff both $\mathcal{L}^f$ and $\mathcal{R}^f$ have finite index.*
2. *f is FOT-definable iff both $\mathcal{L}^f$ and $\mathcal{R}^f$ have finite index and for all $u \in \Sigma^*$, the $\mathcal{L}^f$-class of u and the $\mathcal{R}^f$-class of u are FO-definable languages.*

As shown in [9], if f is MSOT-definable, then the $\mathcal{L}^f$- and $\mathcal{R}^f$-equivalence classes are regular languages that can be effectively represented by automata. Characterization (2) is therefore effective, as it suffices to decide whether all equivalence classes are FO-definable, which is decidable [18].

6 Beyond Functional Finite Word Transformations

In this section, we discuss other transducer-logic connections for non-functional transformations and other structures (infinite words and trees).

6.1 Non-functional Finite Word Transformations

The state-based transducer models (FST, 2FST and SST) can all be extended with non-determinism (leading to the classes NFT, 2NFT, and NSST resp.), and rather define relations from word to words instead of functions. MSO-transducers can, as well, be extended with non-determinism.

[4] We adopt in this paper a slightly different, but equivalent, formalisation as in [9].

Non-deterministic state-based models. We have already seen in Section 4.1 how to extend FST with non-determinism. Similarly, 2FST can also be extended with non-determinism, the transition relation Δ being of type $\Delta \subseteq Q \times (\Sigma \cup \{\dashv, \vdash\}) \times Q \times \{+1, -1\} \times \Sigma^*$. Unlike FST, there can be infinitely many runs on the same input word, leading to infinitely many outputs for that word. For instance, 2NFT can loop an arbitrary number of times between two input positions and non-deterministically decide to apply the halting transition function. Their runs can even be infinite but we consider only finite runs to define the transformation.

SST can, as well, be extended with non-determinism, with transition relation $\Delta \subseteq Q \times \Sigma \times Q$. Unlike 2NFT, there is alway a finite number (exponential in the worst-case) of accepting runs on the same input word.

Non-deterministic MSO-transducers (NMSOT). MSO-transducers can be extended with non-determinism by allowing all the formulas (including the domain formula) to use a finite set of second-order of variables $X_1, \dots, X_n$. Given an input word u, the outputs of u depend on the valuations of these second order variables by subsets of $\mathrm{dom}(u)$. In particular, modulo a valuation of $X_1, \dots, X_n$ by subsets of $\mathrm{dom}(u)$, the transformation becomes functional, and the outputs of u are the set of output words defined for each valuation. For instance, it is possible to define with an NMSO-transducer T_{sub} the tranformation R_{sub} which maps any word u to all its subwords, using one second-order variable X and only one copy, as follows: $\phi_{dom} \equiv \top$, $\phi^1_{pos}(x, X) = x \in X$, $\phi^1_\sigma(x) = L_\sigma(x)$ for all $\sigma \in \Sigma$, and $\phi^{1,1}_{\preceq}(x, y) \equiv x \preceq y$. If one wants to restrict all the subwords to subwords generated by the even positions, it suffices to strengthen the domain formula to $\phi_{dom}(X) \equiv \exists Y \phi_{o/e}(Y, X)$, where $\phi_{o/e}$ has been defined in Example 2.

Transducer-Logic Connections. A transformation R is *finitary* if for all words $u \in \Sigma^*$, $R(u)$ is finite. It is clear that NFT, NSST and NMSOT define finitary transformations. However, 2NFT does not define, in general, finitary transformations. It turns out that 2NFT and NMSOT define incomparable classes of transformations.To capture exactly NMSOT with a two-way device, *Hennie machines* have been introduced in [21]. Hennie machine can rewrite their input tape, but each input position must be visited a constant number of times. On the one-way model side, it has been shown that NSTT corresponds exactly to NMSOT [5].

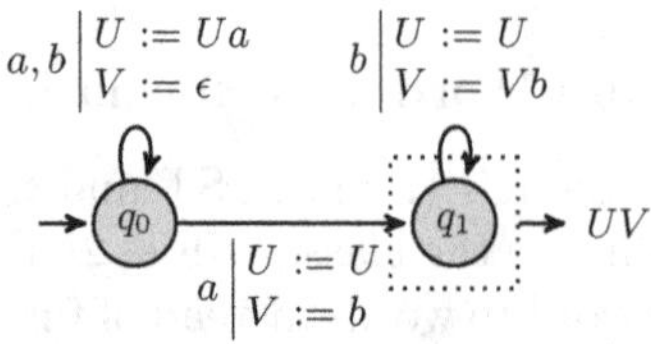

Fig. 7. Example of ω-SST defining f_ω, with output function $O(\{q_1\}) = UV$

6.2 Infinite Word Transformations

An *ω-word* over Σ is a mapping $w : \mathbb{N} \to \Sigma$. The i-th letter of w is $w(i)$. An infinite words w over Σ is either a finite word or an ω-word. The set of ω-words over Σ is denoted by Σ^ω, and the set of infinite words by Σ^∞. Note that $\Sigma^\infty = \Sigma^* \cup \Sigma^\omega$. A transformation R of infinite words over Σ is a binary relation from Σ^ω to Σ^∞ (we assume indeed in this section that the input word is an ω-word).

MSO-transducers can naturally be generalised to define functional infinite word transformations, by seeing an ω-word as a structure over $\mathcal{S}_\Sigma$ whose domain is $\mathbb{N}$. For an MSO-transducer T to be an MSO-transducer of infinite words, we require that for all input words $u \in \Sigma^\omega$, the image $[\![T]\!](u)$ is a structure that corresponds to an infinite word. Although it is a semantical restriction, it is decidable, similarly as in the proof of Proposition 1.

As an example, consider the transformation $f_\omega : \Sigma^\omega \to \Sigma^\omega$, where $\Sigma = \{a, b\}$, that maps any input word of the form uab^ω to $a^{|u|}b^\omega$, where $u \in \Sigma^*$. It is definable by the following MSO-transducer with one copy and $\phi_{dom} \equiv \exists x \cdot L_a(x)$,

$$\phi^1_{pos}(x) \equiv \top \quad \phi^1_a(x) \equiv \exists y \succ x \cdot L_a(y) \quad \phi^1_b(x) \equiv \neg\phi^1_a(x) \quad \phi^{1,1}_{\preceq}(x, y) \equiv x \preceq y$$

(Deterministic) 2FST can be extended to define functional infinite word transformations, by using for instance a Muller acceptance condition (they are called ω-2FST). However, they cannot even define f_ω, because they can never decide locally whether a b letter should be transformed into an a letter or kept unchanged, because it depends on the existence of an a letter in the future. They could use the two-wayness as a kind of look-ahead to check the existence of such an a, but they cannot come back exactly to the position they were coming from, because they get lost, due to the finite state device. One therefore needs to extend ω-2FST with regular look-ahead: each transition of an ω-2FST with regular look-ahead is extended with a finite state automaton over ω-words that checks a property of the (infinite) suffix. Such transition can be triggered only if the suffix belong to the look-ahead automaton. It should be clear that ω-2FST with regular look-ahead strictly extend the expressive power of ω-2FST.

SST have been extended to define functional infinite word transformations, with a Muller accepting condition (called ω-SST). They run on ω-words in a deterministic way, and the (partial) output function O has type $2^Q \to \mathcal{X}^*$. Given a run r over an ω-word u, let P the set of states visited infinitely many times in r. The output of r is defined only if $O(P)$ is defined, as the limit of the sequence of finite words $s_\epsilon s_{r,i}(O(P))$ for $i \to \infty$ (remind that s_ϵ and $s_{r,i}$ have been defined in Section 4.3). In order to ensure the existence of that limit (and to make sure that this limit is an infinite word), syntactic restrictions are put on the SST: if $P \in \mathrm{dom}(O)$ and $O(P) = U_1 \dots U_n$, it is required that on the connected component induced by P in T, the registers $U_1, \dots, U_{n-1}$ are never modified, and the register U_n can only be modified by appending something (i.e. updates are of the form $U_n := U_n\alpha$ for some $\alpha \in (\Sigma \cup \mathcal{X})^*$). The transition monoid of an ω-SST is defined similarly as SST.

As an example, consider the ω-SST of Fig. 7 that defines transformation f_ω. It can loop an arbitrary number of steps in state q_0 while replacing all symbols by a, and storing the current output in register U. Non-deterministically, it guesses the last occurrence of an input a symbol, and from that point on, never modifies register U again, and always append b to register V. Register U is intended to capture the word $a^{|u|}$ in the definition of f_ω, while V captures b^ω. The output is then UV, and it is defined only for the singleton $\{q_1\}$, which enforces that after some time, only b symbols are read.

It has been shown in [3] that MSO-transducers of infinite words correspond exactly to ω-2FST with regular look-ahead, to ω-SST with finite transition monoid, and to ω-SST with copyless transition monoid.

6.3 Tree Transformations

We present a result that establishes a correspondence between MSO-transducers, and a transducer model for functional transformations of finite ranked trees. Recall that ranked trees are ordered trees over a ranked alphabet. Each symbol f of the alphabet has a rank denoted by $r(f)$ and, if some node α is labelled f, this node has exactly $r(f)$ successor nodes, called the children of α (see [13] for a formal definition). We denote by $\mathcal{T}_\Gamma$ the set of ranked trees over a ranked alphabet Γ, and a (ranked) tree transformation is a binary relation over $\mathcal{T}_\Gamma$.

Tree transducers and their connection with term rewriting systems have been deeply studied, see for instance [27]. More recent results on tree transducers can be found in [15]. Like words, ranked trees over Γ can be seen as logical structures over the signature $\mathcal{S} = \{S_1, \dots, S_n, (L_a)_{a\in\Gamma}\}$, where n is the maximum arity in Γ, S_i are binary successor predicates interpreted by pairs of nodes (α, β) such that β is the i-th child of α, and L_a are unary predicates for the node labels. An MSO-transducer over the signature $\mathcal{S}$ defines a functional tree transformation on $\mathcal{T}_\Gamma$, provided the output is a ranked tree structure (which is a decidable property). For instance, consider a ranked alphabet $\Gamma = \{g, a\}$ where g is a binary symbol and a a constant, and the transformation t_{rev} which reverses a tree, i.e., reverses the order relation between the children of any internal node. The transformation t_{rev} is definable by the following one-copy MSO-transducer:

$$\phi_{dom} \equiv \phi^1_{pos}(x) \equiv \top \quad \phi^1_{L_\gamma}(x) \equiv L_\gamma(x) \quad \phi^{1,1}_{S_i}(x,y) \equiv \bigvee_{\gamma\in\Gamma} L_\gamma(x) \wedge \phi_{S_{r(\gamma)-i+1}}(x,y)$$

for $\gamma \in \Gamma$ and $i \in \{1, 2\}$.

Correspondences between MSO-transducers on ranked trees and tree transducers has been first studied in [22,8,23] for attribute grammars and macro tree transducers, and more recently in [4] for streaming tree transducers.

Macro tree transducers extend top-down tree transducers with parameters in which to store partial output trees. They correspond to purely functional programs working on tree structures: states are mutually recursive functions and can carry parameters. Due to lack of space, we do not define formally macro tree transducers. MSO-transducers on ranked trees correspond exactly to (deterministic) macro tree transducers of linear-size increase, i.e. macro tree transducers that define functions whose output tree size depends linearly on the size of the input tree. It is shown in [22,8] that functional MSOT-transformations of ranked trees are definable by macro tree transducers. The other, more difficult direction, which shows that macro tree transducers of linear-size increase are effectively MSOT-definable was proved in [23]. A consequence of this effective correspondence is the decidability of equivalence for linear-size increase macro tree transducers. It was indeed shown in [24] that MSO-transducers of ranked trees have decidable equivalence problem (see [33] for a survey on equivalence problems for tree transducers).

Other connections between MSO-transducers and tree transducers have been obtained, for an extension of streaming string transducers to trees [4], and for some classes of tree walking transducers [15].

Acknowledgments. I warmly thank Jean-François Raskin and Jean-Marc Talbot for reading preliminary versions of this paper, and Sebastian Maneth for his helpful comments.

References

1. Alur, A., Černý, P.: Streaming transducers for algorithmic verification of single-pass list-processing programs. In: POPL, pp. 599–610 (2011)
2. Alur, R., Černý, P.: Expressiveness of streaming string transducers. In: FSTTCS, vol. 8, pp. 1–12 (2010)
3. Alur, R., Filiot, E., Trivedi, A.: Regular transformations of infinite strings. Technical Report MS-CIS-12-05, University of Pennsylvania (2012)
4. Alur, R., D'Antoni, L.: Streaming tree transducers. In: Czumaj, A., Mehlhorn, K., Pitts, A., Wattenhofer, R. (eds.) ICALP 2012, Part II. LNCS, vol. 7392, pp. 42–53. Springer, Heidelberg (2012)
5. Alur, R., Deshmukh, J.V.: Nondeterministic streaming string transducers. In: Aceto, L., Henzinger, M., Sgall, J. (eds.) ICALP 2011, Part II. LNCS, vol. 6756, pp. 1–20. Springer, Heidelberg (2011)
6. Alur, R., Durand-Gasselin, A., Trivedi, A.: From monadic second-order definable string transformations to transducers. In: LICS, pp. 458–467 (2013)
7. Béal, M.-P., Carton, O., Prieur, C., Sakarovitch, J.: Squaring transducers: an efficient procedure for deciding functionality and sequentiality. Theoretical Computer Science 292(1), 45–63 (2003)
8. Bloem, R., Engelfriet, J.: A comparison of tree transductions defined by monadic second order logic and by attribute grammars. J. Comput. Syst. Sci. 61(1), 1–50 (2000)
9. Bojańczyk, M.: Transducers with origin information. In: Esparza, J., Fraigniaud, P., Husfeldt, T., Koutsoupias, E. (eds.) ICALP 2014, Part II. LNCS, vol. 8573, pp. 26–37. Springer, Heidelberg (2014)
10. Büchi, J.R.: Weak second-order arithmetic and finite automata. Zeitschrift für Mathematische Logik und Grundlagen der Mathematik 6(1-6), 66–92 (1960)
11. Büchi, J.R.: On a decision method in restricted second-order arithmetic. In: Int. Congr. for Logic Methodology and Philosophy of Science, pp. 1–11. Standford University Press, Stanford (1962)

12. Church, A.: An unsolvable problem of elementary number theory. Amer. J. Math. 58, 345–363 (1936)
13. Comon-Lundh, H., Dauchet, M., Gilleron, R., Löding, C., Jacquemard, F., Lugiez, D., Tison, S., Tommasi, M.: Tree Automata Techniques and Applications (November 2007)
14. Courcelle, B.: Monadic second-order definable graph transductions: A survey. Theoretical Computer Science 126(1), 53–75 (1994)
15. Courcelle, B., Engelfriet, J.: Graph Structure and Monadic Second-Order Logic - A Language-Theoretic Approach. Encyclopedia of mathematics and its applications, vol. 138. Cambridge University Press (2012)
16. Culik II, K., Karhumäki, J.: The equivalence problem for single-valued two-way transducers (on NPDTOL languages) is decidable. SIAM J. Comput. 16(2), 221–230 (1987)
17. Curry, H.B.: Functionality in Combinatory Logic. Proceedings of the National Academy of Sciences of the United States of America 20, 584–590 (1934)
18. Diekert, V., Gastin, P.: First-order definable languages. In: Logic and Automata: History and Perspectives, Texts in Logic and Games, pp. 261–306. Amsterdam University Press (2008)
19. Ebbinghaus, H.D., Flum, J.: Finite Model Theory. Springer, Berlin (1995)
20. Elgot, C.C.: Decision problems of finite automata design and related arithmetics. Transactions of the American Mathematical Society 98(1), 21–51 (1961)
21. Engelfriet, J., Hoogeboom, H.J.: MSO definable string transductions and two-way finite-state transducers. ACM Trans. Comput. Logic 2, 216–254 (2001)
22. Engelfriet, J., Maneth, S.: Macro tree transducers, attribute grammars, and MSO definable tree translations. Inf. Comput. 154(1), 34–91 (1999)
23. Engelfriet, J., Maneth, S.: Macro tree translations of linear size increase are MSO definable. SIAM J. Comput. 32(4), 950–1006 (2003)
24. Engelfriet, J., Maneth, S.: The equivalence problem for deterministic MSO tree transducers is decidable. Inf. Process. Lett. 100(5), 206–212 (2006)
25. Filiot, E., Gauwin, O., Reynier, P.A., Servais, F.: From two-way to one-way finite state transducers. In: LICS, pp. 468–477 (2013)
26. Filiot, E., Krishna, S.N., Trivedi, A.: First-order definable string transformations. To appear in FSTTCS (2014)
27. Fülöp, Z., Vogler, H.: Syntax-Directed Semantics - Formal Models Based on Tree Transducers. Monographs in Theoretical Computer Science. An EATCS Series. Springer (1998)
28. Griffiths, T.V.: The unsolvability of the equivalence problem for -free nondeterministic generalized machines. Journal of the ACM (1968)
29. Howard, W.: The formulae-as-types notion of construction. In: To H.B.Curry: Essays on Combinatory Logic, Lambda-Calculus and Formalism. Academic Press (1980)
30. Immerman, N.: Languages that capture complexity classes. SIAM Journal on Computing 16(4), 760–778 (1987)
31. Berstel, J.: Transductions and Context-Free Languages. Teubner, Stuttgart (1979)
32. Lautemann, C., McKenzie, P., Schwentick, T., Vollmer, H.: The descriptive complexity approach to LOGCFL. Journal of Computer and System Sciences 62 (2001)
33. Maneth, S.: Equivalence problems for tree transducers: A brief survey. In: AFL, pp. 74–93 (2014)
34. McKenzie, P., Schwentick, T., Therienand, D., Vollmer, H.: The many faces of a translation. Journal of Computer and System Sciences 72 (2006)

35. McNaughton, R.: Testing and generating infinite sequences by a finite automaton. Information and Control 9, 521–530 (1966)
36. Rabin, M.O.: Decidability of second-order theories and automata on infinite trees. Trans. Amer.Math.Soc. 141, 1–35 (1969)
37. Sakarovich, J.: Elements of Automata Theory. Cambridge University Press, Cambridge (2009)
38. Schützenberger, M.P.: On finite monoids having only trivial subgroups. Information and Control 8(2), 190–194 (1965)
39. Shepherdson, J.C.: The reduction of two-way automata to one-way automata. In: Moore, E.F. (ed.) Sequential Machines: Selected Papers. Addison-Wesley (1964)
40. Straubing, H.: Finite Automata, Formal Logic, and Circuit Complexity. Birkhäuser, Boston, Basel and Berlin (1994)
41. Thomas, W.: Languages, automata and logic. In: Salomaa, A., Rozenberg, G. (eds.) Handbook of Formal Languages. Beyond Words, vol. 3. Springer, Berlin (1997)
42. Trakhtenbrot, B.A.: Finite automata and logic of monadic predicates. Dokl. Akad. Nauk SSSR 140, 326–329 (1961) (in Russian)
43. Turing, A.M.: On computable numbers, with an application to the entscheidungsproblem. Procedings of the London Mathematical Society 42(2), 230–265 (1936)
44. Vardi, M.Y., Wilke, T.: Automata: from logics to algorithms. In: Logic and Automata: History and Perspectives (in Honor of Wolfgang Thomas), pp. 629–736 (2008)

Truths about Simpson's Paradox: Saving the Paradox from Falsity

Prasanta S. Bandyopadhyay[1], R. Venkata Raghavan[2], Don Wallace Dcruz[2] and Gordon Brittan Jr.[1]

[1]Department of Philosophy, Montana State University, Bozeman, USA
psb@montana.edu, gbrittan17@gmail.com
[2]Department of Philosophy, University of Hyderabad, Hyderabad, India
{raghavan.rv,don.wallace}@uohyd.ac.in

Abstract. There are three questions associated with Simpson's paradox (SP): (i) Why is SP paradoxical? (ii) What conditions generate SP? and (iii) How to proceed when confronted with SP? An adequate analysis of the paradox starts by distinguishing these three questions. Then, by developing a formal account of SP, and substantiating it with a counter-example to causal accounts, we argue that there are no causal factors at play in answering questions (i) and (ii). Causality enters only in connection with action.

Keywords: Simpson's Paradox, Formal Analysis, Collapsibility Principle, Inference rules, Causal Accounts, Definition of Paradox, First-level-truth, Second-level-truth.

1 Overview

In his recent book, *Saving Truth from Paradox*, Hartry Field discusses the philosophical significance of paradoxes. According to him, "[a]ny resolution of the paradoxes will involve giving up (or at least restricting) some very firmly held principles:... [and] [t]he principles to be given up, are the ones to which the average person simply can't conceive of alternatives. That's why the paradoxes are *paradoxes*." [4, p.17]. Their significance and the firmly held principles which we have to give up in resolving them is a recurring theme in philosophical logic. We will illustrate this in the case of Simpson's paradox (SP), which involves the reversal of the direction of a comparison or the cessation of an association when data from several groups are combined to form a single whole [17]. At least *three* distinct questions are important in understanding the nature of the paradox: (i) Why or in what sense, is SP a paradox? (ii) What are the conditions in which the paradox arises? (iii) How should one proceed when confronted with a typical case of the paradox, hereafter to be called the "what-to-do" question?[1] The three questions are distinct: answering one of them does not entail answers to the

[1] Daniel Hausman was perhaps the first philosopher who drew our attention to the significance of these three types of questions (in an email communication).

M. Banerjee and S.N. Krishna (eds.): ICLA 2015, LNCS 8923, pp. 58–73, 2015.

others. Following these three questions, we distinguish two types of truth about SP: the first-level truth and the second-level truth. The significance of the three questions about the paradox is what we call the "first-level truth", while the significance of the first two questions in unlocking its paradoxical nature and the conditions for its emergence is what we call the "second-level truth." The failure to appreciate the difference between these two levels of truth, we will contend, is the source of its misdiagnosis. Tables 1 and 2 illustrate the two types of SP. The data in both tables represent acceptance and rejection rates of male and female applicants for graduate school in two departments of an imaginary university in some year.

Table 1. Simpson's Paradox (Type I)

Two Groups	Dept 1		Dept 2		Acceptance Rates		Overall Acceptance Rates
	Accept	Reject	Accept	Reject	Dept 1	Dept 2	
Females	180	20	100	200	90%	33%	56%
Males	480	120	10	90	80%	10%	70%

Table 2. Simpson's Paradox (Type II)

Two Groups	Dept 1		Dept 2		Acceptance Rates		Overall Acceptance Rates
	Accept	Reject	Accept	Reject	Dept 1	Dept 2	
Females	90	1410	110	390	6%	22%	10%
Males	20	980	380	2620	2%	13%	10%

Table 1 represents an example of the paradox in which the association in the sub-populations (Dept 1 and Dept 2) with higher acceptance rate for females is *reversed* in the combined population, with overall higher rates for males. Table 2 is an example that shows the paradoxical effect when the association between "gender" and "acceptance rates" in the sub-populations *ceases* to exist in the combined population. Though the acceptance rates for females are higher in each department, in the combined population, those rates cease to be different.

This paper is divided into eight sections. In section two, we will propose our response to the first two questions. Then we will briefly introduce two influential causal accounts of SP proposed independently by Judea Pearl [9] and Peter Spirtes, Clark Glymour and Richard Scheines (hereafter called 'SGS') [15]. In section four, we will produce a counter-example to the causal accounts. The next section will be devoted to the "what-to-do" question. In section six, we evaluate causal accounts (with special attention to Pearl's) of the paradox and compare them with ours. In section seven, we will discuss how our account affects the general notion of paradoxes and their classification while providing a general definition of a paradox. We conclude with some remarks in section eight.

2 Formal Analysis of SP

2.1 Conditions of SP[2]

We begin with an analysis of the paradox in response to question (ii), "what are the conditions in which the paradox arises?" Consider two groups, [A, B], taken to be mutually exclusive and jointly exhaustive. The overall rates for each group are $[\alpha, \beta]$ respectively. Each group is partitioned into categories [1, 2] and the rates within each partition are $[A_1, A_2, B_1, B_2]$. Let's assume that f_1 = the number of females accepted in D_1, F_1 = the total number of females applied to D_1, m_1 = the number of males accepted in D_1, M_1 = the total number of males applied to D_1. Then $A_1 = f_1/F_1$, and $B_1 = m_1/M_1$. Defining f_2, F_2, m_2 and M_2 in a similar way, we get $A_2 = f_2/F_2$ and $B_2 = m_2/M_2$. Likewise, we could understand α and β as representing the overall rates for females and males, respectively. So the terms $\alpha = (f_1+f_2)/(F_1+F_2)$ and $\beta = (m_1+m_2)/(M_1+M_2)$. To help conceptualize these notations in terms of Table 1, we provide their corresponding numerical values: $A_1 = 180/200 = 90\%$, $A_2 = 100/300 = 33\%$, $B_1 = 480/600 = 80\%$, $B_2 = 10/100 = 10\%$, $\alpha = 280/500 = 56\%$, and finally $\beta = 490/700 = 70\%$. Since α, β, A_1, A_2, B_1, and B_2 are rates of some form, they will range between 0 and 1 inclusive. We further stipulate the following definitions where, "$\equiv$" means "is defined as".

$$C_1 \equiv A_1 \geq B_1.$$

$$C_2 \equiv A_2 \geq B_2.$$

$$C_3 \equiv \beta \geq \alpha.$$

$$C \equiv (C_1 \& C_2 \& C_3).$$

In terms of Table 1, these definitions become C_1: 90% > 80%, C_2: 33% > 10%, C_3: 70% > 56% and thus C is satisfied. But C alone is not a sufficient condition for SP. We could have a case where $A_1 = B_1$, $A_2 = B_2$ and $\beta = \alpha$ resulting in no paradox, yet C being satisfied. Hence, we stipulate another definition:

$$C_4 \equiv \theta > 0.$$

where, $$\theta = (A_1 - B_1) + (A_2 - B_2) + (\beta - \alpha).$$

For the data in Table 1, θ equals 10% + 23% + 14%. Again, C_4 alone is not sufficient for SP since we could have a case where $A_1 > B_1$, $B_2 > A_2$ and $\beta > \alpha$ resulting in no paradox (C is violated) and yet C_4 being satisfied.[3] Hence,

[2] Some parts of this section are based on our previous work [1,2].

[3] As a heuristic rule we take A_1 to be that sub-group ratio which is the greater of the two ratios and B_1 as that which is the lesser of the two. In table 1, the ratio of women admitted to department 1 is greater than that of men. Hence, the former will be taken as A_1 and the latter will be taken as B_1. Similarly, since the ratio of women admitted to department 2 is greater than that of men, the former is taken as A_2 and the latter as B_2. This avoids the complexity of taking the absolute value of their difference in the calculation of θ.

a situation is a case of SP if and only if:

$$C \equiv (C_1 \& C_2 \& C_3) \quad \text{(a)}$$

$$C_4 \equiv \theta = (A_1 - B_1) + (A_2 - B_2) + (\beta - \alpha) > 0^4 \quad \text{(b)}$$

Both (a) and (b) are necessary conditions, but they jointly constitute sufficient conditions for generating SP [1].[5] Both conditions for the paradox generate two key theorems which specify the relationship between the two acceptance rates in both sub-populations. These are: 1. $A_1 \neq A_2$, and 2. $B_1 \neq B_2$. Table 3 shows why the condition for Theorem 1 needs to hold. Since $A_1 = A_2$, i.e., 25% = 25%, no paradox results. Similarly, in Table 4, since $B_1 = B_2$, i.e., 25% = 25%, the paradox does not occur. Proofs of these theorems are provided in the appendix.

Table 3. No SP ($A_1 = A_2$)

Two Groups	Dept 1		Dept 2		Acceptance Rates		Overall Acceptance Rates
	Accept	Reject	Accept	Reject	Dept 1	Dept 2	
Females	75	225	75	225	25%	25%	25%
Males	10	90	20	80	10%	20%	15%

Table 4. No SP ($B_1 = B_2$)

Two Groups	Dept 1		Dept 2		Acceptance Rates		Overall Acceptance Rates
	Accept	Reject	Accept	Reject	Dept 1	Dept 2	
Females	10	90	20	80	10%	20%	15%
Males	75	225	75	225	25%	25%	25%

There are four points worth mentioning. First, Clark Glymour [5] would call our account an application of the "Socratic method" in which we provide necessary and sufficient conditions for the analysis of a concept.[6] Second, the characterization of the puzzle in terms of our two conditions captures the paradoxical

[4] See Blyth [3] for similar conditions. However, our conditions and notations are slightly different from his.

[5] See [6], [16]. The latter paper shows that SP reversal involves Boolean disjunction of events in an algebra rather than being restricted to cells of a partition.

[6] Glymour contrasts this method with what he calls the "Euclidean"-method based theories where one could derive interesting consequences from them although Euclidean-method based theories, according to him, are invariably incomplete. It is interesting to note two very different points. First, although Glymour is not fond of the Socratic-method on which, however, a large part of the western philosophical tradition rests, our Socratic-method based logical account at the same time is also able to generate some interesting logical consequences (See [1,2]). Second, it is not only the Greeks who applied this method. In classical Indian philosophical tradition, the Socratic method is also very much prevalent where a definition of a term is evaluated in terms of whether it is able to escape from being both "too narrow" and "too wide."

nature of the data in the examples given, namely, the reversal or the cessation of an association in the overall population; they are in no way ad hoc. Third, the paradox is "structural" in character, in the sense that the reasoning that leads to it is deductive. Consider our examples, which involve simple arithmetic. The overall rates of acceptance for both females and males follow from their rates of acceptance in two departments taken separately. Note that both conditions of the paradox can be defined in terms of the probability theory, which is purely deductive [3]. Fourth, unless someone uses the notion of causation trivially, for example, believes that 2+2 "causes" 4, there is no reason to assume that there are causal intuitions lurking in the background. We will return to the last point in greater detail in the following sections.

2.2 Why is SP "Paradoxical"?

To answer question (i), "why is SP a paradox?" we now provide an explanation of how the paradox arises in people's minds and why it is found perplexing. In other words, what is the reasoning that the "average person" follows that leads him/her to a paradoxical conclusion? For our purposes, we have reconstructed our type I version of SP in terms of its premises and conclusion to show how the paradox arises. However, the point of the reconstruction will be adequately general to be applicable to all types of SP. We introduce a numerical principle called the collapsibility principle (CP) which plays a crucial role in the reconstruction. CP says that relationships between variables that hold in the sub-populations (e.g., the rate of acceptance of females being higher than the rate of acceptance of males in both sub-populations) must hold in the overall population as well (i.e., the rate of acceptance of females must be higher than the rate of acceptance of males in the population). There are two versions of CP corresponding to the two types of SP represented by Tables 1 and 2. The first version of CP (CP1) says that a dataset is collapsible if and only if $[(A_1 > B_1)\&(A_2 > B_2) \rightarrow (\alpha > \beta)]$. The second version of CP (CP2) states that a dataset is collapsible if and only if $[(A_1 = B_1)\&(A_1 = B_2) \rightarrow (\alpha = \beta)]$. That CP1 and CP2 can lead to paradoxical results demonstrates that both versions of the principle are not, in all their applications, true. That is, CP $\rightarrow$ ~SP, whether it is CP1 or CP2, where "$\rightarrow$" is to be construed as the implication sign. If f_1, F_2, m_1, M_2, A_1, A_2, B_1, B_2, α, and β have the same meanings as given in section 2.1, then CP1 takes the following form.

$$\left(\left(\frac{f_1}{F_1} > \frac{m_1}{M_1}\right) \& \left(\frac{f_2}{F_2} > \frac{m_2}{M_2}\right)\right) \rightarrow \left(\frac{f_1 + f_2}{F_1 + F_2} > \frac{m_1 + m_2}{M_1 + M_2}\right)$$

Likewise, CP2 says

$$\left(\left(\frac{f_1}{F_1} = \frac{m_1}{M_1}\right) \& \left(\frac{f_2}{F_2} = \frac{m_2}{M_2}\right)\right) \rightarrow \left(\frac{f_1 + f_2}{F_1 + F_2} = \frac{m_1 + m_2}{M_1 + M_2}\right)$$

As we can see, CP is a numerical inference principle devoid of any causal intuition. Here is the reconstruction of type I version of SP:
(1) Female and male populations are mutually exclusive and jointly exhaustive; one can't be a student of both departments and satisfy the two conditions of SP.
(2) The acceptance rate of females is higher than that of males in Department 1. (observed from data)
(3) The acceptance rate of females is higher than that of males in Department 2. (observed from data)
(4) If (2) and (3) are true, then the acceptance rate for females is higher than that of males overall. (from CP1)
(5) Hence the acceptance rate for females is higher than that of males overall. (from (2), (3) and (4))
(6) However, fewer females are admitted overall. (observed from data)
(7) Overall acceptance rate for females is both higher and lower than that of males. (from (5) and (6))
In our derivation of the paradox, premise (4) plays a crucial role. In type I version of SP, as given in Table 1, CP1 does not hold ($A_1 > B_1$ and $A_2 > B_2$, but $\alpha < \beta$). That CP1 is not generally true is shown by our derivation of a contradiction. The same result can be obtained for Type II version of SP in Table 2 where CP2 has to be given up if the paradox is to be avoided.

Our answer to the first question, (i), then, is simply that humans tend to invoke CP uncritically, as a rule of thumb, and thereby make mistakes in certain cases about proportions and ratios; they find it paradoxical when their usual expectation that CP is applicable across the board, turns out to be incorrect. And the reason we think people invoke CP uncritically, is its remarkable (formal) resemblance with the two inference rules given below.[7]
1. In elementary algebra, the following truth holds for real numbers:

$$x_1 > y_1$$
$$x_2 > y_2$$
$$\therefore (x_1 + x_2) > (y_1 + y_2)$$

While it is correct to substitute $A_1(f_1/F_1)$ for x_1, $B_1(m_1/M_1)$ for y_1, $A_2(f_2/F_2)$ for x_2 and $B_2(m_2/M_2)$ for y_2, people might confuse $(x_1 + x_2)$ and $(y_1 + y_2)$ for $\alpha((f_1 + f_2)/(F_1 + F_2))$ and $\beta((m_1 + m_2)/(M_1 + M_2))$ respectively, leading them to think that CP is also a mathematical truth. Thus, mistakes about proportions and ratios could lead the average person to see a superficial resemblance between CP and the above mathematical truth.
2. In propositional logic, the following rule is valid:

$$P1 \rightarrow Q \quad \text{(A)}$$
$$P2 \rightarrow Q \quad \text{(B)}$$

$$\therefore (P1 \vee P2) \rightarrow Q \quad \text{(C)}$$

[7] We are thankful to Joseph Hanna and John G. Bennett for helpful emails on this point.

In our case, let P1 = "A student applies to Department 1", P2 = "A student applies to Department 2" and Q = "The student has greater chance of being accepted, if the gender of the student is female". Now, (A) partially captures the condition $A_1 > B_1$ whereas (B) partially captures $A_2 > B_2$. (C), which reads, "If a student applied to Department 1 or Department 2 then, the student has greater chance of being accepted if the gender of the student is female" resembles the condition $\alpha > \beta$. We do not suggest that propositional logic can capture the essence of the paradox. The reasoning leading to SP involves probabilistic considerations which, unlike propositional logic, is not truth-functional. For example, the probability of a disjunction is not a function of the probability of its disjuncts. Likewise, SP is a weighted average of probabilities, or, in other words, averages of averages. No such concept of weighted average exists in truth-functional logic. The above comparison of CP with a valid propositional rule no more than suggests why people tend to use CP even in cases where it leads to contradiction.

3 Causal Accounts of SP

3.1 Pearl's Account

Pearl argues that the arithmetical inferences in SP seem counter-intuitive only because we commonly make two incompatible assumptions, that causal relationships are governed by the laws of probability and that causal relationships are more *stable* than probabilistic relationships [9, pp. 180, 25]. Once we reject either of these assumptions, and he opts for rejecting the first, the "paradox" is no longer paradoxical. On the other hand, when we fail to distinguish causal from statistical hypotheses, the paradox results.

Pearl makes two basic points. One, SP is to be understood in causal terms for its correct diagnosis. In the type I version, for example, the effect on "acceptance" (A) of the explanatory variable, "gender" (G), is hopelessly mixed up (or "confounded") with the effects on A of the other variable, "department" (D). We are interested in the direct effect of G on A and not an indirect effect by way of another variable like D. His other point is that causal hypotheses, which support counterfactuals, often cannot be analyzed in statistical terms. Suppose we would like to know Bill Clinton's place in US history had he not met Monica Lewinsky. The counterfactual for the causal hypothesis is "Clinton's status in the US history would be different had he not met Monica Lewinsky" [9, p. 34]. However, there is no statistical model one could construct that would provide the joint occurrence of 'Clinton' and 'no Lewinsky'. There simply are no appropriate data, as there are, for instance, in the fair coin-flipping experiments where the model about flipping a coin and data about it are well known.

3.2 SGS Account

Spirtes, Glymour, and Scheines suggest a subject-matter-neutral automated causal inference engine that provides causal relationships among variables from observational data using information about their probabilistic correlations and assumptions about their causal structure. These assumptions are: 1. Causal Markov Condition (CMC), 2. Faithfulness Condition (FC) and 3. Causal Sufficiency Condition (CSC). According to CMC, a variable X is independent of every other variable (except X's effects) conditional on all of its direct causes. A is a direct cause of X if A exerts a causal influence on X that is not mediated by any other variables in a given graph. The FC says that all the conditional independencies in the graph are only implied by CMC, while CSC states that all common causes of measured variables are explicitly included in the model. Since these theorists are interested in teasing out reliable causal relationships from data, they would like to make sure that those probability distributions are faithful in representing causal relations in them.

One reason for SP being causal, according to this account, is that (for the example given in Table 1) applying to the school has a causal dimension involving causal dependencies between "gender" and "acceptance rate". More female students chose to apply to the departments where rates of acceptance are significantly lower, *causing* their overall rates of acceptance to be lower in the combined population. Similarly, with regard to Simpson's own example in the literature, Spirtes et al. write, "[t]he question is what *causal dependencies* can produce such a table, and that question is properly known as "Simpson's paradox"." [15, p. 40].

4 Counter-Example to the Causal Account

It is not easy to come up with an example which precludes invoking some sort of appeal to "causal intuitions" with regard to SP. But what follows is, we think, such a case. It tests in a crucial way the persuasiveness of the causal accounts.[8]

Table 5. Simpson's Paradox (Marble Example)

Marbles of two sizes	Bag 1		Bag 2		Rates of red Marbles		Overall rates for red marbles
	Red	Blue	Red	Blue	Bag 1	Bag 2	
Big marbles	180	20	100	200	90%	33%	56%
Small Marbles	480	120	10	90	80%	10%	70%

Suppose, as in Table 5, we have two bags of marbles, all of which are either big or small, and red or blue. Suppose in each bag, the proportion of big marbles that are red is greater than the portion of small marbles that are red (Bag 1: 90% > 80% and Bag 2: 33% > 10%). Now suppose we pour all the marbles from both bags into a box. Would we expect the portion of big marbles in the box that are red to be greater than the portion of small marbles in the box that are red? Most of us would be surprised to find that our usual expectation is incorrect.

[8] This counter-example is due to John G. Bennett.

The big marbles in the first bag have a higher ratio of red to blue marbles than do the small marbles; the same is true about the ratio in the second bag. But considering all the marbles together, the small marbles have a higher ratio of reds to blues than the big marbles do (in the combined bag: 70% > 56%).

We argue that this marble example is a case of SP since it has the same mathematical structure as the type I version of SP. There are no causal assumptions made in this example, no possible causal "confounding" and yet it seems paradoxical. We believe this counter-example shows that at least sometimes, there is a purely mathematical mistake about ratios that people customarily make. Some causal theorists might be tempted to contend that even in this example there is confounding between the effects of the marble size on the color with the effects of the bag on the color. However, this confounding is not a causal confounding since one *cannot* say that Bag 1 *has caused* big marbles to become more likely to be red or that Bag 2 has caused big marbles to become more likely to be blue. In short, one must admit that the above counter-example does not involve causal intuitions, yet it is still a case of SP.

5 "What-To-Do" Question and Causal Accounts

In the case of SP, "what-to-do" questions arise when investigators are confronted with choosing between two conflicting statistics. For example, in Table 1, the conflict is between the uncombined statistics of the two departments and their combined statistics. Which one should they use to act? It is evident that many interesting cases of choosing actions arise when we infer causes/patterns from proportions. The standard examples[9] deal with cases in which "what-to-do" questions become preeminent. But it should be clear in what follows that there is no unique response to this sort of question for all cases of the paradox. Consider Table 6 based on data about 80 patients. 40 patients were given the treatment, T, and 40 assigned to a control, ∼T. Patients either recovered, R, or didn't recover, ∼R. There were two types of patients, males (M) and females (∼M).

Table 6. Simpson's Paradox (Medical Example)

Two Groups	M		∼M		Recovery Rates		Overall Recovery Rates
	R	∼R	R	∼R	M	∼M	
T	18	12	2	8	60%	20%	50%
∼T	7	3	9	21	70%	30%	40%

One would think that treatment is preferable to control in the combined statistics, whereas, given the statistics of the sub-population, one gathers the impression that control is better for both men and women. Given a person of unknown gender, would one recommend the control? The standard response is clear: control is better for a person of unknown gender (since Pr(R| ∼T) >

[9] These recommendations are standard because they are agreed upon by philosophers [8], statisticians, and computer scientists [9].

Pr(R|T)). Call this first example 'the medical example'. In the second example, call it 'the agricultural example', we are asked to consider the same data, but now T and ∼T are replaced by the varieties of plants (white [W] or black variety [∼W]), R and ∼R by the yield (high [Y] or low yield [∼Y]) and M and ∼M by the height of plants (tall [T] or short [∼T]).

Table 7. Simpson's Paradox (Agricultural Example)

Two Groups	T		∼T		Yield Rates		Overall Yield Rates
	Y	∼Y	Y	∼Y	T	∼T	
W	18	12	2	8	60%	20%	50%
∼W	7	3	9	21	70%	30%	40%

Given this new interpretation, the overall yield rate suggests that planting the white variety is preferable since it is 10% better overall, although the white variety is 10% worse among both tall and short plants (sub-population statistics). Which statistics should one follow in choosing between which varieties to plant in the future? The standard recommendation is to take the combined statistics and thus recommend the white variety for planting (since Pr(Y|W) > Pr(Y|∼W)), which is in stark contrast with the recommendation given in the medical example. In short, both medical and agricultural examples provide varying responses to the "what-to-do" question. There is no unique response regarding which statistics, subpopulation or whole, to follow in every case of SP. We agree with standard recommendations with a proviso, i.e., we need to use substantial background information, which is largely causal in nature, to answer "what-to-do" questions, as *doing* something means *causing* something to happen.

6 Truths about SP: An Evaluation of Causal Accounts

We argued that to understand the significance of SP as a whole, we need to distinguish three types of questions (first-level truth) as well as divorce the first two questions from the third to show that causality is irrelevant both in unlocking the paradoxical nature of SP and providing conditions for its emergence (second-level truth). Based on our discussion of the causal accounts, one realizes that causal theorists have in fact addressed the "what-to-do" question. We don't deny that causal inference plays a crucial role in choosing the right statistic when confronted with the paradox. Hence we agree with both Pearl and SGS about the third question. However, as far as we know, SGS have not distinguished the three questions about SP, and thereby failed to appreciate the first-level truth about SP. Pearl on the other hand, does distinguish the three questions. But both causal accounts fail to understand the second-level truth about the paradox. Notice that one may, like Pearl, recognize the first-level truth and yet fail to recognize the second-level truth. An examination of his responses to the first two questions will reveal the reason behind this, showing how his causal account falls short of providing an adequate explanation for the first two questions and thereby not being able to appreciate the full significance of SP.

In response to the first question, Pearl draws attention to the distinction between what he calls "Simpson's reversal", which is merely an "arithmetic phenomenon in the calculus of proportions" and "Simpson's paradox" which is "a psychological phenomenon that evokes surprise and disbelief" [10, p. 9]. He thinks that the latter is the result of intuitions guided by causal considerations and the fallacy of equating correlation with causation. While agreeing with him about the fallacy, we pointed out, with the help of the marble counter-example, that fundamentally, SP is devoid of any causal intuitions, although most day-to-day examples of SP can be interpreted causally. We think that human puzzlement about SP stems from the unexpected failure of CP which closely resembles valid inference rules (section 2.2). With respect to the second question, Pearl identifies "scenarios" in which one can expect a reversal. A scenario, according to him, is "a process by which data is generated" [10, p. 10]. The causal calculus/models which represent these causal scenarios are different from our formal conditions which have been derived from the structure of the paradox (section 2.1). So our conditions capture *all* cases of SP regardless of the causal process involved and provide a more general account than either of the causal accounts.

7 Re-evaluating the Place of SP in Paradox Literature

Logicians tend to hold different views concerning what paradoxes are. Whether SP is a paradox depends on how one defines and slices paradoxes. Priest [11], for example, may not consider SP to be a paradox as it is neither a set-theoretic paradox such as Russell's nor a semantic one like the Liar Paradox. But, under Sainsbury's construal, SP could be regarded as a paradox since he understands a paradox as "an apparently unacceptable conclusion derived by apparently acceptable reasoning from apparently acceptable premises." [14, p. 1]. However, this might not furnish a genuine rationale for what makes paradoxes paradoxical since one might worry what an "apparently acceptable reasoning" is. In this regard, we find a better explanation in W.V.Quine, who both defines and provides a general rationale for the apparently paradoxical nature of paradoxes. A paradox, according to him, is "just any conclusion that at first sounds absurd but that has an argument to sustain it" [12, p. 1]. SP can be treated as a paradox in this Quinean sense.

Two points are to be noted here. First, Quine's use of the word "absurd" could be ambiguous since it lends itself to two interpretations: a) psychological confusion and b) logical contradiction. Our analysis of SP suggests that SP "sounds absurd" under both interpretations. Given the logical reconstruction of SP (section 2.2), we see how it leads to a self-contradictory conclusion. And, given our response to question (i), we find that people tend to apply CP across the board and their psychological confusion results when they find out that CP, in fact, cannot be so applied. Second, our research shows that the sharp distinction Quine draws between "veridical paradox" and "falsidical paradox" does not necessarily hold about SP. Distinguishing between these two varieties while justifying each, he writes, "[a] veridical paradox packs a surprise, but the surprise quickly dissipates itself as we ponder the *proof*. A falsidical paradox packs a surprise, but it is seen as a false alarm when we solve the *underlying fallacy*." [12, p. 9, emphasis is ours]. He argues that Gödel's discovery and other paradoxes in set theory are veridical paradoxes. We think that SP can be seen as a case of veridical paradox as soon as we realize that the population data in all tables follow necessarily from the sub-population data along with the proofs we provided for the paradox to hold. To explore whether SP could fall under the category of a falsidical paradox, consider Quine's own example of the latter. According to him, paradoxes of Zeno are instances of falsidical paradoxes since they rest on the fallacious assumption that "an infinite succession of intervals must add up to an infinite interval." Once we note this, it becomes clear that the initial surprise about them was unwarranted. The same reasoning can be offered for SP being a falsidical paradox. Our analysis shows that the surprise SP packs rests on holding the dubious assumption that CP is unconditionally applicable. Once we realize this, the paradoxical nature of SP disappears. So, the unique feature of SP is that it is a paradox in both veridical and falsidical senses. Therefore, there need not be a sharp distinction between these two types of paradoxes as Quine once argued.

Two issues emerge from the preceding discussion. First, we rely on Quine's definition of a paradox and how it fares with regard to SP; As we will see in a moment Roy Sorensen thinks that Quine's definition is flawed as, according to him, it is neither necessary nor sufficient [13]. Second, whether it is possible to advance a definition of a paradox which could include all types of paradoxes including SP and the Liar paradox under its banner. The rest of this section will be devoted to addressing these two issues.

Sorensen's method is to turn the definition of a paradox against what he takes to be Quine's own "paradox" of radical translation. Quine sets out his "paradox" by first assuming the possibility of a "radical translation" situation, in which neither speaker knows a word of the other's language. Consider a group of linguists interested in understanding what the native speakers' utterances mean. Suppose the speakers utter "gavagai." The linguists observe the speakers, hear what they utter, observe the conditions under which they utter a word or sentence, and determine what they are looking at or pointing out when they utter. Armed with such information, let's assume these linguists make a hypothesis that "gavagai" means "rabbit". In the same way, it is possible that another group of linguists having the same evidence as the first group translates "gavagai" as "undetached rabbit part." Which one is the correct translation of "gavagai"? Based on this thought experiment, Quine contends that radical translation is not possible as meaning, here understood as referent, is indeterminate or at least undermined by the totality of empirical evidence that is available. There is no way to know whether the translation of "gavagai" as "rabbit" or "undetached rabbit part" is the correct hypothesis. But the conclusion seems absurd; at least most of the time, we know what others in our language group (or outside it) are referring to when they utter sounds. Sorensen rejects Quine's construal of paradoxes by pointing out that, " 'What is the translation of 'Gavagai'?' has infinitely many rival answers. According to Quine, the problem is that infinitely many of these are equally good answers. Quine's paradox of radical translation is a counterexample to his own definition of paradox. In addition to showing that absurdity is inessential to paradox, the paradox of radical translation shows that the paradox can be free of arguments and conclusions. 'What is the translation of 'Gavagai'?' has answers obtained by translation, not conclusions derived by arguments."[13, p. 560].

Even though we agree with Sorensen that actual, in contrast to merely apparent, absurdity is not necessary for understanding the nature of paradoxes, we disagree with his claim that it is not helpful to construe paradoxes in terms of an argument consisting of premises and a conclusion. What Sorensen misses in his criticism of Quine is that while a paradox need not present itself in canonical forms, their canonical forms are useful tools in understanding them, just as the canonical form of an argument (with numbered premises and designated conclusion) is a useful tool for discussing arguments that, in real life, do not always present themselves in that way. To force the paradox into the canonical form, suppressed premises must be revealed and hidden assumptions made explicit. If the radical translation claims are paradoxical, they can be fitted into the canonical forms, though there may be different ways to do that. Here's one version in our favored canonical form:

(1) A correct translation of one natural language into another is one that is entirely compatible with all the facts about usage.
(2) If two translations translate a given term in one language into incompatible terms in another language, one of the translations is not correct.
(3) There are two correct translations of the native language word "gavagai" into English; one translates it as "rabbit" and the other translates it as "undetached rabbit part."
(4) "Rabbit" and "undetached rabbit part" are incompatible terms in English (in the sense that they do not have the same referent).
(5) The native language and English are natural languages.

Contrary to Sorensen, we find that it is possible to exhibit the paradox of radical translation in terms of an argument with premises and a conclusion, revealing the assumption on which it rests. At the same time, we agree with Sorensen in a different way when he holds that a paradox need not have a genuinely absurd conclusion. We tend to think that "sounding absurd" lends a psychological air to the issue of a paradox. In light of these two considerations, we propose a general definition of a paradox. A paradox is an (apparently) inconsistent set of sentences each of which seems to be true.[10] The word "apparently" in this account, as in Quine's, is to allow for cases that depend on fallacious arguments, as in the well-known "proofs" for 1=2. Another advantage of this account is that one might make several arguments from a set of inconsistent sentences, but one

[10] We owe this definition to John G. Bennett. Lycan [7] has also provided a similar definition.

would probably not want to call them distinct paradoxes. Any paradox worth the name, including SP, should obey this definition. Simplifying our reconstruction of SP as a paradox in section 2.2, we provide a rough schema for SP with the (optional) false premise marked by an asterisk, for the type 1 version of the paradox.

(1) Sub-population 1 has a positive correlation between two variables.
(2) Sub-population 2 has a positive correlation between two variables.
(*3) If each sub-population in a partition of a larger population exhibits a positive correlation between two variables, then the population as a whole will also exhibit that same positive correlation between the same two variables
(4) Overall population has a negative correlation between the same two variables.

If *3 is included, the set is inconsistent, since premise *3 is false. If *3 is not included, the set seems to be inconsistent, but is not. Whether to analyze the paradox one way or the other may depend on the example and the context. We think that our definition is adequately general to include even the Liar paradox. Call "this sentence is false" the liar sentence. The following provides a canonical reconstruction of the Liar paradox with two premises and a conclusion.

(1) The liar sentence is true.
(2) The liar sentence is false.
(3) A sentence is either true or false, but not both.

In this section, among other issues, we both discussed and evaluated different views on paradoxes. As a result, we are able to provide a general framework to understand paradoxes while showing that both SP and the Liar paradox satisfy it even though the former has an apparently contradictory conclusion while the latter has a genuinely contradictory one.

8 Conclusion

Unraveling paradoxes is crucial to philosophers of logic as they challenge our deeply held intuitions in a fundamental way. While addressing SP, we distinguished three types of questions. We showed that answering one does not necessarily lead to the answers of the rest. Although, admittedly, the "what-to-do" question is the most important insofar as the practical side of SP is concerned, some causal theorists have overlooked the need to distinguish these three questions, thus failing to appreciate the first-level truth about the paradox. Even if they recognize this first-level truth, the importance of the "what-to-do" question drives them to assume that the causal calculus needed to address this question is the correct way to unlock the riddle about the paradox. We, however, showed that the truth about the paradoxical nature of SP and conditions for its emergence need to be isolated from the "what-to-do" question. This failure on the part of the causal theorists leads to their failure in appreciating the second-level truth about the paradox. Pivoting on the question "why is SP paradoxical?", we provide a general framework for understanding any paradox. Our analysis of

SP also highlights the significant role played by CP in generating the paradoxical result. Such principles are what Field would suggest we jettison to escape paradoxes.

Acknowledgments. We wish to thank Prajit Basu, Abhijit Dasgupta, S. G. Kulkarni, Vineet Nair, Davin Nelson and members of the Research Scholar Group of the University of Hyderabad, Philosophy Department (where an earlier version of the paper was presented) and three anonymous referees for their very helpful comments. We are indebted to John G. Bennett for several key email communications concerning the paradox.

References

1. Bandyopadhyay, P.S., Nelson, D., Greenwood, M., Brittan, G., Berwald, J.: The logic of Simpson's paradox. Synthese 181, 185–208 (2011)
2. Bandyopadhyay, P.S., Greenwood, M., Dcruz, Don Wallace F., Venkata Raghavan, R.: Simpson's Paradox and Causality. Am Philos Quart. (forthcoming)
3. Blyth, C.: On Simpson's Paradox and the Sure-Thing Principle. J. Am. Stat. Assoc. 67(338), 364–366 (1972)
4. Field, H.: Saving Truth from Paradox. Oxford University Press, Oxford (2009)
5. Glymour, C.: Critical Notice. James Woodward's Making Things Happen: A Theory of Causal Explanation. Brit. J. Philos. Sci. 55, 779–790 (2004)
6. Good, I., Mitttal, Y.: The amalgamation and geometry of two-by-two contingency tables. Annals of Statistics 15(2), 694–711 (1988)
7. Lycan, W.G.: What, exactly, is a paradox? Analysis 70(4), 615–622 (2010)
8. Meek, C., Glymour, C.: Conditioning and Intervening. Brit. J. Philos. Sci. 45, 1001–1021 (1994)
9. Pearl, J.: Causality. Cambridge University Press, Cambridge (2009)
10. Pearl, J.: Comment: Understanding Simpson's Paradox. Am. Stat. 68(1), 8–13 (2014)
11. Priest, G.: The logic of paradox. J. Philos. Logic. 8(1), 219–241 (1979)
12. Quine, W.V.O.: The Ways of Paradox and Other Essays. Revised and Enlarged. Harvard University Press, Cambridge (1976)
13. Sorensen, R.: A Brief History of the Paradox. Oxford University Press (2003)
14. Sainsbury, R.M.: Paradoxes, 3rd edn. Cambridge University Press, Cambridge (2009)
15. Spirtes, P., Glymour, C., Scheines, R.: Causation, Prediction, and Search. MIT Press, Cambridge (2000)
16. Wheeler, G.: Two Puzzles concerning measures of uncertainty and the positive Boolean connectives. In: Neves, J., Santos, M.F., Machado, J.M. (eds.) EPIA 2007. LNCS (LNAI), vol. 4874, pp. 170–180. Springer, Heidelberg (2007)
17. Yule, G.U.: Notes on the Theory of Association of Attributes in Statistics. Biometrika 2(2), 121–134 (1903)

Appendix[11]

For proving theorems 1 and 2 we firstly assume that the conditions of SP (arrived at in section 2.1) are satisfied. That is,

$$C \equiv (C_1 \& C_2 \& C_3)$$
$$\theta = (A_1 - B_1) + (A_2 - B_2) + (\beta - \alpha) > 0$$

Further, we stipulate the following definitions:

$$\text{a} = \text{(members of A in partition 1)/(total members of A)}$$
$$\text{b} = \text{(members of B in partition 1)/(total members of B)}$$
$$\alpha = aA_1 + A_2(1-a)$$
$$\beta = bB_1 + B_2(1-b)$$

A_1, A_2, B_1 and B_2 have the same meanings defined in section 2.1. We have defined α and β differently than what we had done in section 2.1 only to ease the proofs of the following theorems; otherwise the two sets of definitions are mathematically equivalent. To take an example, in Table 1 (Type I SP), $A_1 = 180/200$, $A_2 = 100/300$, $B_1 = 480/600$, $B_2 = 10/100$, $a = 200/500$, $b = 600/700$. Hence, $\alpha = (180/500) + (100/500) = 280/500 = 56\%$ and $\beta = (480/700) + (10/700) = 490/700 = 70\%$.

Theorem 1. *Simpson's paradox results only if $A_1 \neq A_2$.*

Proof: Let us assume that $A_1 = A_2$. Then, $\alpha = aA_1 + A_2(1-a) = aA_1 + A_2 - aA_2 = A_1 = A_2$. Given this, there are three possible scenarios. (I) $B_1 > B_2$, or (II) $B_1 < B_2$ or (III) $B_1 = B_2$.

(I) If $B_1 > B_2$, then $[B_1 b + B_1(1-b)] > [B_1 b + B_2(1-b)]$. Therefore, $B_1 > \beta$. Yet, if $A_1 \geq B_1$, and $\alpha = A_1$, then $\alpha > \beta$, which contradicts the assumption that $\beta \geq \alpha$. Therefore, if $A_1 = A_2$, then it can't be that $B_1 > B_2$.

(II) If $B_1 < B_2$, then $[B_1 b + B_2(1-b) < [B_2 b + B_2(1-b)] = B_2$. Therefore, $\beta < B_2$. Yet, $A_2 \geq B_2$, $A_1 \geq B_2$, and $\alpha \geq B_2 > \beta$. This contradicts the assumption that $\beta \geq \alpha$. Therefore, if $A_1 = A_2$, then it can't be the case that $B_1 < B_2$.

(III) If $B_1 = B_2$, then $\beta = bB_1 + B_2(1-b) = bB_1 + B_1(1-b) = B_1$. Given that $A_1 \geq B_1$, $A_1 = \alpha$, and $B_1 = \beta$, then $\alpha \geq \beta$. Yet, by assumption, $\beta \geq \alpha$. Therefore, $\beta = \alpha$. Since $A_1 = A_2 = \alpha$, and $B_1 = B_2 = \beta$, it must be that $A_1 = B_1$, $A_2 = B_2$, and $\alpha = \beta$. That $\alpha = \beta$ contradicts the assumption that out case is paradoxical, characterized by the reversal which we don't find here. Therefore, if $A_1 = A_2$, it can't be the case that $B_1 = B_2$. Therefore, $A_1 \neq A_2$. Without $A_1 \neq A_2$, Simpson's paradox cannot occur.

Theorem 2. *Simpson's paradox arises only if $B_1 \neq B_2$.*

Proof: Let us assume that $B_1 = B_2$. Then $\beta = bB_1 + B_2(1-b) = bB_1 + B_1(1-b) = B_1 = B_2$. Given that $A_1 > A_2$, it is true that $[aA_1 + A_2(1-a)] > [aA_2 + A_2(1-a)]$. Given that $A_1 > A_2$, it follows that $[aA_1 + A_2(1-a)] > [aA_2 + A_2(1-a)]$. Therefore, $\alpha > A_2$. Yet, $A_2 \geq B_2 = \beta$. So $\alpha > \beta$, which contradicts the assumption. Therefore, $B_1 \neq B_2$. Without $B_1 \neq B_2$, Simpson's paradox cannot occur.

[11] We are indebted to Davin Nelson for the following proofs.

Some Instances of Graded Consequence in the Context of Interval-Valued Semantics

Soma Dutta[1], Benjamín R.C. Bedregal[2], and Mihir Kr. Chakraborty[3]

[1] MIMUW, University of Warsaw, Poland
somadutta9@gmail.com
[2] UFRN, Natal, Brazil
bedregal@dimap.ufrn.br
[3] Jadavpur University, Kolkata, India
mihirc4@gmail.com

Abstract. This paper proposes some instances of graded consequence relation where the object language formulae are interpreted by sub-intervals of [0, 1]. These instances represent different attitudes of decision making that may be called conservative, liberal, and moderate.

Keywords: Graded consequence, Interval semantics, Imprecise reasoning.

1 Introduction

The theory of graded consequence (GCT) [6, 7] was introduced as a general meta-theory where for any set of formulae X and formula α, that a formula α follows from X is a matter of grade. Let us explain two main features of the theory of graded consequence.

(i) Classically, $X \models \alpha$ iff for all states of affair T_i if $X \subseteq T_i$ then $\alpha \in T_i$. Formally this meta-linguistic sentence turns out to be $\forall T_i\{(X \subseteq T_i) \rightarrow \alpha \in T_i\}$, where $X \subseteq T_i$ is again a meta-level sentence representing $\forall_{x \in F}(x \in X \rightarrow x \in T_i)$. In graded context, T_is are fuzzy sets assigning values to the object level formulae; and the meta-linguistic connective $\rightarrow$ and quantifier $\forall$, are computed by a fuzzy implication and the lattice 'infimum' operator respectively. Thus the sentence, 'α (semantically) follows from X' becomes graded; the grade is denoted by $gr(X \mid\approx \alpha)$. It is to be noted that following and extending [20], $\{T_i\}_{i \in I}$ is taken to be any arbitrary collection of fuzzy sets over formulae; that is considering the whole collection of T_i's is not a necessity here. In [7], a complete residuated lattice is considered for interpreting the meta-linguistic entities of the notion of graded consequence. So, given any collection of fuzzy sets $\{T_i\}_{i \in I}$, the meta-linguistic sentence viz., $X \mid\approx \alpha$ gets the value,

$$gr(X \mid\approx \alpha) = \inf_i\{inf_{x \in X} T_i(x) \rightarrow_f T_i(\alpha)\},$$

where $\rightarrow_f$ is the residuum operator of the complete residuated lattice.

(ii) GCT, thus, proposes a meta-theoretic set up where derivation is a graded notion. As a part of the programme of building the meta-theory, some of its areas of concern are (a) axiomatizing the notion of consequence ($\mid\sim$) (b) defining its semantic counterpart ($\mid\approx$), (c) studying their interrelations, and also (d) studying other meta-logical notions and their interrelations.

M. Banerjee and S.N. Krishna (eds.): ICLA 2015, LNCS 8923, pp. 74–87, 2015.

A graded consequence relation [7] is characterized as a fuzzy relation $|\!\sim$ from $P(F)$, the power set of formulae, to F, the set of formulae, satisfying the following axioms. These axioms are generalizations of that of the classical notion of consequence [21, 17].
(GC1) If $\alpha \in X$ then $gr(X |\!\sim \alpha) = 1$.
(GC2) If $X \subseteq Y$ then $gr(X |\!\sim \alpha) \leq gr(Y |\!\sim \alpha)$.
(GC3) $inf_{\beta \in Z}\, gr(X |\!\sim \beta) * gr(X \cup Z |\!\sim \alpha) \leq gr(X |\!\sim \alpha)$.

For each set of formulae X and formula α, $gr(X |\!\sim \alpha)$, read as the degree to which α is a consequence of X, is a member of the underlying set of the complete residuated lattice. The monoidal operation $*$, of the residuated lattice, computes meta-level conjunction. In [7] a representation theorem is proved establishing the soundness-completeness like connection between $|\!\sim$ and $|\!\approx$. There are several other papers [8, 9, 13, 14] where GCT is developed considering other meta-logical notions, their interrelations, axiomatic counterpart of graded consequence, its proof theory, and GCT in the context of fuzzy sets of premises too. It may be mentioned that some other researchers have also dealt with similar ideas and of them some contributed towards generalization of the above mentioned notion of GCT [5, 18, 19].

Development of GCT, to date, assumes a semantic base which is an arbitrary collection $\{T_i\}_{i\in I}$ of fuzzy sets assigning single values to the object language formulae. Each T_i may be counted as an expert whose opinion, i.e., values assigned to the object language formulae, forms the initial context or the database. Based on the collective database of $\{T_i\}_{i\in I}$ decisions are made. The decision maker wants to decide whether a particular formula α is a consequence of a set X of formulae, which is a matter of grade in GCT. In this paper we shall consider interval-valued semantics for the notion of graded consequence. More specifically, we shall consider that experts are allowed to assign an interval to the object language formulae, and then based on the mechanism of GCT the value to which *a formula is a consequence of a set of formulae* will be computed.

There are plenty of instances where it is impossible to claim *precisely* that *an imprecise concept applies to an object to a specific degree*. As a result when an imprecise concept is quantized by a single value, the inherent impreciseness of the concept is somewhat lost. Assigning an interval-value, to some extent, manages to retain the uncertainty *of understading an imprecise concept* as it only attaches a set of possible interpretations to the concept. In this regard let us quote a few lines from [10]. "*IVFS theory emerged from the observation that in a lot of cases no objective procedure is available to select the crisp membership degrees of elements in a fuzzy set. It was suggested to alleviate that problem by allowing to specify only an interval ... to which the actual membership degree is assumed to belong*. Thus interval mathematics and its application in the context of imprecise reasoning is quite significant. GCT provides a general set up for imprecise reasoning. So, developing GCT in the context of interval-valued semantics is meaningful both from the angle of theory building and applications. In this paper we shall present three different attitudes of decision making based on GCT. The information coming from different sources, which may be counted as the collection of T_i's, as well as the attitude (conservative, liberal, moderate) of the decision maker play roles in the process of decision making and in the final conclusion. Keeping this practical motivation in mind we here propose three different notions for deriving conclusion which satisfy the graded consequence axioms [7]. In each of these cases, the object language formulae are interpreted by closed sub-intervals of [0, 1], but the notion of consequence is made single-valued. This value assignment is done taking either the left-hand end point or the right-hand end point or some value in between from the final interval that is computed as an outcome. It is not completely unrealistic to think that experts i.e., T_i's are entitled to assign a range of values, but the decision maker is constrained to conclude a single value, and such a practice of precisification in final result prevails in the literature of fuzzy set theory, especially in the area of application of the theory. The meta-linguistic notions, e.g. consequence, consistency, inconsistency, could also get interval-values, and this direction of research will be taken up in our future work.

2 Interval Mathematics: Some Basic Notions

Assigning a specific grade to an imprecise sentence often pushes us into a situation where from a range of possible values we are to choose a single one for the sake of the mathematical ease of computation. Lifting the whole mathematics of fuzzy set theory in the context of interval-valued fuzzy set theory, researchers [1–4, 10, 12, 11, 15, 16] to a great extent could manage to resolve this problem. In this section we present some part of the development [2–4, 10, 12, 11, 16] according to the purpose of this paper.

Let us consider $\mathcal{U} = \{[a,b]/0 \leq a \leq b \leq 1\}$ along with two order relations $\leq_I$ and $\subseteq$, defined by: $[x_1,x_2] \leq_I [y_1,y_2]$ iff $x_1 \leq y_1$ and $x_2 \leq y_2$ and

$[x_1,x_2] \subseteq [y_1,y_2]$ iff $y_1 \leq x_1 \leq x_2 \leq y_2$.

$(\mathcal{U},\leq_I)$ forms a complete lattice, and $(\mathcal{U},\subseteq)$ forms a poset. Let $\bigwedge$ be the lattice meet corresponding to the order relation $\leq_I$.

Definition 2.1 [2]. An interval t-norm is a mapping $T : \mathcal{U} \times \mathcal{U} \mapsto \mathcal{U}$ such that T is commutative, associative, monotonic with respect to $\leq_I$ and $\subseteq$, and $[1,1]$ is the identity element with respect to T.

Definition 2.2. [16] Let T be an interval t-norm. T is called t-representable if there exists t-norms t_1,t_2 on $[0,1]$ such that $T([x_1,x_2],[y_1,y_2]) = [t_1(x_1,y_1),t_2(x_2,y_2)]$.

Definition 2.3. [16] For any t-norm $*$ on $[0,1]$ and $a \in [0,1]$, $T_{*,a}$ is defined below.
$T_{*,a}([x_1,x_2],[y_1,y_2]) = [x_1 * y_1, max((x_2 * y_2) * a, x_1 * y_2, x_2 * y_1)]$.

In [11] it has been shown that for any t-norm $*$ on $[0,1]$ and any $a \in [0,1]$, $T_{*,a}$ is an interval t-norm. Moreover, for $a = 1$, $T_{*,a}$ becomes a t-representable t-norm [16]; i.e. $T_{*,1}([x_1,x_2],[y_1,y_2]) = [x_1 * y_1, x_2 * y_2]$. For the purpose of this paper we shall consider such an $T_{*,1}$, and denote this interval t-norm based on $*$ as $*_I$.

Definition 2.4. Given $\rightarrow$, the residuum of $*$ on $[0,1]$, and $1 \in [0,1]$, $\rightarrow_I$: $\mathcal{U} \times \mathcal{U} \mapsto \mathcal{U}$ is defined by: $[x_1,x_2] \rightarrow_I [y_1,y_2] = [min\{x_1 \rightarrow y_1, x_2 \rightarrow y_2\}, min\{(x_2 * 1) \rightarrow y_2, x_1 \rightarrow y_2\}]$.
$= [min\{x_1 \rightarrow y_1, x_2 \rightarrow y_2\}, min\{x_2 \rightarrow y_2, x_1 \rightarrow y_2\}]$.

In [16] it is shown that $\rightarrow_I$ is an interval fuzzy implication with the adjoint pair $(*_I,\rightarrow_I)$ on $\mathcal{U}$. For $I_1,I_2,I' \in \mathcal{U}$, the following properties of $(*_I,\rightarrow_I)$ are of particular interest here.

(i) If $I_1 \leq_I I_2$ then $I_2 \to_I I \leq_I I_1 \to_I I$
(ii) If $I_1 \leq_I I_2$ then $I \to_I I_1 \leq_I I \to_I I_2$
(iii) $I \to_I I' \geq_I I'$
(iv) $[1,1] \to_I I = I$
(v) If $I \leq_I I'$ then $I \to_I I' = [1,1]$
(vi) $(I_1 \to_I I_2) *_I ((I_1 \wedge I_2) \to_I I) \leq_I (I_1 \to_I I)$.
(vii) $\bigwedge_k (I' \to_I I_k) = (I' \to_I \bigwedge_k I_k)$
(viii) $I_1 *_I I_2 \leq_I I_3$ iff $I_2 \leq_I I_1 \to_I I_3$.

3 GCT in the Context of Interval-Valued Semantics

In this section we propose a few different definitions for the semantic notion of graded consequence. These definitions incorporate different decision making attitudes from practical perspectives, and when the semantics for the object language formulae is restricted to the degenerate intervals, each of the notions yields the original notion of graded consequence [7, 8].

3.1 Graded Consequence: Form (Σ)

Definition 3.1. Given a collection of interval-valued fuzzy sets, say $\{T_i\}_{i\in I}$, the grade of $X \mid\approx \alpha$, i.e. $gr(X \mid\approx \alpha) = l([\bigwedge_{i\in I}\{\bigwedge_{x\in X} T_i(x) \to_I T_i(\alpha)\}])$, $\quad\ldots(\Sigma)$
where $l([.])$ represents the left-hand end point of an interval; that is, $l([x_1,x_2]) = x_1$.

The similarity and the differences between the notions of $gr(X \mid\approx \alpha)$, given in Section 1 and form (Σ), are quite visible from their respective forms. According to (Σ), to find out the degree to which α follows from X one has to first find out the truth-interval assignment to the formulae by a set of experts $\{T_i\}_{i\in I}$. Then, the left-hand end point of the least interval-value assigned to $\bigwedge_{x\in X} T_i(x) \to_I T_i(\alpha)$ (if every member of X is true then α is true) needs to be computed. In order to stick to a single value for the notion of derivation, in this case, the left-hand end point of the resultant interval is taken.

One might think that value for (Σ) would be the same if before computing $\to_I$ with the component intervals, the left-hand end point of the concerned intervals are taken out, and the corresponding implication operation for single-valued case is applied. In order to show that (Σ) is not the same as computing $\inf_i[l(\wedge_{x\in X} T_i(x)) \to l(T_i(\alpha)]$, let us consider $l([.3,.7] \to_I [.2,.3])$, where $\to_I$ is defined in terms of the ordinary Łukasiewicz implication following the Definition 2.4. Then it can be checked that $l([.3,.7] \to_I [.2,.3])$ $= .6$, and $l([.3,.7]) \to_Ł l([.2,.3]) = .9$.

Lemma 3.1. $\inf_i l(\{[x_i,y_i]\}_{i\in K}) = l(\bigwedge_i \{[x_i,y_i]\}_{i\in K})$.

Lemma 3.2. $\bigwedge_i I_i *_I \bigwedge_i I'_i \leq_I \bigwedge_i (I_i *_I I'_i)$, where I_i, I'_i are intervals.

Lemma 3.3. $\bigwedge_i \bigwedge_j I_{ij} = \bigwedge_j \bigwedge_i I_{ij}$.

Lemma 3.4. If $*_I$ is a t-representable t-norm with respect to an ordinary t-norm $*$ then $I_1 *_I I_2 \leq_I I_3$ implies $l(I_1) * l(I_2) \leq l(I_3)$.

Representation Theorems with Respect to (Σ)

Theorem 3.1. For any $\{T_i\}_{i\in I}$, $|\!\approx$ in the sense of (Σ) is a graded consequence relation.

Proof. (GC1) For $\alpha \in X$, $\bigwedge_{x\in X} T_i(x) \leq_I T_i(\alpha)$. So, using (v) we have $gr(X \mid\!\approx \alpha) = 1$.
(GC2) For $X \subseteq Y$, $\bigwedge_{x\in Y} T_i(x) \leq_I \bigwedge_{x\in X} T_i(x)$. Hence by (i) GC2 is immediate.

(GC3) $\inf_{\beta\in Z} gr(X \mid\!\approx \beta) * gr(X\cup Z \mid\!\approx \alpha)$
$= \inf_{\beta\in Z} l[\bigwedge_{i\in I}\{\bigwedge_{x\in X} T_i(x) \to_I T_i(\beta)\}] * l[\bigwedge_{i\in I}\{\bigwedge_{x\in X\cup Z} T_i(x) \to_I T_i(\alpha)\}]$
$= l[\bigwedge_{\beta\in Z}\{\bigwedge_{i\in I}(\bigwedge_{x\in X} T_i(x) \to_I T_i(\beta))\}] * l[\bigwedge_{i\in I}\{\bigwedge_{x\in X\cup Z} T_i(x) \to_I T_i(\alpha)\}]$ (Lemma 3.1)
$= l[\bigwedge_{i\in I}\{\bigwedge_{\beta\in Z}(\bigwedge_{x\in X} T_i(x) \to_I T_i(\beta))\}] * l[\bigwedge_{i\in I}\{\bigwedge_{x\in X\cup Z} T_i(x) \to_I T_i(\alpha)\}]$ (Lemma 3.3)
$= l[\bigwedge_{i\in I}(\bigwedge_{x\in X} T_i(x) \to_I \bigwedge_{\beta\in Z} T_i(\beta))\}] * l[\bigwedge_{i\in I}\{\bigwedge_{x\in X\cup Z} T_i(x) \to_I T_i(\alpha)\}]$ (by (vii))...(1)

Now for each T_i, $\{\bigwedge_{x\in X} T_i(x) \to_I \bigwedge_{\beta\in Z} T_i(\beta)\} *_I \{\bigwedge_{x\in X\cup Z} T_i(x) \to_I T_i(\alpha)\}$
$= \{\bigwedge_{x\in X} T_i(x) \to_I \bigwedge_{\beta\in Z} T_i(\beta)\} *_I \{[(\bigwedge_{x\in X} T_i(x)) \bigwedge (\bigwedge_{x\in Z} T_i(x))] \to_I T_i(\alpha)\}$
$\leq_I \{\bigwedge_{x\in X} T_i(x) \to_I T_i(\alpha)\}$. (by (vi)).

Therefore, $\bigwedge_{i\in I}[\{\bigwedge_{x\in X} T_i(x) \to_I \bigwedge_{\beta\in Z} T_i(\beta)\} *_I \{\bigwedge_{x\in X\cup Z} T_i(x) \to_I T_i(\alpha)\}]$

$$\leq_I \bigwedge_{i\in I}[\bigwedge_{x\in X} T_i(x) \to_I T_i(\alpha)]. \quad \ldots(2)$$

Also, $\bigwedge_{i\in I}[\{\bigwedge_{x\in X} T_i(x) \to_I \bigwedge_{\beta\in Z} T_i(\beta)\}] *_I \bigwedge_{i\in I}[\{\bigwedge_{x\in X\cup Z} T_i(x) \to_I T_i(\alpha)\}]$
$\leq_I \bigwedge_{i\in I}[\{\bigwedge_{x\in X} T_i(x) \to_I \bigwedge_{\beta\in Z} T_i(\beta)\} *_I \{\bigwedge_{x\in X\cup Z} T_i(x) \to_i T_i(\alpha)\}]$ (Lemma 3.2)
$\leq_I \bigwedge_{i\in I}[\bigwedge_{x\in X} T_i(x) \to_I T_i(\alpha)]$ (by (2))
$l[\bigwedge_{i\in I}(\bigwedge_{x\in X} T_i(x) \to_I \bigwedge_{\beta\in Z} T_i(\beta))\}] * l[\bigwedge_{i\in I}\{\bigwedge_{x\in X\cup Z} T_i(x) \to_I T_i(\alpha)\}]$
$\leq l[\bigwedge_{i\in I}\{\bigwedge_{x\in X} T_i(x) \to_I T_i(\alpha)\}]$ (Lemma 3.4 as $*_I$ is t-representable) ...(3)

Hence from (1) and (3) we have, $\inf_{\beta\in Z} gr(X \mid\!\approx \beta) * gr(X\cup Z \mid\!\approx \alpha) \leq gr(X \mid\!\approx \alpha)$.

Theorem 3.2. For any graded consequence relation $|\!\sim$ there is a collection of interval-valued fuzzy sets such that $|\!\approx$ generated in the sense of (Σ) coincides with $|\!\sim$.

Proof. Given $|\!\sim$, a graded consequence relation, let us consider the collection $\{T_X\}_{X\in P(F)}$ of interval-valued fuzzy sets over formulae such that $T_X(\alpha) = [gr(X \mid\!\sim \alpha)]_{BIR}$, where $[x]_{BIR}$ represents the best interval representation [3] of x, i.e. the interval $[x,x]$.

We want to prove $gr(X \mid\!\sim \alpha) = l[\bigwedge_{Y\in P(F)}\{\bigwedge_{x\in X} T_Y(x) \to_I T_Y(\alpha)\}]$.
By (GC3) and Lemma 3.1, we have $l(T_Y(\beta)) \geq l(T_{X\cup Y}(\beta)) * l(\wedge_{\alpha\in X} T_Y(\alpha))$.
As T_Y's are degenerate intervals by proposition 4.1 of [16] every true identity expressible in [0, 1] is expressible in $\mathcal{U}$. So, $T_{X\cup Y}(\beta) *_I \wedge_{\alpha\in X} T_Y(\alpha) \leq_I T_Y(\beta)$. Then following the proof in [7] we can show, $T_X(\beta) \leq_I \wedge_{Y\in P(F)}[\wedge_{\alpha\in X} T_Y(\alpha) \to_I T_Y(\beta)] \ldots$(i)
$\wedge_{Y\in P(F)}[\wedge_{\alpha\in X} T_Y(\alpha) \to_I T_Y(\beta)] \leq_I \wedge_{X\subseteq Y\in P(F)}[\wedge_{\alpha\in X} T_Y(\alpha) \to_I T_Y(\beta)] \ldots$(ii)
Now for $X \subseteq Y$, by (GC2), $T_X(\alpha) \leq_I T_Y(\alpha)$. Then by (GC1) we have
$[1,1] = \wedge_{\alpha\in X} T_X(\alpha) \leq_I \wedge_{\alpha\in X} T_Y(\alpha)$. i.e. $\wedge_{\alpha\in X} T_Y(\alpha) \to_I T_Y(\beta) = T_Y(\beta)$.
Hence (ii) becomes $\wedge_{Y\in P(F)}[\wedge_{\alpha\in X} T_Y(\alpha) \to_I T_Y(\beta)] \leq_I \wedge_{X\subseteq Y\in P(F)} T_Y(\beta) = T_X(\beta) \ldots$(iii)
Combining (i) and (iii) we have $gr(X \mid\!\sim \beta) = l[\wedge_{Y\in P(F)}\{\wedge_{\alpha\in X} T_Y(\alpha) \to_I T_Y(\beta)\}]$.

Remark 1. In this new context the meta-level algebraic structure may be viewed as $\langle \mathcal{U}, *_I, \to_I, [0,0], [1,1], l\rangle$, a complete residuated lattice with a function $l : \mathcal{U} \mapsto [0,1]$. The structure $\langle \mathcal{U}, *_I, \to_I, [0,0], [1,1], l\rangle$ is formed out of a complete residuated lattice $([0,1], *, \to, 0, 1)$. Specifically, the adjoint pair $(*_I, \to_I)$ is defined in terms of the adjoint pair $(*, \to)$.

Definition (Σ) endorses the minimum truth value assignment ensuring the limit, below which none of the assignments for 'if every member of X is true then α is true' i.e., $\bigwedge_{x\in X} T_i(x) \rightarrow_I T_i(\alpha)$, considering expert's opinion, lie. As an instance let, out of two experts the value assignment to $\bigwedge_{x\in X} T_i(x) \rightarrow_I T_i(\alpha)$ following one's opinion is $[.1,.3]$ and that of the other is $[.7,.9]$. Then following (Σ) the value for 'α follows from X' only considers the left-hand end point of the least interval i.e. $[.1,.3]$. So, this value assignment does not take care of the consensus of all. It only emphasises that the lower bound of everyone's point of agreeing is .1, no matter whether someone really has marked a high grade.

3.2 Extension of $\subseteq$ as a Lattice Order Relation

Let us now explore a method of assigning the value to 'α follows from X' in such a way that it takes care of every individual's opinion. That is, we are looking for an interval which lie in the intersection of everyone's opinion. For this we need a complete lattice structure with respect to the order relation $\subseteq$ on $\mathcal{U}$. Let us extend the partially ordered relation $\subseteq$ into a lattice order by the following definition.

Definition 3.2. $\subseteq_e$ is a binary relation on $\mathcal{U}$ defined as below.
$[x_1,x_2] \subseteq_e [y_1,y_2]$ if $y_1 \leq x_1 < x_2 \leq y_2$,
$[x_1,x_1] \subseteq_e [y_1,y_2]$ if $x_1 \leq y_2$.

Proposition 3.1. If $[x_1,x_2] \subseteq [y_1,y_2]$ then $[x_1,x_2] \subseteq_e [y_1,y_2]$.

Note 3.1. The converse of the above proposition does not hold. For the intervals $[.2,.2]$ and $[.3,.7]$, $[.2,.2] \subseteq_e [.3,.7]$, but $[.2,.2] \not\subseteq [.3,.7]$. Also to be noted that for two intervals $[x_1,x_2]$ and $[y_1,y_2]$, $[x_1,x_2] \subseteq_e [y_1,y_2]$ does not hold if $x_1 < x_2$ and $y_1 = y_2$. Let $[x_1,x_2] \subseteq_e [y_1,y_2]$ be such that $x_1 = x_2 < y_1 \leq y_2$ holds. This pair of intervals are called non-overlapping intervals under the relation $\subseteq_e$; other pairs of intervals under the relation $\subseteq_e$ are known as overlapping intervals under the relation $\subseteq_e$.

Proposition 3.2. $(\mathcal{U},\subseteq_e)$ forms a poset.

Proposition 3.3. $(\mathcal{U},\subseteq_e)$ is a lattice where the greatest lower bound, say $\bigcap$, and the least upper bound, say $\bigcup$ are defined as follows.

$$[x_1,x_2]\bigcap[y_1 y_2] = [\max(x_1,y_1),\min(x_2,y_2)] \text{ if } \max(x_1,y_1) \leq \min(x_2,y_2)$$
$$= [\min(x_2,y_2),\min(x_2,y_2)], \text{ otherwise.}$$
$$[x_1,x_2]\bigcup[y_1 y_2] = [\max(x_2,y_2),\max(x_2,y_2)], \text{ if } x_1 = x_2,\ y_1 = y_2.$$
$$= [\max(x_1,y_1),\max(x_2,y_2)], \text{ if } x_1 = x_2 < y_1 < y_2$$
$$= [\min(x_1,y_1),\max(x_2,y_2)], \text{ otherwise.}$$
$$\text{(i.e. either } x_1 = x_2,\ y_1 < y_2,\ y_1 \leq x_1, \text{ or } x_1 < x_2,\ y_1 < y_2)$$

Proposition 3.4. $(\mathcal{U},\subseteq_e)$ forms a complete lattice.
Proof For arbitrary collection $\{[x_i,y_i]\}_{i\in K}$, $\bigcap_{i\in K}[x_i,y_i] = [\sup_i x_i, \inf_i y_i]$ if $\sup_i x_i \leq \inf_i y_i$
$= [\inf_i y_i, \inf_i y_i]$, otherwise.

$\bigcup_{i\in K}[x_i,y_i] = [\inf_i x_i, \sup_i y_i]$, if $\inf_i x_i < \inf_i y_i \leq \sup_i y_i$
$= [\sup_i x_i, \sup_i y_i]$, if $\inf_i x_i = \inf_i y_i \leq \sup_i y_i$.
Rest is straightforward as being a closed set, $[0,1]$ contains infimum and supremum of x_i's and y_i's.

3.3 Graded Consequence: Form (Σ')

Now let us define an alternative definition for semantic graded consequence relation which takes care of the common consensus zone.

Definition 3.3. Given any collection of interval-valued fuzzy sets, say $\{T_i\}_{i\in I}$,
$gr(X \mid\approx \alpha) = r([\bigcap_{i\in I}\{\bigwedge_{x\in X} T_i(x) \to_I T_i(\alpha)\}])$, where $r([x_1,x_2]) = x_2$. ...(Σ')

So, in this definition for graded semantic consequence the value of 'α follows from X' is the right-hand end point of the common interval-value assigned to the sentence 'whenever every member of X is true α is also true' taking care of every expert's opinion; that is, this method counts the maximum truth-value assignment where all the experts agree.

Lemma 3.5. $\inf_i r([x_i,y_i]_{i\in K}) = r(\bigcap_{i\in K}[x_i,y_i])$.

Lemma 3.6. $\bigcap_l \bigcap_k I_{lk} = \bigcap_k \bigcap_l I_{lk}$ for each $I_{lk} \in \mathcal{U}$.

Lemma 3.7. If $*_I$ is a t-representable t-norm with respect to an ordinary t-norm $*$ then $I_1 *_I I_2 \leq_I I_3$ implies $r(I_1) * r(I_2) \leq r(I_3)$.

Lemma 3.8. If $\{[x_i^1,x_i^2]\}_{i\in K}$ and $\{[y_i^1,y_i^2]\}_{i\in K}$ are two collections of intervals such that $[x_i^1,x_i^2] \leq_I [y_i^1,y_i^2]$ for each $i \in K$, then $\bigcap_{i\in K}[x_i^1,x_i^2] \leq_I \bigcap_{i\in K}[y_i^1,y_i^2]$.

Proof. $x_i^1 \leq y_i^1$ and $x_i^2 \leq y_i^2$ for each $i \in K$.
Hence, $\sup\{x_i^1\}_{i\in K} \leq \sup\{y_i^1\}_{i\in K}$ and $\inf\{x_i^2\}_{i\in K} \leq \inf\{y_i^2\}_{i\in K}$. ...(1)
As $x_i^1 \leq x_i^2$ for $i \in K$, there are two possibilities - (i) $\sup_i x_i^1 \leq \inf_i x_i^2$, (ii) $\sup_i x_i^1 > \inf_i x_i^2$.
(i) $\sup_i x_i^1 \leq \inf_i x_i^2 \leq x_i^2 \leq y_i^2$ for each $i \in K$. So, $\sup_i x_i^1 \leq \inf_i x_i^2 \leq \inf_i y_i^2$.
Here again two subcases arise. (ia) $\sup_i y_i^1 \leq \inf_i y_i^2$ and (ib) $\sup_i y_i^1 > \inf_i y_i^2$.
(ia) If $\sup_i y_i^1 \leq \inf_i y_i^2$, then $\bigcap_{i\in K}[y_i^1,y_i^2] = [\sup_i y_i^1, \inf_i y_i^2]$.
Hence inequalities of (1) ensure that $\bigcap_{i\in K}[x_i^1,x_i^2] \leq_I \bigcap_{i\in K}[y_i^1,y_i^2]$.
(ib) If $\sup_i y_i^1 > \inf_i y_i^2$, then $\bigcap_{i\in K}[y_i^1,y_i^2] = [\inf_i y_i^2, \inf_i y_i^2]$.
Again from (i) $\sup_i x_i^1 \leq \inf_i x_i^2 \leq \inf_i y_i^2$ implies $\bigcap_{i\in K}[x_i^1,x_i^2] \leq_I \bigcap_{i\in K}[y_i^1,y_i^2]$.
(ii) $\inf_i x_i^2 < \sup_i x_i^1$ implies $\inf_i x_i^2 < \sup_i x_i^1 \leq \sup_i y_i^1$.
So, $\bigcap_{i\in K}[x_i^1,x_i^2] = [\inf_i x_i^2, \inf_i x_i^2] \leq_I \bigcap_{i\in K}[y_i^1,y_i^2]$ since $\inf_i x_i^2 \leq \sup_i y_i^1$ and $\inf_i x_i^2 \leq \inf_i y_i^2$.

Lemma 3.9. For any collection $\{I_j\}_{j\in J}$ of intervals, $\bigwedge_{j\in J} I_j \leq_I \bigcap_{j\in J} I_j$.

Lemma 3.10. For any collection $\{I_j\}_{j\in J}$ of intervals, $\bigcap_{j\in J} I_j \subseteq_e \bigwedge_{j\in J} I_j$.

Corollary 3.1. $r[\bigcap_{j\in J}\bigcap_{l\in L} I_{lj}] \leq r[\bigcap_{j\in J}\bigwedge_{l\in L} I_{lj}]$.

Representation Theorems with Respect to (Σ')

Theorem 3.3. For any $\{T_i\}_{i\in I}$, $|\approx$ in the sense of (Σ') is a graded consequence relation.

Proof. (GC1) is proved as in Theorem 3.1. (GC2) is obtained using (i) and Lemma 3.8.

(GC3) $\inf_{\beta\in Z} gr(X \mid\approx \beta) * gr(X\cup Z \mid\approx \alpha)$
$= \inf_{\beta\in Z} r[\bigcap_{i\in I}\{\bigwedge_{x\in X} T_i(x) \rightarrow_I T_i(\beta)\}] * r[\bigcap_{i\in I}\{\bigwedge_{x\in X\cup Z} T_i(x) \rightarrow_I T_i(\alpha)\}]$
$= r[\bigcap_{\beta\in Z}\{\bigcap_{i\in I}(\bigwedge_{x\in X} T_i(x) \rightarrow_I T_i(\beta))\}] * r[\bigcap_{i\in I}\{\bigwedge_{x\in X\cup Z} T_i(x) \rightarrow_I T_i(\alpha)\}]$ (Lemma 3.5)
$= r[\bigcap_{i\in I}\{\bigcap_{\beta\in Z}(\bigwedge_{x\in X} T_i(x) \rightarrow_I T_i(\beta))\}] * r[\bigcap_{i\in I}\{\bigwedge_{x\in X\cup Z} T_i(x) \rightarrow_I T_i(\alpha)\}]$ (Lemma 3.6)
$\leq r[\bigcap_{i\in I}\{\bigwedge_{\beta\in Z}(\bigwedge_{x\in X} T_i(x) \rightarrow_I T_i(\beta))\}] * r[\bigcap_{i\in I}\{\bigwedge_{x\in X\cup Z} T_i(x) \rightarrow_I T_i(\alpha)\}]$ (Corollary 3.1)
$= r[\bigcap_{i\in I}(\bigwedge_{x\in X} T_i(x) \rightarrow_I \bigwedge_{\beta\in Z} T_i(\beta))\}] * r[\bigcap_{i\in I}\{\bigwedge_{x\in X\cup Z} T_i(x) \rightarrow_I T_i(\alpha)\}]$ (by (vii))...(1)
Also we obtain, $r[\bigcap_{i\in I}[\{\bigwedge_{x\in X} T_i(x) \rightarrow_I \bigwedge_{\beta\in Z} T_i(\beta)\} *_I \{\bigwedge_{x\in X\cup Z} T_i(x) \rightarrow_I T_i(\alpha)\}]]$
$\leq r[\bigcap_{i\in I}\{\bigwedge_{x\in X} T_i(x) \rightarrow_I T_i(\alpha)\}]$ (by (vi) and Lemma 3.8). ...(2)

Now $\bigwedge_{i\in I}(\bigwedge_{x\in X} T_i(x) \rightarrow_I \bigwedge_{\beta\in Z} T_i(\beta))\} *_I \bigwedge_{i\in I}\{\bigwedge_{x\in X\cup Z} T_i(x) \rightarrow_I T_i(\alpha)\}$
$\leq_I \bigwedge_{i\in I}[\{\bigwedge_{x\in X} T_i(x) \rightarrow_I \bigwedge_{\beta\in Z} T_i(\beta)\} *_I \{\bigwedge_{x\in X\cup Z} T_i(x) \rightarrow_I T_i(\alpha)\}]$ (Lemma 3.2)
$\leq_I \bigcap_{i\in I}[\{\bigwedge_{x\in X} T_i(x) \rightarrow_I \bigwedge_{\beta\in Z} T_i(\beta)\} *_I \{\bigwedge_{x\in X\cup Z} T_i(x) \rightarrow_I T_i(\alpha)\}]$. (Lemma 3.9)
i.e., $r[\bigwedge_{i\in I}(\bigwedge_{x\in X} T_i(x) \rightarrow_I \bigwedge_{\beta\in Z} T_i(\beta))\}] * r[\bigwedge_{i\in I}\{\bigwedge_{x\in X\cup Z} T_i(x) \rightarrow_I T_i(\alpha)\}]$
$\leq r[\bigcap_{i\in I}[\{\bigwedge_{x\in X} T_i(x) \rightarrow_I \bigwedge_{\beta\in Z} T_i(\beta)\} *_I \{\bigwedge_{x\in X\cup Z} T_i(x) \rightarrow_I T_i(\alpha)\}]]$. (Lemma 3.7)...(3)

$\bigcap_{i\in I}(\bigwedge_{x\in X} T_i(x) \rightarrow_I \bigwedge_{\beta\in Z} T_i(\beta))\} \subseteq_e \bigwedge_{i\in I}(\bigwedge_{x\in X} T_i(x) \rightarrow_I \bigwedge_{\beta\in Z} T_i(\beta))\}$. (Lemma 3.10)
So, $r[\bigcap_{i\in I}(\bigwedge_{x\in X} T_i(x) \rightarrow_I \bigwedge_{\beta\in Z} T_i(\beta))\}] \leq r[\bigwedge_{i\in I}(\bigwedge_{x\in X} T_i(x) \rightarrow_I \bigwedge_{\beta\in Z} T_i(\beta))\}]$.
Thus we have, $r[\bigcap_{i\in I}(\bigwedge_{x\in X} T_i(x) \rightarrow_I \bigwedge_{\beta\in Z} T_i(\beta))\}] * r[\bigcap_{i\in I}\{\bigwedge_{x\in X\cup Z} T_i(x) \rightarrow_I T_i(\alpha)\}]$
$\leq r[\bigwedge_{i\in I}(\bigwedge_{x\in X} T_i(x) \rightarrow_I \bigwedge_{\beta\in Z} T_i(\beta))\}] * r[\bigwedge_{i\in I}\{\bigwedge_{x\in X\cup Z} T_i(x) \rightarrow_I T_i(\alpha)\}]$
$\leq r[\bigcap_{i\in I}[\{\bigwedge_{x\in X} T_i(x) \rightarrow_I \bigwedge_{\beta\in Z} T_i(\beta)\} *_I \{\bigwedge_{x\in X\cup Z} T_i(x) \rightarrow_I T_i(\alpha)\}]]$ (by 3)
$\leq r[\bigcap_{i\in I}\{\bigwedge_{x\in X} T_i(x) \rightarrow_I T_i(\alpha)\}]$ (by (2)).

Hence (GC3) is proved.

Theorem 3.4. For any graded consequence relation $|\sim$ there is a collection of interval-valued fuzzy sets such that $|\approx$ generated in the sense of (Σ') coincides with $|\sim$.
Proof. Given a graded consequence relation $|\sim$, let $\{T_X\}_{X\in P(F)}$ be such that $T_X(\alpha) = [gr(X \mid\sim \alpha)]_{BIR}$, and we want to show $gr(X \mid\sim \alpha) = r[\bigcap_{Y\in P(F)}\{\bigwedge_{x\in X} T_Y(x) \rightarrow_i T_Y(\alpha)\}]$.
Arguing as Theorem 3.2 we have, $T_X(\beta) \leq_I \wedge_{Y\in P(F)}[\wedge_{\alpha\in X} T_Y(\alpha) \rightarrow_I T_Y(\beta)]$
$\subseteq_e \cap_{Y\in P(F)}[\wedge_{\alpha\in X} T_Y(\alpha) \rightarrow_I T_Y(\beta)]$ (Lemma 3.9)
$r(T_X(\beta)) \leq r(\wedge_{Y\in P(F)}[\wedge_{\alpha\in X} T_Y(\alpha) \rightarrow_I T_Y(\beta)]) \leq r(\cap_{Y\in P(F)}[\wedge_{\alpha\in X} T_Y(\alpha) \rightarrow_I T_Y(\beta)])$...(1)
$\cap_{Y\in P(F)}[\wedge_{\alpha\in X} T_Y(\alpha) \rightarrow_I T_Y(\beta)] \subseteq_e \cap_{X\subseteq Y\in P(F)}[\wedge_{\alpha\in X} T_Y(\alpha) \rightarrow_I T_Y(\beta)]$ ...(2)
Following the proof of Theorem 3.2, for $X \subseteq Y$, we have $\wedge_{\alpha\in X} T_Y(\alpha) \rightarrow_I T_Y(\beta) = T_Y(\beta)$.
From (2) and Lemma 3.10, $\cap_{Y\in P(F)}[\wedge_{\alpha\in X} T_Y(\alpha) \rightarrow_I T_Y(\beta)] \subseteq_e \wedge_{X\subseteq Y} T_Y(\beta) = T_X(\beta)$.
Thus, $r(T_X(\beta)) = r(\cap_{Y\in P(F)}[\wedge_{\alpha\in X} T_Y(\alpha) \rightarrow_I T_Y(\beta)]) = gr(X \mid\sim \alpha)$.

Here the same structure $\langle \mathcal{U}, *_I, \rightarrow_I, [0,0], [1,1], r\rangle$, as mentioned in Remark 1, is taken; only the differences are: (i) $\mathcal{U}$ is endowed with both the lattice order relations $\leq_I$ and $\subseteq_e$, and (ii) a function $r : \mathcal{U} \mapsto [0,1]$, different from l, is considered here.

Let us present, below, a diagram to visualize the beauty and purpose of dealing with two lattice structures over the same domain. We consider a linear scale D and intervals over D. We consider $D = \{3,5,7,9\}$ and $I_D = \{[a,b] : a \leq b \text{ and } a,b \in D\}$.

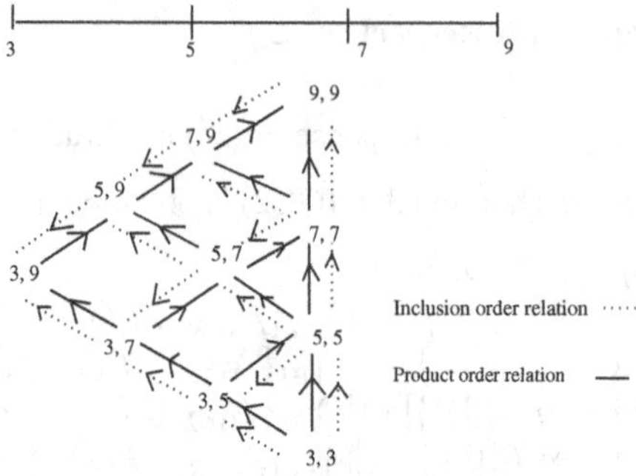

The diagram represents two lattice structures with respect to $\leq_I$ (product order relation) and $\subseteq_e$ (inclusion order relation) over I_D. Let $\{T_1, T_2, T_3\}$ be the set of experts who assign interval-value from I_D to every formula. Assuming the ordinary t-norm $\wedge$ on D, one can immediately obtain the corresponding residuum $\rightarrow$ on D, and hence $\wedge_{I_D}$, $\rightarrow_{I_D}$ on I_D can be constructed. Given a set of formulae X and a formula α, to compute $gr(X \mid\approx \alpha)$ in the sense of both (Σ) and (Σ'), first for each T_i (i = 1, 2, 3), the value $\bigwedge_{x \in X} T_i(x) \rightarrow_{I_D} T_i(\alpha)$ needs to be computed. Let the respective values corresponding to T_1, T_2, T_3 be $[3,5]$, $[5,7]$, $[7,7]$. Then, for (Σ), the least interval $[3,5]$ will be selected and the left-hand end point 3 would be counted as the grade of $X \mid\approx \alpha$. Following (Σ'), $[5,5]$ will be chosen as the interval included in all the intervals in the sense of $\subseteq_e$, and its right-hand end point 5 would be counted as the grade of $X \mid\approx \alpha$. The first method pulls down one of the experts high opinion, which is here 7, drastically to 3; whereas the second admits some room for adjustment between different opinions, and pulls down the value to 5. Thus, (Σ') provides a good sense of respecting individual's opinion.

3.4 Graded Consequence: Form (Σ'')

Among the above two forms of graded consequence, (Σ) is based on a *conservative* attitude as it choses the left-hand end point of the interval lying below each expert's (T_i) opinion for computing the value of $\wedge_{x \in X} T_i(x) \rightarrow_I T_i(\alpha)$. On the other hand, form (Σ') admits very *liberal* attitude as it takes the right-hand end point of the interval-value which lies at the common consensus zone of the values for $\wedge_{x \in X} T_i(x) \rightarrow_I T_i(\alpha)$ taking care of every expert's opinion. Both of these reflect two extremities of decision making attitude. Below we would look for an approach where considering each expert's opinion first an interval for $\wedge_{x \in X} T_i(x) \rightarrow_I T_i(\alpha)$ is assigned. Then, a number of times the assigned interval can be revised; the number being stipulated by different constraints. Finally among these iterations for the values of $\wedge_{x \in X} T_i(x) \rightarrow_I T_i(\alpha)$ one would be chosen, and from all such revised interval-values the common zone will be selected. This idea of iterative revision of an interval-value assignment is taken care of in the following series of definitions. Finally Theorem 3.8 of this section throws light on the fact that the forms generated from iterative-revisions (Σ_1'', Σ_2'') retain a place between the two extreme attitudes of decision making.

Definition 3.4. $I_{[a,b]}$ is a collection of iterated revisions $[x^i, y^i]$'s of $[a,b]$, given by: $I_{[a,b]} = \{[x^i, y^i] : x^0 = a, y^0 = b, [x^i, y^i] \subseteq_e [x^{i-1}, y^{i-1}]$ and $[x^{i-1}, y^{i-1}]$ is non-degenerate$\}$.

Let $I^j_{[a,b]}(\subseteq I_{[a,b]})$ be a set containing iterated revisions of $[a,b]$ upto j-th iterations.

Definition 3.5. C_I is a choice function over $\{I^j_{[a,b]} : j \geq 0, [a,b] \in \mathcal{U}\}$ such that (i) $C_I(I^j_{[a,b]}) \in I^j_{[a,b]}$ and (ii) $[a,b] \leq_I [c,d]$ implies $C_I(I^j_{[a,b]}) \leq_I C_I(I^j_{[c,d]})$.

Let us present one such case of iterative-revision of intervals below.

Definition 3.6. For $\varepsilon > 0$, $\underline{C}_\varepsilon$: $\mathcal{U} \mapsto \mathcal{U}$ such that $\underline{C}_\varepsilon([a,b]) = [a+\varepsilon, b]$ if $a+\varepsilon < b$
$= [b,b]$, otherwise.
$\overline{C}_\varepsilon$: $\mathcal{U} \mapsto \mathcal{U}$ such that $\overline{C}_\varepsilon([a,b]) = [a, b-\varepsilon]$ if $a < b-\varepsilon$
$= [a,a]$, otherwise.
$\mathcal{C}_\varepsilon$: $\mathcal{U} \mapsto \mathcal{U}$ such that $\mathcal{C}_\varepsilon([a,b]) = \underline{C}_\varepsilon([a,b]) \cap \overline{C}_\varepsilon([a,b])$.

Let us choose an arbitrarily fixed number n. Given $[a,b]$, fixing $\varepsilon \geq \frac{b-a}{n}$ and applying $\mathcal{C}_\varepsilon$ finite number of times on $[a,b]$, one instance of $I_{[a,b]}$, we call $I^\varepsilon_{[a,b]}$, can be obtained in the following way.

Definition 3.7. $I^\varepsilon_{[a,b]} = \{\mathcal{C}^i_\varepsilon([a,b]) : i \geq 0, \mathcal{C}^0_\varepsilon([a,b]) = [a,b], \mathcal{C}^{i-1}_\varepsilon([a,b])$ is non-degenerate$\}$.

Note 3.2. Taking $\varepsilon = \frac{1}{2}$ we have $I^{\frac{1}{2}}_{[.3,.7]} = \{[.3,.7],[.3,.3]\}$, $I^{\frac{1}{2}}_{[.1,.9]} = \{[.1,.9], [.4,.4]\}$. It is to be noted that as $I^{\frac{1}{2}}_{[.3,.7]}$ contains only two iterations $I^{\frac{1}{2}}_{[.3,.7]} = I^{\frac{1}{2},j}_{[.3,.7]} = \{[.3,.7],[.3,.3]\}$ for any number of iterations $j \geq 2$, where $I^{\frac{1}{2},j}_{[.3,.7]}$ contains intervals upto j-th iterations.

Definition 3.8. $C_I(I^{\varepsilon,j}_{[a,b]}) = \cap\, I^{\varepsilon,j}_{[a,b]}$, where j denotes the number of iterations.

Note 3.3. As I^ε is obtained by finitely many iterations, $C_I(I^\varepsilon_{[a,b]}) = \cap\, I^\varepsilon_{[a,b]}$, which is a degenerate interval. Clearly, C_I (of Definition 3.8) satisfies condition (i) of the Definition 3.5 as $C_I(I^{\varepsilon,j}_{[a,b]}) = \mathcal{C}^j_\varepsilon([a,b]) \in I^\varepsilon_{[a,b]}$. To check that C_I also satisfies condition (ii) of Definition 3.5 we need to prove a series of results below.

Proposition 3.5. $\mathcal{C}_\varepsilon([a,b]) \subseteq_e [a,b]$.

Proof. Two cases arise. (i) $a+\varepsilon < b$ i.e. $a < b-\varepsilon$ (ii) otherwise.
For all these cases the result is straightforward from the definitions of $\subseteq_e$ and $\cap$.

Theorem 3.5. $[a,b] \leq_I [c,d]$ implies $\mathcal{C}_\varepsilon([a,b]) \leq_I \mathcal{C}_\varepsilon([c,d])$.

Proof. Let $[a,b] \leq_I [c,d]$, then $a \leq c$, $b \leq d$, and $a \leq b$, $c \leq d$.
Now (i) $\underline{C}_\varepsilon([a,b]) = [a+\varepsilon, b]$ or (ii) $\underline{C}_\varepsilon([a,b]) = [b,b]$.
(i) Let $\underline{C}_\varepsilon([a,b]) = [a+\varepsilon, b]$ i.e. $a+\varepsilon < b \leq d$. Also $a+\varepsilon \leq c+\varepsilon$.
Now either $c+\varepsilon < d$ or $d \leq c+\varepsilon$. Both the cases yield $\underline{C}_\varepsilon([a,b]) \leq_I \underline{C}_\varepsilon([c,d])$, as for $c+\varepsilon < d$, $\underline{C}_\varepsilon([c,d]) = [c+\varepsilon, d]$. i.e., $[a+\varepsilon, b] \leq_I [c+\varepsilon, d]$, and for $d \leq c+\varepsilon$, $\underline{C}_\varepsilon([c,d]) = [d,d]$ i.e., $[a+\varepsilon, b] \leq_I [d,d]$, since $a+\varepsilon < b \leq d$.
(ii) Let $\underline{C}_\varepsilon([a,b]) = [b,b]$, i.e. $b \leq a+\varepsilon \leq c+\varepsilon$ and also $b \leq d$.

Now either $c+\varepsilon<d$ or $d\leq c+\varepsilon$. If $c+\varepsilon<d$, $\underline{C}_\varepsilon([c,d])=[c+\varepsilon,d]$. So, $[b,b]\leq_I[c+\varepsilon,d]$. And if $d\leq c+\varepsilon$, then $\underline{C}_\varepsilon([c,d])=[d,d]$ i.e., $[b,b]\leq_I[c+\varepsilon,d]$. Hence $[a,b]\leq_I[c,d]$ implies $\underline{C}_\varepsilon([a,b])\leq_I\underline{C}_\varepsilon([c,d])$.
Now we shall check the case for $\overline{C}_\varepsilon$. Let $[a,b]\leq_I[c,d]$.
There are two cases. (a) $\overline{C}_\varepsilon([a,b])=[a,b-\varepsilon]$ (b) $\overline{C}_\varepsilon([a,b])=[a,a]$.
(a) If $\overline{C}_\varepsilon([a,b])=[a,b-\varepsilon]$, then $a<b-\varepsilon<b\leq d$ and $b-\varepsilon\leq d-\varepsilon$.
Now either $c<d-\varepsilon$ or $d-\varepsilon\leq c$.
If $c<d-\varepsilon$, $\overline{C}_\varepsilon([c,d])=[c,d-\varepsilon]$. i.e., $[a,b-\varepsilon]\leq_I[c,d-\varepsilon]$ as $a\leq c$, $b\leq d$.
If $d-\varepsilon\leq c$, $\overline{C}_\varepsilon([c,d])=[c,c]$, So, $[a,b-\varepsilon]\leq_I[c,c]$ as $a\leq c$, $b-\varepsilon\leq d-\varepsilon\leq c$.
(b) Let $\overline{C}_\varepsilon([a,b])=[a,a]$. Then $b-\varepsilon\leq a\leq b\leq d$, $a\leq c$.
Now either $c<d-\varepsilon$ or $d-\varepsilon\leq c$.
If $c<d-\varepsilon$, $\overline{C}_\varepsilon([c,d])=[c,d-\varepsilon]$. i.e., $[a,a]\leq_I[c,d-\varepsilon]$ since $a\leq c<d-\varepsilon$.
And if $d-\varepsilon\leq c$, $\overline{C}_\varepsilon([c,d])=[c,c]$. So, $[a,a]\leq_I[c,c]$.
Therefore, $[a,b]\leq_I[c,d]$ implies $\overline{C}_\varepsilon([a,b])\leq_I\overline{C}_\varepsilon([c,d])$.
Hence, $C_\varepsilon([a,b])=\underline{C}_\varepsilon([a,b])\cap\overline{C}_\varepsilon([a,b])\leq_I\underline{C}_\varepsilon([c,d])\cap\overline{C}_\varepsilon([c,d])=C_\varepsilon([c,d])$.

Corollary 3.2. $[a,b]\leq_I[c,d]$ implies $C_I(I^{\varepsilon,j}_{[a,b]})\leq_I C_I(I^{\varepsilon,j}_{[c,d]})$.

Theorem 3.6. $[a,b]\leq_I[c,d]$ implies $C_I(I^{\varepsilon}_{[a,b]})\leq_I C_I(I^{\varepsilon}_{[c,d]})$.

Proof. Let $C^i_\varepsilon([a,b])$, $C^j_\varepsilon([c,d])$ be degenerate. Then either (i) $i=j$ or (ii) $i\leq j$, or (iii) $j\leq i$. (i) If $i=j$, by Theorem 3.5, $C_I(I^\varepsilon_{[a,b]})=C^i_\varepsilon([a,b])\leq_I C^i_\varepsilon([c,d])=C_I(I^\varepsilon_{[c,d]})$. (ii) By Theorem 3.5, $l(C_I(I^\varepsilon_{[a,b]}))=l(C^i_\varepsilon([a,b]))\leq l(C^i_\varepsilon([c,d]))\leq l(C^j_\varepsilon([c,d]))=l(C_I(I^\varepsilon_{[c,d]}))$. That is, $C_I(I^\varepsilon_{[a,b]})\leq_I C_I(I^\varepsilon_{[c,d]})$. (iii) If $j\leq i$, $r(C_I(I^\varepsilon_{[c,d]}))=r(C^j_\varepsilon([c,d]))\geq r(C^j_\varepsilon([a,b]))$. By Note 3.2, $r(C^j_\varepsilon([a,b]))$= $r(C^i_\varepsilon([a,b]))=r(C_I(I_{[a,b]}))$ i.e., $C_I(I^\varepsilon_{[a,b]})\leq_I C_I(I^\varepsilon_{[c,d]})$.

Thus we have shown existence of a C_I which satisfies conditions of Definition 3.5. Now we come back to the general context of the Definition 3.5 for C_I.

Note 3.4. $C_I(I^j_{[a,b]})$, $C_I(I_{[a,b]})\subseteq_e[a,b]$, and $C_I(I^j_{[a,b]})$, $C_I(I_{[a,b]})$, $[a,b]$ are mutually overlapping pairs of intervals. And for I^ε, $C_I(I^\varepsilon_{[a,b]})\subseteq_e C_I(I^{\varepsilon,j}_{[a,b]})\subseteq_e[a,b]$

We now propose different notions of $|\!\approx$ based on the notion of iterative revision.

Definition 3.9. Given any collection of interval-valued fuzzy sets $\{T_i\}_{i\in I}$,
(Σ_1'') $gr(X\,|\!\approx^{(\Sigma_1'')}\alpha)=r(\cap_{i\in I}\{C_I(I^j_{[\bigwedge_{x\in X}T_i(x)\rightarrow_I T_i(\alpha)]})\})$ for an arbitrarily fixed $j\geq 0$,
(Σ_2'') $gr(X\,|\!\approx^{(\Sigma_2'')}\alpha)=r(\cap_{i\in I}\{C_I(I^\varepsilon_{[\bigwedge_{x\in X}T_i(x)\rightarrow_I T_i(\alpha)]})\})=l(\cap_{i\in I}\{C_I(I^\varepsilon_{[\bigwedge_{x\in X}T_i(x)\rightarrow_I T_i(\alpha)]})\})$.

Theorem 3.7. For any graded consequence relation $|\!\sim$, there is a collection interval-valued fuzzy sets such that $|\!\approx$ generated in the sense of (Σ_1''), (Σ_2'') coincide with $|\!\sim$.

Proof. Given $|\!\sim$, a graded consequence relation, let us consider $\{T_X\}_{X\in P(F)}$ such that T_X (α)=$[gr(X\,|\!\sim\alpha)]_{BIR}$. We want to prove $gr(X|\!\sim\alpha)$=$r[\bigcap_{Y\in P(F)}\{C_I(I^i_{[\bigwedge_{x\in X}T_Y(x)\rightarrow_I T_Y(\alpha)]})\}]$

considering (Σ_1''), and $gr(X \mid\sim \alpha) = r[\bigcap_{Y \in P(F)}\{C_I(I^{\varepsilon}_{[\bigwedge_{x\in X} T_Y(x) \to_I T_Y(\alpha)]})\}]$ considering (Σ_2''). As for $\alpha \in F$ and $Y \subseteq F$, $T_Y(\alpha)$ is a degenerate interval, $\bigwedge_{x\in X} T_Y(x) \to_I T_Y(\alpha)$ is a degenerate interval. Hence by the construction of I, I^{ε},

$C_I(I^i_{[\bigwedge_{x\in X} T_Y(x) \to_I T_Y(\alpha)]}) = \bigwedge_{x\in X} T_Y(x) \to_I T_Y(\alpha) = C_I(I^{\varepsilon}_{[\bigwedge_{x\in X} T_Y(x) \to_I T_Y(\alpha)]})$.
i.e. $r(\cap_{Y\in P(F)}\{C_I(I^i_{[\bigwedge_{x\in X} T_i(x) \to_I T_i(\alpha)]})\}) = r(\cap_{Y\in P(F)}\{C_I(I^{\varepsilon}_{[\bigwedge_{x\in X} T_i(x) \to_I T_i(\alpha)]})\})$
$= r[\bigcap_{Y\in P(F)}[\bigwedge_{x\in X} T_Y(x) \to_I T_Y(\alpha)]$
Rest follows from the Theorem 3.4.

Let us distinguish the notions of graded semantic consequence by superscribing $|\approx$ with their respective forms. So, we have $|\approx^{(\Sigma)}$, $|\approx^{(\Sigma')}$, and $|\approx^{(\Sigma_n'')}$, $n = 1, 2$.

Theorem 3.8. $gr(X |\approx^{(\Sigma)} \alpha) \le gr(X |\approx^{(\Sigma_1'')} \alpha) \le gr(X |\approx^{(\Sigma')} \alpha)$ and $gr(X |\approx^{(\Sigma)} \alpha) \le gr(X |\approx^{(\Sigma_2'')} \alpha) \le gr(X |\approx^{(\Sigma')} \alpha)$.

Proof. Using Lemma 3.9 we have, $l(\wedge_{i\in I}[\wedge_{x\in X} T_i(x) \to_I T_i(\alpha)])$
$\le l(\cap_{i\in I}[\wedge_{x\in X} T_i(x) \to_I T_i(\alpha)]) \le r(\cap_{i\in I}[\wedge_{x\in X} T_i(x) \to_I T_i(\alpha)])$ ...(i)
Also $\cap_{i\in I} C_I(I_{[\wedge_{x\in X} T_i(x) \to_I T_i(\alpha)]})$, $\cap_{i\in I} C_I(I^j_{[\wedge_{x\in X} T_i(x) \to_I T_i(\alpha)]}) \subseteq_e \cap_{i\in I}[\wedge_{x\in X} T_i(x) \to_I T_i(\alpha)]$ and they are overlapping. So, $l(\cap_{i\in I}[\wedge_{x\in X} T_i(x) \to_I T_i(\alpha)]) \le l(\cap_{i\in I} C_I(I^j_{[\wedge_{x\in X} T_i(x) \to_I T_i(\alpha)]}))$
$\le r(\cap_{i\in I} C_I(I^j_{[\wedge_{x\in X} T_i(x) \to_I T_i(\alpha)]})) \le r(\cap_{i\in I}[\wedge_{x\in X} T_i(x) \to_I T_i(\alpha)])$ ...(ii)
(i) and (ii) imply, $gr(X |\approx^{(\Sigma)} \alpha) \le gr(X |\approx^{(\Sigma_1'')} \alpha) \le gr(X |\approx^{(\Sigma')} \alpha)$.
Also following Note 3.4, $l(\cap_{i\in I}[\wedge_{x\in X} T_i(x) \to_I T_i(\alpha)]) \le r(\cap_{i\in I} C_I(I^{\varepsilon}_{[\wedge_{x\in X} T_i(x) \to_I T_i(\alpha)]}))$
$\le r(\cap_{i\in I}[\wedge_{x\in X} T_i(x) \to_I T_i(\alpha)])$ ...(iii).
Hence, combining (i) and (iii), $gr(X |\approx^{(\Sigma)} \alpha) \le gr(X |\approx^{(\Sigma_2'')} \alpha) \le gr(X |\approx^{(\Sigma')} \alpha)$.

Note 3.5. If for (Σ_1'') $I^{\varepsilon,j}$ (of Definition 3.8) is chosen instead of I^j then $gr(X |\approx^{(\Sigma)} \alpha) \le gr(X |\approx^{(\Sigma_2'')} \alpha) \le gr(X |\approx^{(\Sigma_1'')} \alpha) \le gr(X |\approx^{(\Sigma')} \alpha)$.

Theorem 3.9. Given $\{T_i\}_{i\in I}$, $|\approx$ in the sense of (Σ_n'') $(n = 1, 2)$, satisfies (GC1), (GC2), and a variant of (GC3).

Proof. (GC1) is immediate. If $X \subseteq Y$, $C_I(I^j_{[\wedge_{x\in X} T_i(x) \to_I T_i(\alpha)]}) \le_I C_I(I^j_{[\wedge_{x\in Y} T_i(x) \to_I T_i(\alpha)]})$ and $C_I(I^{\varepsilon}_{[\wedge_{x\in X} T_i(x) \to_I T_i(\alpha)]}) \le_I C_I(I^{\varepsilon}_{[\wedge_{x\in Y} T_i(x) \to_I T_i(\alpha)]})$ are obtained by condition (ii) of Definition 3.5 and Theorem 3.6 respectively. Hence by Lemma 3.8, GC2 holds for (Σ_1''), (Σ_2''). Now for each of (Σ_n''), $n = 1, 2$, we prove that a variant form of (GC3), i.e.
$\inf_{\beta\in Z} gr(X |\approx^{\Sigma_n''} \beta) * gr(X \cup Z |\approx^{\Sigma_n''} \alpha) \le gr(X |\approx^{\Sigma'} \alpha)$ holds.
$\inf_{\beta\in Z} gr(X |\approx^{\Sigma_2''} \beta) * gr(X \cup Z |\approx^{\Sigma_2''} \alpha)$
$= \inf_{\beta\in Z} [r\{\cap_{i\in I} C_I(I^{\varepsilon}_{[\wedge_{x\in X} T_i(x) \to_I T_i(\beta)]})\}] * r[\cap_{i\in I} C_I(I^{\varepsilon}_{[\wedge_{x\in X\cup Z} T_i(x) \to_I T_i(\alpha)]})\}]$.
$= r[\cap_{\beta\in Z}\{\cap_{i\in I} C_I(I^{\varepsilon}_{[\wedge_{x\in X} T_i(x) \to_I T_i(\beta)]})\}] * r[\cap_{i\in I} C_I(I^{\varepsilon}_{[\wedge_{x\in X\cup Z} T_i(x) \to_I T_i(\alpha)]})\}]$. (Lemma 3.5)
$= r[\cap_{i\in I}\{\cap_{\beta\in Z} C_I(I^{\varepsilon}_{[\wedge_{x\in X} T_i(x) \to_I T_i(\beta)]})\}] * r[\cap_{i\in I} C_I(I^{\varepsilon}_{[\wedge_{x\in X\cup Z} T_i(x) \to_I T_i(\alpha)]})\}]$ (Lemma 3.6) ...(i)
Using Note 3.4, $\cap_{\beta\in Z} C_I(I^{\varepsilon}_{[\wedge_{x\in X} T_i(x) \to_I T_i(\beta)]}) \subseteq_e \cap_{\beta\in Z} \{\wedge_{x\in X} T_i(x) \to_I T_i(\beta)\}$
$\subseteq_e \wedge_{\beta\in Z} \{\wedge_{x\in X} T_i(x) \to_I T_i(\beta)\}$ (Lemma 3.10).
Hence, $r(\cap_{i\in I}[\cap_{\beta\in Z} C_I(I^{\varepsilon}_{[\wedge_{x\in X} T_i(x) \to_I T_i(\beta)]})]) \le r(\cap_{i\in I}[\wedge_{\beta\in Z} \{\wedge_{x\in X} T_i(x) \to_I T_i(\beta)\}])$.
Similarly, $r(\cap_{i\in I}[C_I(I^{\varepsilon}_{[\wedge_{x\in X\cup Z} T_i(x) \to_I T_i(\alpha)]})]) \le r(\cap_{i\in I}[\{\wedge_{x\in X\cup Z} T_i(x) \to_I T_i(\alpha)\}])$.

(i) becomes, $\inf_{\beta \in Z} gr(X \mid\approx^{\Sigma''_2} \beta) * gr(X \cup Z \mid\approx^{\Sigma''_2} \alpha)$

$$\leq r(\cap_{i\in I}[\wedge_{\beta\in Z}\{\wedge_{x\in X} T_i(x) \rightarrow_I T_i(\beta)\}]) * r(\cap_{i\in I}[\{\wedge_{x\in X\cup Z} T_i(x) \rightarrow_I T_i(\alpha)\}])$$
$$= r(\cap_{i\in I}[\{\wedge_{x\in X} T_i(x) \rightarrow_I \wedge_{\beta\in Z} T_i(\beta)\}]) * r(\cap_{i\in I}[\{\wedge_{x\in X\cup Z} T_i(x) \rightarrow_I T_i(\alpha)\}])$$

Now following the steps below the inequality (1) of the proof of Theorem 3.3 we get: $\inf_{\beta\in Z} gr(X \mid\approx^{\Sigma''_2} \beta) * gr(X \cup Z \mid\approx^{\Sigma''_2} \alpha) \leq r[\cap_{i\in I}\{\wedge_{x\in X} T_i(x) \rightarrow_I T_i(\alpha)\}] = gr(X \mid\approx^{\Sigma'} \alpha)$. Similar is the argument for $\inf_{\beta\in Z} gr(X \mid\approx^{\Sigma''_1} \beta) * gr(X \cup Z \mid\approx^{\Sigma''_1} \alpha) \leq gr(X \mid\approx^{\Sigma'} \alpha)$.

4 Conclusion

In this paper we have studied some possible ways of obtaining the notion of graded consequence incorporating interval semantics for the object language only. One natural direction is to extend the idea when both object language formulae and the notion of consequence assume intervals. This step is yet to be developed in our further work, and the attempt made in this paper would work as a basis for this future plan.

From the development made in this paper we observe that, in the context of interval semantics, GCT is simultaneously exploiting two different lattice structures ($\leq_I$, $\subseteq_e$) over the set of sub-intervals of $[0,1]$. This adds an important dimension. Having endowed with both the notions of *interval lying below a set of intervals* ($\leq_I$) and *interval lying in the intersection of a set of intervals* ($\subseteq_e$) we manage to address different attitudes of decision making. Given a databse based on a set of experts opinion, different notions of $\mid\approx$ are introduced to address the following aspects of decision making. (i) (Σ) proposes a set up where the interval lying below all the experts' opinion would be counted. (ii) (Σ') proposes a set up where the interval lying in the common consensus zone would be counted. (iii) (Σ''_1) proposes a set up where the interval, considering expert's opinion, can be revised equally (finitely) many times, and then the interval lying in the common consensus zone would be counted. (iv) (Σ''_2) proposes a set up where the interval, taking care of expert's opinion, are revised (following a specific method viz., I^{ε}) till they reach a concrete value, and then the common consensus zone is considered. We call the approach (i) as *conservative*, (ii) as liberal, and (iii), (iv) as moderate. Thus this study provides a theoretical framework where a decision maker having some of the above attitudes derives, with certain degree, a decision from a set of imprecise information.

Acknowledgements. The authors acknowledge the valuable comments of the anonymous reviewers. The first author of this paper also acknowledges the support obtained from The Institute of Mathematical Sciences, Chennai, India, during the initial phase of preparation of this paper; the final preparation of this paper has been carried out during the tenure of an ERCIM 'Alain Bensoussan' fellowship.

References

1. Alcalde, C., Burusco, A., Fuentes-Gonzlez, R.: A constructive method for the definition of interval-valued fuzzy implication operators. Fuzzy Sets and Systems 153(2), 211–227 (2005)
2. Bedregal, B.R.C., Takahashi, A.: The best interval representations of t-norms and automorphisms. Fuzzy Sets and Systems 157, 3220–3230 (2006)
3. Bedregal, B.R.C., et al.: On interval fuzzy S-implications. Information Sciences 180, 1373–1389 (2010)
4. Bedregal, B.R.C., Santiago, R.H.N.: Interval representations, Łukasiewicz implicators and Smets-Magrez axioms. Information Sciences 221, 192–200 (2013)
5. Castro, J.L., Trillas, E., Cubillo, S.: On consequence in approximate reasoning. Journal of Applied Non-Classical Logics 4(1), 91–103 (1994)
6. Chakraborty, M.K.: Use of fuzzy set theory in introducing graded consequence in multiple valued logic. In: Gupta, M.M., Yamakawa, T. (eds.) Fuzzy Logic in Knowledge-Based Systems, Decision and Control, pp. 247–257. Elsevier Science Publishers, B.V(North Holland) (1988)
7. Chakraborty, M.K.: Graded Consequence: further studies. Journal of Applied Non-Classical Logics 5(2), 127–137 (1995)
8. Chakraborty, M.K., Basu, S.: Graded Consequence and some Metalogical Notions Generalized. Fundamenta Informaticae 32, 299–311 (1997)
9. Chakraborty, M.K., Dutta, S.: Graded Consequence Revisited. Fuzzy Sets and Systems 161, 1885–1905 (2010)
10. Cornelis, C., Deschrijver, G., Kerre, E.E.: Implication in intuitionistic fuzzy and interval-valued fuzzy set theory: construction, classification, application. International Journal of Approximate Reasoning 35, 55–95 (2004)
11. Deschrijver, G., Kerre, E.E.: Classes of intuitionistic fuzzy t-norms satisfying the residuation principle. Int. J. Uncertainty, Fuzziness and Knowledge-Based Systems 11, 691–709 (2003)
12. Deschrijver, G., Cornelis, C.: Representability in interval-valued fuzzy set theory. International Journal of Uncertainty, Fuzziness and Knowledge-Based Systems 15(3), 345–361 (2007)
13. Dutta, S., Basu, S., Chakraborty, M.K.: Many-valued logics, fuzzy logics and graded consequence: a comparative appraisal. In: Lodaya, K. (ed.) ICLA 2013. LNCS, vol. 7750, pp. 197–209. Springer, Heidelberg (2013)
14. Dutta, S., Chakraborty, M.K.: Graded consequence with fuzzy set of premises. Fundamenta Informaticae 133, 1–18 (2014)
15. Li, D., Li, Y.: Algebraic structures of interval-valued fuzzy (S, N)-implications. Int. J. Approx. Reasoning 53(6), 892–900 (2012)
16. Gasse, B.V., et al.: On the properties of a generalized class of t-norms in interval-valued fuzzy logics. New Mathematics and Natural Computation 2(1), 29–41 (2006)
17. Gentzen, G.: Investigations into Logical Deductions In: the collected papers of Gentzen, G., Szabo, M.E. (ed.), pp. 68–131. North Holland Publications, Amsterdam (1969)
18. Gerla, G.: Fuzzy Logic: Mathematical Tools for Approximate Reasoning. Kluwer Academic Publishers (2001)
19. Rodríguez, R.O., Esteva, F., Garcia, P., Godo, L.: On implicative closure operators in approximate reasoning. International Journal of Approximate Reasoning 33, 159–184 (2003)
20. Shoesmith, D.J., Smiley, T.J.: Multiple Conclusion Logic. Cambridge University Press (1978)
21. Tarski, A.: Methodology of Deductive Sciences. In Logic, Semantics, Metamathematics, pp. 60–109. Clavendon Press (1956)

Neighborhood Contingency Logic

Jie Fan[1] and Hans van Ditmarsch[2]

[1] Department of Philosophy, Peking University
fanjie@pku.edu.cn
[2] LORIA, CNRS — Université de Lorraine
hans.van-ditmarsch@loria.fr

Abstract. A formula is contingent, if it is possibly true and possibly false; a formula is non-contingent, if it is not contingent, i.e., if it is necessarily true or necessarily false. In this paper, we propose a neighborhood semantics for contingency logic, in which the interpretation of the non-contingency operator is consistent with its philosophical intuition. Based on this semantics, we compare the relative expressivity of contingency logic and modal logic on various classes of neighborhood models, and investigate the frame definability of contingency logic. We present a decidable axiomatization for classical contingency logic (the obvious counterpart of classical modal logic), and demonstrate that for contingency logic, neighborhood semantics can be seen as an extension of Kripke semantics.

1 Introduction

Like necessity and possibility, contingency is a very important notion in philosophical logic. This notion goes back to Aristotle, who develops a logic of statements about contingency [2]. As first defined in [11], a formula is *contingent*, if it is possibly true and possibly false; otherwise, it is *non-contingent*, i.e., if it is necessarily true or necessarily false. (Non-)Contingency also arose in the area of epistemic logic but with different terminology: ignorance [8, 14] and 'knowing whether' [6]; 'a formula φ is non-contingent' there means 'the agent knows whether φ', and 'φ is contingent' there means 'the agent is ignorant about φ'.

Non-contingency can be defined with necessity, namely as $\Delta\varphi =_{df} \Box\varphi \vee \Box\neg\varphi$. But necessity is *not* always definable in terms of non-contingency [4, 5, 11]. Moreover, a known difficulty for contingency logic is the absence of axioms characterizing Kripke frame properties, which makes it hard to find axiomatizations of contingency logic over various classes of Kripke frames (refer to [5] and the reference therein). As shown in [6], contingency logic is not normal, since $\Delta(\varphi \to \psi) \to (\Delta\varphi \to \Delta\psi)$ is invalid. This suggests that it may be interesting to investigate neighborhood semantics for contingency logic.

Neighborhood semantics was proposed independently by Scott and Montague in 1970 [10, 13]. Since it was introduced, neighborhood semantics has become a standard semantics tool for studying non-normal modal logics [3]. To our knowledge, Steinsvold can be said to have been the first to have explored neighborhood semantics for contingency logic [14]. He gave a topological semantics for

M. Banerjee and S.N. Krishna (eds.): ICLA 2015, LNCS 8923, pp. 88–99, 2015.

the logic of ignorance (the epistemic counterpart of contingency), in which non-contingency operator is interpreted essentially as $\Box\varphi \vee \Box\neg\varphi$. His topological models correspond to **S4** Kripke models. He did not refer to the tradition on contingency logic in his work. In this paper, we present a neighborhood semantics for contingency logic on a much wider range of model classes.

In Section 2, we define contingency logic and a neighborhood semantics for it. Sections 3, 4, and 5 contain our main contributions. In Section 3 we compare the relative expressivity of contingency logic and modal logic over various classes of neighborhood models, and investigate the frame definability of contingency logic. Section 4 completely axiomatizes a decidable contingency logic over the class of all neighborhood frames. This logic is called *classical contingency logic*, which is also characterized by another class of neighborhood frames. Section 5 deals with the relationship between neighborhood semantics and Kripke semantics for contingency logic. We conclude with some future work in Section 6.

2 Syntax and Neighborhood Semantics

Let us first recall the language of contingency logic, which is a fragment of the following logical language with both the necessity operator and the non-contingency operator as primitive modalities.

Definition 1 (Languages CML, ML and CL). *Given a set* $\boldsymbol{P}$ *of propositional variables, the logical language* ***CML*** *is defined recursively as:*

$$\varphi ::= \top \mid p \mid \neg\varphi \mid (\varphi \wedge \varphi) \mid \Delta\varphi \mid \Box\varphi$$

where $p \in \boldsymbol{P}$.

Without the construct $\Delta\varphi$*, we obtain the* language **ML** of modal logic*; without the construct* $\Box\varphi$*, we obtain the* language **CL** of contingency logic.

We always omit the parentheses whenever convenient. Formula $\Delta\varphi$ is read as "it is non-contingent that φ", and $\Box\varphi$ is read as "it is necessary that φ". Other operators are defined as usual; in particular, $\nabla\varphi$ is defined as $\neg\Delta\varphi$, for which we read "it is contingent that φ." Note that ∇ is defined as the negation, rather than the dual, of Δ, although we will see from the neighborhood semantics below that $\neg\Delta\varphi$ is equivalent to $\neg\Delta\neg\varphi$. In this paper, we will mainly focus on the language **CL** which has Δ as the only primitive modality.

Definition 2 (Neighborhood Structure). *A* neighborhood model *is a tuple* $\mathcal{M} = \langle S, N, V\rangle$*, where* S *is a nonempty set of possible worlds called* the domain, *N is a neighborhood function from S to $\mathcal{P}(\mathcal{P}(S))$, V is a valuation function assigning a set of worlds $V(p) \subseteq S$ to each $p \in \boldsymbol{P}$. Given a world $s \in S$, the pair $(\mathcal{M}, s)$ is a* pointed model*; we will omit these parentheses whenever convenient. We also write $s \in \mathcal{M}$ to denote $s \in S$. A neighborhood frame is a neighborhood model without valuation. Sometimes we write model and frame without 'neighborhood'.*

Definition 3 (Properties of Neighborhoods). *Let $\mathcal{F} = \langle S, N \rangle$ be a neighborhood frame, $\mathcal{M}$ be a neighborhood model based on $\mathcal{F}$. Let $s \in S$ and let $X, Y \subseteq S$. We define various properties of neighborhoods.*

- *(n): $N(s)$* contains the unit, *if $S \in N(s)$.*
- *(r): $N(s)$* contains its core, *if $\bigcap N(s) \in N(s)$.*
- *(i): $N(s)$* is closed under intersections, *if $X, Y \in N(s)$ implies $X \cap Y \in N(s)$.*
- *(s): $N(s)$ is* supplemented, *or* closed under supersets, *if $X \in N(s)$ and $X \subseteq Y \subseteq S$ implies $Y \in N(s)$.*
- *(c): $N(s)$ is* closed under complements, *if $X \in N(s)$ implies $S \backslash X \in N(s)$.*
- *(d): $X \in N(s)$ implies $S \backslash X \notin N(s)$.*
- *(t): $X \in N(s)$ implies $s \in X$.*
- *(b): $s \in X$ implies $\{u \in S \mid S \backslash X \notin N(u)\} \in N(s)$.*
- *(4): $X \in N(s)$ implies $\{u \in S \mid X \in N(u)\} \in N(s)$.*
- *(5): $X \notin N(s)$ implies $\{u \in S \mid X \notin N(u)\} \in N(s)$.*

The function N satisfies such a property, if $N(s)$ has the property for all $s \in S$; $\mathcal{F}$ has a property, if N has. Frame $\mathcal{F}$ is augmented, *if $\mathcal{F}$ satisfies (r) and (s). Model $\mathcal{M}$ is said to have a property, if $\mathcal{F}$ has it.*

The conditions (d), (t), (b), (4), (5) are the neighborhood structural correspondents of the relational properties of seriality, reflexivity, symmetry, transitivity, and Euclidicity that are, as known, characterized by the axioms D, T, B, 4, and 5, respectively. This also explains their names.

In order to interpret the non-contingency operator Δ in the neighborhood setting, we let ourselves be inspired by neighborhood semantics for ordinary $\Box$ modality. We recall a formula is non-contingent, if it is necessarily true or necessarily false. We therefore propose to interpret Δ such that $\Delta\varphi \leftrightarrow \Box\varphi \vee \Box\neg\varphi$ is valid using neighborhood semantics for $\Box$. The same idea, but in the setting of topological semantics, can be found in [14, Def. 3.2].

Definition 4 (Neighborhood Semantics). *Let $\mathcal{M} = \langle S, N, V \rangle$ be a neighborhood model and $s \in S$, the neighborhood semantics of* **CML** *is defined as follows:*

$\mathcal{M}, s \vDash \top$	*iff true*
$\mathcal{M}, s \vDash p$	*iff* $s \in V(p)$
$\mathcal{M}, s \vDash \neg\varphi$	*iff* $\mathcal{M}, s \nvDash \varphi$
$\mathcal{M}, s \vDash \varphi \wedge \psi$	*iff* $\mathcal{M}, s \vDash \varphi$ *and* $\mathcal{M}, s \vDash \psi$
$\mathcal{M}, s \vDash \Delta\varphi$	*iff* $\varphi^{\mathcal{M}} \in N(s)$ *or* $(\neg\varphi)^{\mathcal{M}} \in N(s)$
$\mathcal{M}, s \vDash \Box\varphi$	*iff* $\varphi^{\mathcal{M}} \in N(s)$

where $\varphi^{\mathcal{M}}$ denotes the truth set *of φ in $\mathcal{M}$, that is, $\varphi^{\mathcal{M}} = \{s \in S \mid \mathcal{M}, s \vDash \varphi\}$.*

We say a formula φ is true *in $(\mathcal{M}, s)$, if $\mathcal{M}, s \vDash \varphi$, and sometime write $s \vDash \varphi$ simply when $\mathcal{M}$ is clear from the context; we say φ is* valid, *if $\mathcal{M}, s \vDash \varphi$ for all $\mathcal{M}$ and all $s \in \mathcal{M}$. We say φ is* satisfiable, *if there is a model $(\mathcal{M}, s)$ such that $\mathcal{M}, s \vDash \varphi$. Given a set of formulas Γ, we say Γ is true in $(\mathcal{M}, s)$, notation: $\mathcal{M}, s \vDash \Gamma$, if for all $\psi \in \Gamma$, $\mathcal{M}, s \vDash \psi$; Γ* entails *φ over a class of frames $\mathbb{F}$, notation: $\Gamma \vDash_{\mathbb{F}} \varphi$, if for all $\mathcal{F} \in \mathbb{F}$, all $\mathcal{M}$ based on $\mathcal{F}$ and all $s \in \mathcal{M}$, $\mathcal{M}, s \vDash \Gamma$*

implies $\mathcal{M}, s \vDash \varphi$. *Formulas* $\varphi, \psi \in$ ***CML*** *are* equivalent *over a class of models* $\mathbb{M}$, *notation:* $\mathbb{M} \vDash \varphi \leftrightarrow \psi$, *if for all* $(\mathcal{M}, s) \in \mathbb{M}$, $\mathcal{M}, s \vDash \varphi$ *iff* $\mathcal{M}, s \vDash \psi$.

Intuitively, $N(s)$ means the set of propositions which are necessary at s. Thus $\Box\varphi$ says that the proposition expressed by φ is necessary at s, and $\Delta\varphi$ says that at least one of the proposition expressed by φ and its denial is necessary at s.

Under this semantics, it is not hard to get the following validities and invalidities, which were all shown to be valid under Kripke semantics [6].

Proposition 1 (Validities and Invalidities).

- $\vDash \Delta\varphi \leftrightarrow \Box\varphi \vee \Box\neg\varphi$, $\vDash \Delta\varphi \leftrightarrow \Delta\neg\varphi$, $\vDash \varphi \leftrightarrow \psi$ *implies* $\vDash \Delta\varphi \leftrightarrow \Delta\psi$
- $\nvDash \Delta(\varphi \to \psi) \wedge \Delta(\neg\varphi \to \psi) \to \Delta\psi$, $\nvDash \Delta\varphi \to \Delta(\varphi \to \psi) \vee \Delta(\neg\varphi \to \chi)$
- $\vDash \varphi$ *does* not *imply* $\vDash \Delta\varphi$

3 Expressivity and Frame Definability

In this section, we first compare the relative expressivity of contingency logic **CL** and modal logic **ML** on various neighborhood models, and then we give some negative results for frame correspondence of contingency logic.

3.1 Expressivity

Definition 5 (Expressivity). *Let L_1 and L_2 be two logical languages that are interpreted in the same class $\mathbb{M}$ of models,*

- L_2 *is* at least as expressive as L_1, *notation:* $L_1 \preceq L_2$, *if for every formula* $\varphi_1 \in L_1$ *there is an equivalent formula* $\varphi_2 \in L_2$ *over* $\mathbb{M}$.
- L_1 *and* L_2 *are* equally expressive, *notation:* $L_1 \equiv L_2$, *if* $L_1 \preceq L_2$ *and* $L_2 \preceq L_1$.
- L_1 *is* less expressive than L_2, *notation:* $L_1 \prec L_2$, *if* $L_1 \preceq L_2$ *and* $L_2 \npreceq L_1$.

Proposition 2. ***CL*** *is less expressive than* ***ML*** *on the class of all models, models satisfying* (*r*) *or* (*i*) *or* (*s*) *or* (*d*).

Proof. Since $\vDash \Delta\varphi \leftrightarrow \Box\varphi \vee \Box\neg\varphi$, **CL** $\preceq$ **ML**. To show **CL** $\prec$ **ML**, consider the following neighborhood models $\mathcal{M} = \langle S, N, V\rangle$ and $\mathcal{M}' = \langle S', N', V'\rangle$:

- $S = \{s, t\}$, $S' = \{s', t'\}$
- $N(s) = \{\{t\}, \{s, t\}\}$, $N(t) = \emptyset$, $N'(s') = \{\{t'\}, \{s', t'\}\}$, $N'(t') = \emptyset$
- $V(p) = \{s\}$, $V'(p) = \{s', t'\}$

The two models can be visualized as follows. An arrow from a world s to a set X means that $X \in N(s)$.

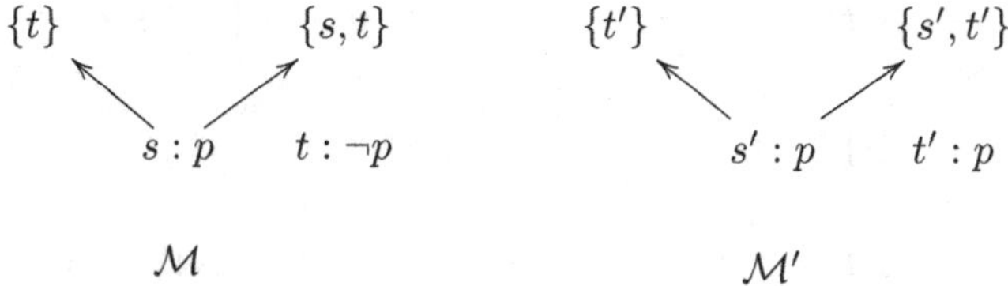

Observe that $\mathcal{M}$ and $\mathcal{M}'$ both satisfy (r), (i), (s), and (d).

We only need to show **ML** $\not\preceq$ **CL**. First, $(\mathcal{M}, s)$ and $(\mathcal{M}', s')$ are distinguished by an **ML**-formula $\Box p$: since $p^{\mathcal{M}} = \{s\} \notin N(s)$, we have $\mathcal{M}, s \nvDash \Box p$; since $p^{\mathcal{M}'} = \{s', t'\} \in N'(s')$, we get $\mathcal{M}', s' \vDash \Box p$. Therefore $\Box p$ can distinguish $(\mathcal{M}, s)$ and $(\mathcal{M}', s')$.

However, $(\mathcal{M}, s)$ and $(\mathcal{M}', s')$ cannot be distinguished by a **CL**-formula. That is, we have that $(\star)$: for all $\varphi \in$ **CL**, $\mathcal{M}, s \vDash \varphi$ iff $\mathcal{M}', s' \vDash \varphi$.

The proof of $(\star)$ proceeds by induction on φ. The only non-trivial case is $\Delta\varphi$. By semantics, we have the following equivalences:

$$\begin{aligned}
\mathcal{M}, s \vDash \Delta\varphi &\iff \varphi^{\mathcal{M}} \in N(s) \text{ or } (\neg\varphi)^{\mathcal{M}} \in N(s) \\
&\iff \varphi^{\mathcal{M}} \in \{\{t\}, \{s,t\}\} \text{ or } (\neg\varphi)^{\mathcal{M}} \in \{\{t\}, \{s,t\}\} \\
&\iff \varphi^{\mathcal{M}} = \{t\} \text{ or } \varphi^{\mathcal{M}} = \{s,t\} \text{ or } (\neg\varphi)^{\mathcal{M}} = \{t\} \text{ or } (\neg\varphi)^{\mathcal{M}} = \{s,t\} \\
&\iff \varphi^{\mathcal{M}} = \{t\} \text{ or } \varphi^{\mathcal{M}} = \{s,t\} \text{ or } \varphi^{\mathcal{M}} = \{s\} \text{ or } \varphi^{\mathcal{M}} = \emptyset \\
&\iff \text{true}
\end{aligned}$$

$$\begin{aligned}
\mathcal{M}', s' \vDash \Delta\varphi &\iff \varphi^{\mathcal{M}'} \in N'(s) \text{ or } (\neg\varphi)^{\mathcal{M}'} \in N'(s) \\
&\iff \varphi^{\mathcal{M}'} \in \{\{t'\}, \{s',t'\}\} \text{ or } (\neg\varphi)^{\mathcal{M}'} \in \{\{t'\}, \{s',t'\}\} \\
&\iff \varphi^{\mathcal{M}'} = \{t'\} \text{ or } \varphi^{\mathcal{M}'} = \{s',t'\} \text{ or } (\neg\varphi)^{\mathcal{M}'} = \{t'\} \\
&\qquad \text{or } (\neg\varphi)^{\mathcal{M}'} = \{s',t'\} \\
&\iff \varphi^{\mathcal{M}'} = \{t'\} \text{ or } \varphi^{\mathcal{M}'} = \{s',t'\} \text{ or } \varphi^{\mathcal{M}'} = \{s'\} \text{ or } \varphi^{\mathcal{M}'} = \emptyset \\
&\iff \text{true}
\end{aligned}$$

In either case, the penultimate line of the proof merely states that φ can be interpreted on the related model: its denotation must necessarily be one of *all* possible subsets of the domain. Notice that the above proof of the inductive case $\Delta\varphi$ does not use the induction hypothesis. We conclude that $\mathcal{M}, s \vDash \Delta\varphi$ iff $\mathcal{M}', s' \vDash \Delta\varphi$.

Proposition 3. ***CL*** *is less expressive than* ***ML*** *on the class of models satisfying* (n) *or* (b).

Proof. Consider the following models $\mathcal{M} = \langle S, N, V\rangle$ and $\mathcal{M}' = \langle S', N', V'\rangle$:

- $S = \{s, t\}$, $S' = \{s', t'\}$
- $N(s) = \{\emptyset, \{t\}, \{s,t\}\}$, $N(t) = \{\emptyset, \{s\}, \{t\}, \{s,t\}\}$, $N'(s') = \{\emptyset, \{t'\}, \{s',t'\}\}$, $N'(t') = \{\emptyset, \{s'\}, \{t'\}, \{s',t'\}\}$
- $V(p) = \{s\}$, $V'(p) = \{s', t'\}$

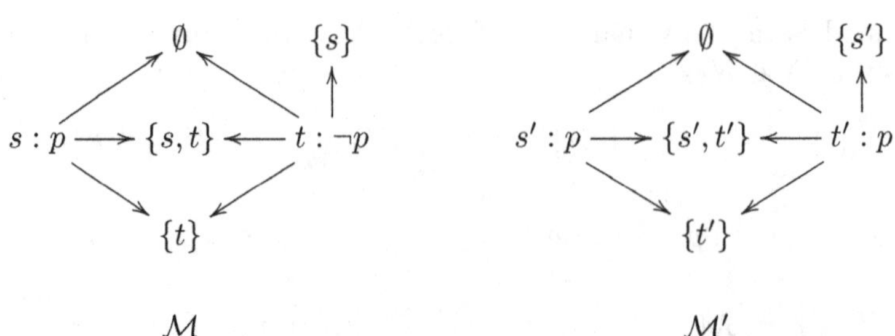

We claim that $\mathcal{M}$ and $\mathcal{M}'$ satisfy (n) and (b). Below we just show the claim for model $\mathcal{M}$; the proof for the case $\mathcal{M}'$ is similar.

For (n): It is clear that $S \in N(s)$ and $S \in N(t)$.

For (b): As $N(t)$ consists of all subsets of the domain, it is clear that (b) holds for $N(t)$. As for $N(s)$, if $s \in X$, then X must be either $\{s\}$ or $\{s,t\}$. Then $S \backslash X$ must be either $\{t\}$ or $\emptyset$, respectively. But $\emptyset, \{t\} \in N(s)$, so $\{u \in \mathcal{M} \mid S \backslash X \notin N(u)\} = \emptyset \in N(s)$.

We continue by demonstrating the expressivity result. Notice that $\{s\} \notin N(s)$ and $\{s', t'\} \in N'(s')$. Similar to the proof of Prop. 2, we show that $s \nvDash \Box p$ but $s' \vDash \Box p$. Therefore $(\mathcal{M}, s)$ and $(\mathcal{M}', s')$ are distinguished by an **ML**-formula $\Box p$.

Moreover, also similar to the proof of Prop. 2, we can show that: for all $\varphi \in$ **CL**, $\mathcal{M}, s \vDash \varphi$ iff $\mathcal{M}', s' \vDash \varphi$. Therefore $(\mathcal{M}, s)$ and $(\mathcal{M}', s')$ cannot be distinguished by a **CL**-formula.

Proposition 4. ***CL*** *is less expressive than* ***ML*** *on the class of models satisfying* (4) *or* (5).

Proof. Consider the following models $\mathcal{M} = \langle S, N, V \rangle$ and $\mathcal{M}' = \langle S', N', V' \rangle$:

- $S = \{s, t\}$, $S' = \{s', t'\}$
- $N(s) = \{\emptyset, \{s\}, \{s,t\}\}$, $N(t) = \{\{t\}\}$, $N'(s') = \{\{s'\}, \{s', t'\}\}$, $N'(t') = \{\{t'\}, \emptyset\}$
- $V(p) = \emptyset$, $V'(p) = \emptyset$

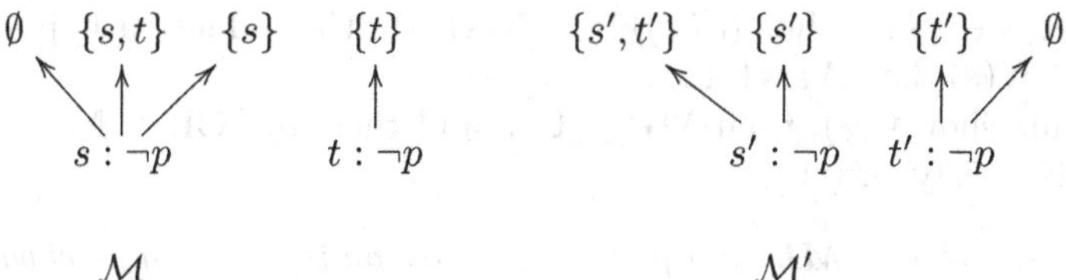

Based on the three observations below, we obtain the statement.

- $\mathcal{M}$ and $\mathcal{M}'$ satisfy (4) and (5). Below we just show the claim for model $\mathcal{M}$; the proof for the case $\mathcal{M}'$ is similar.
  - For (4): suppose that $X \in N(s)$. Then $X = \emptyset$ or $X = \{s\}$ or $X = \{s, t\}$. Observe that $\{u \mid X \in N(u)\} = \{s\} \in N(s)$. Similarly, we can show that (4) applies to $N(t)$.
  - For (5): suppose that $X \notin N(s)$. Then $X = \{t\}$. Observe that $\{u \mid X \notin N(u)\} = \{s\} \in N(s)$. Similarly, we can show that (5) applies to $N(t)$.
- $(\mathcal{M}, s)$ and $(\mathcal{M}', s')$ are distinguished by an **ML**-formula $\Box p$: since $p^{\mathcal{M}} = \emptyset \in N(s)$, we have $\mathcal{M}, s \vDash \Box p$; since $p^{\mathcal{M}'} = \emptyset \notin N'(s')$, we get $\mathcal{M}', s' \nvDash \Box p$. Therefore $\Box p$ can distinguish $(\mathcal{M}, s)$ and $(\mathcal{M}', s')$.
- $(\mathcal{M}, s)$ and $(\mathcal{M}', s')$ cannot be distinguished by any **CL**-formulas. That is, for all $\varphi \in$ **CL**, $\mathcal{M}, s \vDash \varphi$ iff $\mathcal{M}', s' \vDash \varphi$. The proof is similar to the corresponding part of Prop. 2.

However, on the class of neighborhood models satisfying (t), **CL** and **ML** are equally expressive. This is as expected, as the results on Kripke semantics are similar: once frames are reflexive, modal necessity is definable from contingency [11]. Nevertheless, it is surprising that the same story goes with the class of models satisfying (c).

Proposition 5. ***CL*** *and* ***ML*** *are equally expressive on the class of neighborhood models satisfying* (t).

Proof. Define a translation t from **CL** to **ML**, and a translation t' from **ML** to **CL**, such that each translation preserves propositional variables and Boolean connectives, and $t(\Delta\varphi) = \Box t(\varphi) \vee \Box\neg t(\varphi)$ and $t'(\Box\varphi) = \Delta t'(\varphi) \wedge t'(\varphi)$.

Due to Prop. 1, by induction on $\varphi \in \mathbf{CL}$, we can show that $\vDash \varphi \leftrightarrow t(\varphi)$. Therefore $\mathbf{CL} \preceq \mathbf{ML}$.

Moreover, by induction, we show $(\star)$: for any $\varphi \in \mathbf{ML}$, we have $\mathbb{M}_t \vDash \varphi \leftrightarrow t'(\varphi)$, where $\mathbb{M}_t$ is the class of models satisfying (t). Given any $\mathcal{M} = \langle S, N, V\rangle \in \mathbb{M}_t$ and any $s \in \mathcal{M}$. We only need to show the case for $\Box\varphi$, i.e., $\mathcal{M}, s \vDash \Box\varphi$ iff $\mathcal{M}, s \vDash \Delta t'(\varphi) \wedge t'(\varphi)$.

First, suppose that $\mathcal{M}, s \vDash \Box\varphi$, then $\varphi^{\mathcal{M}} \in N(s)$, by the induction hypothesis, $(t'(\varphi))^{\mathcal{M}} \in N(s)$, then $(t'(\varphi))^{\mathcal{M}} \in N(s)$ or $(\neg t'(\varphi))^{\mathcal{M}} \in N(s)$, i.e., $\mathcal{M}, s \vDash \Delta t'(\varphi)$. From $(t'(\varphi))^{\mathcal{M}} \in N(s)$ and condition (t), it follows that $s \in (t'(\varphi))^{\mathcal{M}}$, that is, $\mathcal{M}, s \vDash t'(\varphi)$, therefore $\mathcal{M}, s \vDash \Delta t'(\varphi) \wedge t'(\varphi)$.

Conversely, suppose that $\mathcal{M}, s \vDash \Delta t'(\varphi) \wedge t'(\varphi)$. From $\mathcal{M}, s \vDash \Delta t'(\varphi)$, it follows that $(t'(\varphi))^{\mathcal{M}} \in N(s)$ or $(\neg t'(\varphi))^{\mathcal{M}} \in N(s)$. If $(\neg t'(\varphi))^{\mathcal{M}} \in N(s)$, since $\mathcal{M}$ satisfies (t), it follows that $s \in (\neg t'(\varphi))^{\mathcal{M}}$, i.e., $\mathcal{M}, s \vDash \neg t'(\varphi)$, contrary to the supposition $\mathcal{M}, s \vDash t'(\varphi)$, thus $(t'(\varphi))^{\mathcal{M}} \in N(s)$. By the induction hypothesis, we have $\varphi^{\mathcal{M}} \in N(s)$, i.e., $\mathcal{M}, s \vDash \Box\varphi$.

We have thus shown $(\star)$, then $\mathbf{ML} \preceq \mathbf{CL}$, and therefore $\mathbf{CL} \equiv \mathbf{ML}$ on the class of models satisfying (t).

Proposition 6. ***CL*** *and* ***ML*** *are equally expressive on the class of neighborhood models satisfying* (c).

Proof. Define t and tr, respectively, as t and t' in the proof of Prop. 5, except that $tr(\Box\varphi) = \Delta tr(\varphi)$.

Similar to the corresponding proof in Prop. 5, we can show that $\mathbf{CL} \preceq \mathbf{ML}$.

Besides, by induction, we show $(*)$: for any $\varphi \in \mathbf{ML}$, we have $\mathbb{M}_c \vDash \varphi \leftrightarrow tr(\varphi)$, where $\mathbb{M}_c$ is the class of models satisfying (c). Given any $\mathcal{M} = \langle S, N, V\rangle \in \mathbb{M}_c$ and any $s \in \mathcal{M}$. We only need to show the case for $\Box\varphi$, i.e., $\mathcal{M}, s \vDash \Box\varphi$ iff $\mathcal{M}, s \vDash \Delta tr(\varphi)$.

'Only if' can be shown easily due to $\vDash \Delta\varphi \leftrightarrow \Box\varphi \vee \Box\neg\varphi$.

Now suppose that $\mathcal{M}, s \vDash \Delta tr(\varphi)$, that is, $(tr(\varphi))^{\mathcal{M}} \in N(s)$ or $(\neg tr(\varphi))^{\mathcal{M}} \in N(s)$. If $(\neg tr(\varphi))^{\mathcal{M}} \in N(s)$, since $\mathcal{M}$ satisfies (c), it follows that $(tr(\varphi))^{\mathcal{M}} \in N(s)$, thus we always have $(tr(\varphi))^{\mathcal{M}} \in N(s)$. By the induction hypothesis, we have $\varphi^{\mathcal{M}} \in N(s)$, i.e., $\mathcal{M}, s \vDash \Box\varphi$.

We have thus shown $(*)$, then $\mathbf{ML} \preceq \mathbf{CL}$, and therefore $\mathbf{CL} \equiv \mathbf{ML}$ on the class of models satisfying (c).

3.2 Frame Definability

Formulas from standard modal logic can be used to capture properties of neighborhood frames, e.g., $\Box p \to \Diamond p$ corresponds to the property (d) [12, Section 2.4]. It is therefore remarkable that there is no such correspondence in **CL** for most of the properties of neighborhood frames. Similarly, many properties of Kripke frames do not correspond to frame axioms in contingency logic [16].

Definition 6 (Frame Definability). *Let Γ be a set of **CL**-formulas and $\mathbb{F}$ a class of frames. We say that Γ defines $\mathbb{F}$ if for all frames $\mathcal{F}$, $\mathcal{F}$ is in $\mathbb{F}$ if and only if $\mathcal{F} \vDash \Gamma$. In this case we also say Γ defines the property of $\mathbb{F}$. If Γ is a singleton (e.g. $\{\varphi\}$), we usually write $\mathcal{F} \vDash \varphi$ rather than $\mathcal{F} \vDash \{\varphi\}$. A class of frames (or the corresponding frame property) is definable in **CL** if there is a set of **CL**-formulas that defines it.*

Proposition 7. *The frame properties (n), (r), (i), (s), (c), (d), (t), (b), (4), and (5) are not definable in **CL**.*

Proof. Consider the following frames $\mathcal{F}_1 = \langle S_1, N_1 \rangle$, $\mathcal{F}_2 = \langle S_2, N_2 \rangle$, $\mathcal{F}_3 = \langle S_3, N_3 \rangle$ and $\mathcal{F}_4 = \langle S_4, N_4 \rangle$:

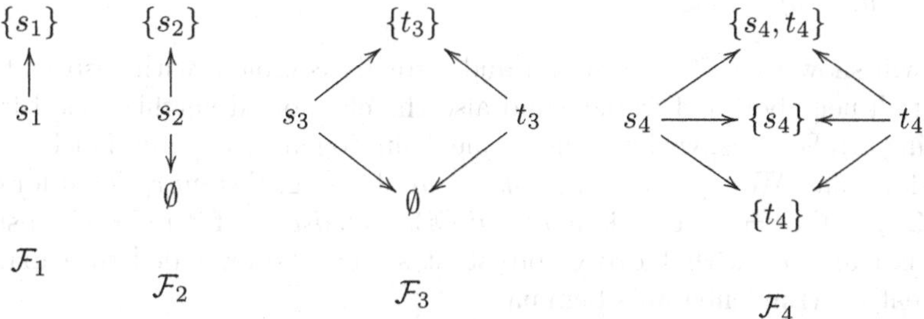

Observe that $\mathcal{F}_1$ satisfies (d) and (t) but $\mathcal{F}_2$ does not satisfy either (d) or (t); $\mathcal{F}_2$ satisfies all of (n), (s), (c), (b), (4), and (5), while $\mathcal{F}_3$ does not satisfy any of those; $\mathcal{F}_3$ satisfies (r) and (i), while $\mathcal{F}_4$ does not satisfy (r) or (i). For instance, $\{s_2\} \in N_2(s_2)$ but $\emptyset = S_2 \backslash \{s_2\} \in N_2(s_2)$, thus $\mathcal{F}_2$ does not satisfy (d); $t_3 \in \{t_3\}$, but $\{u \in S_3 \mid \{s_3\} \notin N_3(u)\} = \{s_3, t_3\} \notin N_3(t_3)$, thus $\mathcal{F}_3$ does not satisfy (b); $\bigcap N_4(s_4) = \emptyset \notin N_4(s_4)$, thus $\mathcal{F}_4$ does not satisfy (r).

We next show that: for any $\varphi \in$ **CL**, $\mathcal{F}_1 \vDash \varphi$ iff $\mathcal{F}_2 \vDash \varphi$ iff $\mathcal{F}_3 \vDash \varphi$ iff $\mathcal{F}_4 \vDash \varphi$.

Suppose that $\mathcal{F}_1 \nvDash \varphi$, then there exists $\mathcal{M}_1 = \langle \mathcal{F}_1, V_1 \rangle$ such that $\mathcal{M}_1, s_1 \nvDash \varphi$. Define a valuation V_2 on $\mathcal{F}_2$ as $p \in V_2(s_2)$ iff $p \in V_1(s_1)$ for all $p \in \mathbf{P}$. By induction on φ, we can show that $\mathcal{M}_1, s_1 \vDash \varphi$ iff $\mathcal{M}_2, s_2 \vDash \varphi$, where the only non-trivial case $\Delta\varphi$ is proved as the proof of Prop. 2. From this, it follows that $\mathcal{M}_2, s_2 \nvDash \varphi$, therefore $\mathcal{F}_2 \nvDash \varphi$. The converse is similar. Therefore $\mathcal{F}_1 \vDash \varphi$ iff $\mathcal{F}_2 \vDash \varphi$.

Similarly, we can show $\mathcal{F}_2 \vDash \varphi$ iff $\mathcal{F}_3 \vDash \varphi$, and $\mathcal{F}_3 \vDash \varphi$ iff $\mathcal{F}_4 \vDash \varphi$.

If (d) were to be defined by a set of **CL**-formulas, say Γ, then since $\mathcal{F}_1$ satisfies (d), we have $\mathcal{F}_1 \vDash \Gamma$. Then we should also have $\mathcal{F}_2 \vDash \Gamma$, i.e., $\mathcal{F}_2$ satisfies (d), contradiction. Thus (d) is not definable in **CL**. The undefinability of other properties in question can be proved similarly.

4 Classical Contingency Logic

In this section, we present an axiomatization for classical contingency logic, which is sound and strongly complete with respect to both the class of all neighborhood frames, and also the class of all neighborhood frames satisfying (c). And we obtain decidability for this logic.

Definition 7 (Proof System $\mathbb{CCL}$). *The* proof system $\mathbb{CCL}$ *is the set of* ***CL****-formulas including axioms TAUT and* Δ*Equ, and closed under the rule RE*Δ.

$$\begin{array}{ll} \textit{TAUT} & \textit{all instances of tautologies} \\ \Delta \textit{Equ} & \Delta\varphi \leftrightarrow \Delta\neg\varphi \\ \textit{RE}\Delta & \dfrac{\varphi \leftrightarrow \psi}{\Delta\varphi \leftrightarrow \Delta\psi} \end{array}$$

A derivation of φ from Γ, *notation:* $\Gamma \vdash \varphi$, *is a finite sequence of* ***CL****-formulas such that each formula is either the instantiation of an axiom, or an element in* Γ, *or follows from the prior formulas in the sequence by an inference rule. A* derivation of φ *is a derivation of* φ *from the empty set* $\emptyset$. *We write* $\vdash \varphi$ *if there is a derivation of* φ *in* $\mathbb{CCL}$.

We will show that $\mathbb{CCL}$ is sound and strongly complete with respect to the class of all neighborhood frames, and also the class of all neighborhood frames satisfying (c). For this, we introduce some definition and notation. Let Γ be a set of **CL**-formulas. We say Γ is $\mathbb{CCL}$*-consistent*, if $\Gamma \nvdash \bot$; Γ is *maximal*, if for every $\varphi \in \mathbf{CL}$, $\varphi \in \Gamma$ or $\neg\varphi \in \Gamma$; Γ is *maximal* $\mathbb{CCL}$*-consistent*, if it is $\mathbb{CCL}$-consistent and maximal. Recall that every consistent set can be extended to a maximal consistent set (Lindenbaum's Lemma).

Definition 8 (Soundness, Strong Completeness). *Let S be a logical system, and let $\mathbb{F}$ be a class of frames.*

- *S is* sound *with respect to $\mathbb{F}$, if for any φ, $\vdash_S \varphi$ implies $\mathbb{F} \vDash \varphi$.*
- *S is* strongly complete *with respect to $\mathbb{F}$, if for any set of formulas Γ and any φ, $\Gamma \vDash_{\mathbb{F}} \varphi$ implies $\Gamma \vdash_S \varphi$.*

Now we construct the canonical model of $\mathbb{CCL}$.

Definition 9 (Canonical Model). *The* canonical neighborhood model *of* $\mathbb{CCL}$ *is the tuple $\mathcal{M}^c = \langle S^c, N^c, V^c\rangle$, such that*

- $S^c = \{s \mid s \textit{ is a maximal } \mathbb{CCL}\textit{-consistent set}\}$
- $N^c(s) = \{|\varphi| \mid \Delta\varphi \in s\}$
- $V^c(p) = \{s \mid s \in |p|\}$

where $|\varphi| = \{s \in S^c \mid \varphi \in s\}$ is the proof set *of φ in* $\mathbb{CCL}$.

Lemma 1 (Truth Lemma). *For any $s \in S^c$ and formula φ, $\mathcal{M}^c, s \vDash \varphi$ iff $\varphi \in s$. That is to say, $\varphi^{\mathcal{M}^c} = |\varphi|$.*

Proof. By induction on φ. The base cases and Boolean cases are straightforward. The non-trivial case is $\Delta\varphi$.

$$\begin{array}{lll} \mathcal{M}^c, s \vDash \Delta\varphi & \Longleftrightarrow_{\text{semantics}} & \varphi^{\mathcal{M}^c} \in N^c(s) \text{ or } (\neg\varphi)^{\mathcal{M}^c} \in N^c(s) \\ & \Longleftrightarrow_{\text{IH}} & |\varphi| \in N^c(s) \text{ or } |\neg\varphi| \in N^c(s) \\ & \Longleftrightarrow_{\text{Def. } N^c} & \Delta\varphi \in s \text{ or } \Delta\neg\varphi \in s \\ & \Longleftrightarrow_{\Delta\text{Equ}} & \Delta\varphi \in s \end{array}$$

We also need to prove that N^c is well defined.

Lemma 2. *If* $|\varphi| \in N^c(s)$ *and* $|\varphi| = |\psi|$, *then* $\Delta\psi \in s$.

Proof. Assume that the conditions hold, to show that $\Delta\psi \in s$. From $|\varphi| \in N^c(s)$, it follows that $\Delta\varphi \in s$. By $|\varphi| = |\psi|$, we have $\vdash \varphi \leftrightarrow \psi$: otherwise, $\neg(\varphi \leftrightarrow \psi)$ would be consistent, then by Lindenbaum's Lemma, there exists $s \in S^c$ such that $\neg(\varphi \leftrightarrow \psi) \in s$, thus $\varphi \in s \not\Leftrightarrow \psi \in s$, contrary to $|\varphi| = |\psi|$. By REΔ we get $\vdash \Delta\varphi \leftrightarrow \Delta\psi$, thus $\Delta\psi \in s$.

Theorem 1. $\mathbb{CCL}$ *is sound and strongly complete with respect to the class of all neighborhood frames.*

Proof. Soundness is clear from Prop. 1. For completeness, suppose that $\Gamma \nvdash \varphi$, then $\Gamma \cup \{\neg\varphi\}$ is consistent. By Lindenbaum's Lemma, there exists $s \in S^c$ such that $\Gamma \cup \{\neg\varphi\} \subseteq s$. By Lemma 1, $\Gamma \nvDash \varphi$, as desired.

Thm. 1 indicates that $\mathbb{CCL}$ is the smallest contingency logic under neighborhood semantics; this is why we call $\mathbb{CCL}$ classical contingency logic. Surprisingly, the same logic is also characterized by the class of frames satisfying (c).

Theorem 2. $\mathbb{CCL}$ *is sound and strongly complete with respect to the class of frames satisfying* (c).

Proof. Due to Thm. 1, it suffices to show that for each $s \in S^c$, $N^c(s)$ satisfies (c).

Let $s \in S^c$. Assume that $X \in N^c(s)$. By definition of N^c, $X = |\varphi| \in N^c(s)$ for some φ, and then $\Delta\varphi \in s$. By **ΔEqu**, we have $\Delta\neg\varphi \in s$. Using definition of N^c again, we obtain that $|\neg\varphi| \in N^c(s)$, i.e., $S\backslash|\varphi| \in N^c(s)$, thus $S\backslash X \in N^c(s)$.

Due to $\vDash \Delta\varphi \leftrightarrow \Box\varphi \vee \Box\neg\varphi$, and the decidability of classical modal logic [3], the logic $\mathbb{CCL}$ is also decidable.

Proposition 8 (Decidability of $\mathbb{CCL}$). *The logic* $\mathbb{CCL}$ *is decidable.*

5 Neighborhood Semantics and Kripke Semantics

This section deals with the relationship between neighborhood semantics and Kripke semantics for contingency logic. The result is: there is a one-to-one correspondence between augmented neighborhood frames and Kripke frames for **CL**. Thus neighborhood semantics for **CL** can be seen as an extension of Kripke semantics for **CL**.

Proposition 9. *For every Kripke model $\mathcal{M}^K = \langle S, R, V\rangle$ there exists an augmented neighborhood model $\mathcal{M}^N = \langle S, N, V\rangle$ such that for all $s \in S$ and all $\varphi \in$ **CL**, $\mathcal{M}^K, s \vDash \varphi \iff \mathcal{M}^N, s \vDash \varphi$, i.e., $\varphi^{\mathcal{M}^K} = \varphi^{\mathcal{M}^N}$.*

Proof. Let $\mathcal{M}^K = \langle S, R, V\rangle$ be a Kripke model. Define a neighborhood model $\mathcal{M}^N = \langle S, N, V\rangle$ by stipulating $N(s) = \{X \subseteq S \mid R(s) \subseteq X\}$ for each $s \in S$, where $R(s) = \{t \in S \mid sRt\}$.

We first show that $\mathcal{M}^N$ is augmented: if $X \in N(s)$ and $X \subseteq Y$, then $R(s) \subseteq X \subseteq Y$, thus $Y \in N(s)$. Given any $X \in N(s)$, we have $R(s) \subseteq X$. From this follows that $R(s) \subseteq \bigcap\{X \mid X \in N(s)\} = \bigcap N(s)$, therefore $\cap N(s) \in N(s)$.

Next, we show that for all $s \in S$ and all $\varphi \in \mathbf{CL}$, $\mathcal{M}^K, s \vDash \varphi \iff \mathcal{M}^N, s \vDash \varphi$. The proof proceeds by induction on φ. The only non-trivial case is $\Delta\varphi$.

$$\begin{array}{lll} \mathcal{M}^K, s \vDash \Delta\varphi & \iff_{\text{Kripke semantics}} & R(s) \subseteq \varphi^{\mathcal{M}^K} \text{ or } R(s) \subseteq (\neg\varphi)^{\mathcal{M}^K} \\ & \iff_{\text{IH}} & R(s) \subseteq \varphi^{\mathcal{M}^N} \text{ or } R(s) \subseteq (\neg\varphi)^{\mathcal{M}^N} \\ & \iff_{\text{Def. } N} & \varphi^{\mathcal{M}^N} \in N(s) \text{ or } (\neg\varphi)^{\mathcal{M}^N} \in N(s) \\ & \iff_{\text{Nb. semantics}} & \mathcal{M}^N, s \vDash \Delta\varphi \end{array}$$

Proposition 10. *For every augmented neighborhood model $\mathcal{M}^N = \langle S, N, V\rangle$, there exists a Kripke model $\mathcal{M}^K = \langle S, R, V\rangle$ such that for all $s \in S$ and all $\varphi \in$ **CL**, $\mathcal{M}^N, s \vDash \varphi \iff \mathcal{M}^K, s \vDash \varphi$, i.e., $\varphi^{\mathcal{M}^N} = \varphi^{\mathcal{M}^K}$.*

Proof. Let $\mathcal{M}^N = \langle S, N, V\rangle$ be an augmented neighborhood model. Define a Kripke model $\mathcal{M}^K = \langle S, R, V\rangle$ by stipulating $R(s) = \bigcap N(s)$ for each $s \in S$.

By induction on φ. The only non-trivial case is $\Delta\varphi$.

$$\begin{array}{lll} \mathcal{M}^N, s \vDash \Delta\varphi & \iff_{\text{Nh. semantics}} & \varphi^{\mathcal{M}^N} \in N(s) \text{ or } (\neg\varphi)^{\mathcal{M}^N} \in N(s) \\ & \iff_{\text{augmentation of } \mathcal{M}^N} & \bigcap N(s) \subseteq \varphi^{\mathcal{M}^N} \text{ or } \bigcap N(s) \subseteq (\neg\varphi)^{\mathcal{M}^N} \\ & \iff_{\text{Def. } R, \text{ IH}} & R(s) \subseteq \varphi^{\mathcal{M}^K} \text{ or } R(s) \subseteq (\neg\varphi)^{\mathcal{M}^K} \\ & \iff_{\text{Kripke semantics}} & \mathcal{M}^K, s \vDash \Delta\varphi \end{array}$$

The above two propositions have nice applications. For instance, by Prop. 9, we get that $\Delta(\varphi \wedge \psi) \to \Delta\varphi$ is not valid on the class of augmented neighborhood frames, because it is not valid on the class of Kripke frames. For another example, consider the axiomatization $\mathbb{CL}$ of contingency logic, which is known to be complete under Kripke semantics [6]. By Prop. 10, $\mathbb{CL}$ is sound on the class of augmented frames, because it is sound on the class of Kripke frames.

6 Conclusion

We proposed to interpret contingency logic on neighborhood models. We showed that contingency logic is less expressive than modal logic on various classes of neighborhood models, but equally expressive on other classes of models, and that most standard properties of neighborhood frames, which are definable in modal logic, are undefinable in contingency logic. We further proposed a decidable axiomatization for classical contingency logic and provided a correspondence between neighborhood semantics for contingency logic on augmented frames and Kripke semantics for that logic.

For further research we leave the question whether the axiomatization $\mathbb{CL}$ is strongly complete with respect to the class of augmented neighbourhood frames (see Section 5 for soundness). We further wish to investigate the suitable notion of bisimulation on neighborhood models for **CL**, and to characterize **CL** as a fragment within modal logic and within first-order logic, where [1, 7] may help. Finally, we wish to advance the investigation of dynamics on neighborhood models and in substructural logics, as initiated in studies like [9, 15].

Acknowledgements. Jie Fan thanks support of the Major Project 12&ZD119 of National Social Science Foundation of China. Hans van Ditmarsch is also affiliated to IMSc, Chennai, as research associate. He acknowledges support from ERC project EPS 313360. We thank Yanjing Wang for useful discussions and the ICLA reviewers for their encouraging enthusiasm about our work.

References

1. Areces, C., Figueira, D.: Which semantics for neighbourhood semantics? In: IJCAI, pp. 671–676 (2009)
2. Brogan, A.: Aristotle's logic of statements about contingency. Mind 76(301), 49–61 (1967)
3. Chellas, B.F.: Modal Logic: An Introduction. Cambridge University Press (1980)
4. Cresswell, M.: Necessity and contingency. Studia Logica 47, 145–149 (1988)
5. Fan, J., Wang, Y., van Ditmarsch, H.: Almost necessary. Advances in Modal Logic 10, 178–196 (2014)
6. Fan, J., Wang, Y., van Ditmarsch, H.: Contingency and knowing whether (to appear, 2014)
7. Hansen, H.H., Kupke, C., Pacuit, E.: Neighbourhood structures: Bisimilarity and basic model theory. Logical Methods in Computer Science 5(2), 1–38 (2009)
8. van der Hoek, W., Lomuscio, A.: A logic for ignorance. Electronic Notes in Theoretical Computer Science 85(2), 117–133 (2004)
9. Ma, M., Sano, K.: How to update neighborhood models. In: Grossi, D., Roy, O., Huang, H. (eds.) LORI. LNCS, vol. 8196, pp. 204–217. Springer, Heidelberg (2013)
10. Montague, R.: Universal grammar. Theoria 36, 373–398 (1970)
11. Montgomery, H., Routley, R.: Contingency and non-contingency bases for normal modal logics. Logique et Analyse 9, 318–328 (1966)
12. Pacuit, E.: Neighborhood semantics for modal logic: An introduction. ESSLLI Lecture (2007), `http://web.pacuit.org/papers/nbhdesslli.pdf`
13. Scott, D.: Advice on modal logic. In: Philosophical Problems in Logic: Some Recent Developments, pp. 143–173 (1970)
14. Steinsvold, C.: A note on logics of ignorance and borders. Notre Dame Journal of Formal Logic 49(4), 385–392 (2008)
15. Wheeler, G.: AGM belief revision in monotone modal logics. In: Proc. of LPAR 2010 (2010)
16. Zolin, E.: Completeness and definability in the logic of noncontingency. Notre Dame Journal of Formal Logic 40(4), 533–547 (1999)

Hierarchies in Inclusion Logic with Lax Semantics

Miika Hannula

University of Helsinki, Department of Mathematics and Statistics,
P.O. Box 68, 00014 Helsinki, Finland
miika.hannula@helsinki.fi

Abstract. We study the expressive power of fragments of inclusion logic under the so-called lax team semantics. The fragments are defined either by restricting the number of universal quantifiers or the arity of inclusion atoms in formulae. In case of universal quantifiers, the corresponding hierarchy collapses at the first level. Arity hierarchy is shown to be strict by relating the question to the study of arity hierarchies in fixed-point logics.

1 Introduction

In this article we study the expressive power of inclusion logic (FO($\subseteq$)) [1] in the lax team semantics setting. Inclusion logic is a variant of dependence logic (FO(=(...))) [2] which extends first-order logic with dependence atoms

$$=(x_1, \ldots, x_n)$$

expressing that the values of x_n depend functionally on the values of $x_1, \ldots, x_{n-1}$. Inclusion logic, instead, extends first-order logic with inclusion atoms

$$\boldsymbol{x} \subseteq \boldsymbol{y}$$

which express that the set of values of $\boldsymbol{x}$ is included in the set of the values of $\boldsymbol{y}$. We study the expressive power of two syntactic fragments of inclusion logic under the lax team semantics. These two fragments, FO($\subseteq$)($k\forall$) and FO($\subseteq$)(k-inc), are defined by restricting the number of universal quantifiers or the arity of inclusion atoms to k, respectively. We will show that FO($\subseteq$)($k\forall$) captures FO($\subseteq$) already with $k = 1$ and that the fragments FO($\subseteq$)(k-inc) give rise to an infinite, strict expressivity hierarchy.

Since the introduction of dependence logic in 2007, many interesting variants of it have been introduced. One reason for this orientation is the semantical framework that is being used. Team semantics, introduced by Hodges in 1997 [3], provides a natural way to extend first-order logic with many different kinds of dependency notions. Although many of these notions have been extensively studied in database theory since the 70s, with team semantics the novelty comes from the fact that also interpretations for logical connectives and quantifiers are provided.

M. Banerjee and S.N. Krishna (eds.): ICLA 2015, LNCS 8923, pp. 100–118, 2015.

In expressive power FO(=(...)) is equivalent to existential second-order logic (ESO) [2]. For some variants of FO(=(...)), the correspondence with ESO does not hold or it can depend on which version of team semantics is being used. For instance, FO($\subseteq$) corresponds in expressive power to ESO if we use the so-called "strict team semantics" [4]. Under the lax team semantics, FO($\subseteq$) corresponds to greatest fixed-point logic (GFP) [5] which captures PTIME over finite ordered models. In FO(=(...)) no separation between the strict and the lax version of team semantics exists since dependence atoms satisfy the so-called downward closure property. In the following we briefly list some complexity theoretical aspects of FO(=(...)) and its variants.

- FO(=(...)) extended with the so-called intuitionistic implication $\rightarrow$ (introduced in [6]) increases the expressive power of FO(=(...)) to full second-order logic [7].
- The model checking problem of FO(=(...)), and many of its variants, was recently shown to be NEXPTIME-complete. Moreover, for any variant of FO(=(...)) whose atoms are PTIME-computable, the corresponding model checking problem is contained in NEXPTIME [8].
- The non-classical interpretation of disjunction in FO(=(...)) has the effect that the model checking problems of $\phi_1 := {=}(x,y) \vee {=}(u,v)$ and $\phi_2 := {=}(x,y) \vee {=}(u,v) \vee {=}(u,v)$ are already NL-complete and NP-complete, respectively [9].

This article pursues the line of study taken in [10,4] where syntactical fragments of dependence and independence logic (FO($\perp_c$)) were investigated, respectively. FO($\perp_c$) extends first-order logic by conditional independence atoms

$$\boldsymbol{y} \perp_{\boldsymbol{x}} \boldsymbol{z}$$

with the informal meaning that the values of $\boldsymbol{y}$ and $\boldsymbol{z}$ are independent of each other, given any value of $\boldsymbol{x}$. As FO($\subseteq$), also FO($\perp_c$) does not have downward closure and is sensitive to the choice between the lax and the strict version of team semantics. For a set of atoms $\mathcal{C}$, we use FO($\mathcal{C}$) (omitting the set parentheses of $\mathcal{C}$) to denote the logic obtained by adding the atoms of $\mathcal{C}$ to first-order logic. FO($\mathcal{C}$)($k\forall$) denotes the sentences of FO($\mathcal{C}$) in which at most k variables are universally quantified. In [10] it was shown that

$$\mathrm{FO}(=(\ldots))(k\forall) \leq \mathrm{ESO}_f(k\forall) \leq \mathrm{FO}(=(\ldots))(2k\forall)$$

where $\mathrm{ESO}_f(k\forall)$ denotes the Skolem normal form ESO sentences in which at most k universally quantified first-order variables appear. In [4] it was shown that (under the lax team semantics)

- FO($\perp$)(2$\forall$) = FO($\perp$) and
- FO($\perp, \subseteq$)(1$\forall$) = FO($\perp, \subseteq$)

where FO($\perp$) is the sublogic of FO($\perp_c$) allowing only so-called pure independence atoms $\boldsymbol{x} \perp \boldsymbol{y}$. Moreover, it is known that FO($\perp$) is equivalent in expressive power to FO($\perp, \subseteq$) and FO($\perp_c$) [1,11].

Also different arity fragments were defined. By $\mathrm{FO}(\mathcal{C})(k\text{-dep})$ we denote the sentences of $\mathrm{FO}(\mathcal{C})$ in which only dependence atoms of the form $=\!(x_1,\ldots,x_{n+1})$ with $n \leq k$ may appear. $\mathrm{FO}(\mathcal{C})(k\text{-ind})$ denotes the sentences of $\mathrm{FO}(\mathcal{C})$ in which only independence atoms containing at most $k+1$ different variables may appear. It was shown in [10,4] that (under the lax team semantics)

$$\mathrm{ESO}(k\text{-ary}) = \mathrm{FO}(=\!(\ldots))(k\text{-dep}) = \mathrm{FO}(\perp_c)(k\text{-ind})$$

where $\mathrm{ESO}(k\text{-ary})$ denote the sentences of ESO in which the quantified functions and relations have arity at most k. This yields a strict arity hierarchy for both $\mathrm{FO}(=\!(\ldots))$ and $\mathrm{FO}(\perp_c)$ since the property "R is even" is definable in $\mathrm{ESO}(k\text{-ary})$ but not in $\mathrm{ESO}(k-1\text{-ary})$, for k-ary R [12].

The main contribution of this article is to show that arity fragments of inclusion logic also give rise to a strict expressivity hierarchy. We let $\mathrm{FO}(\mathcal{C})(k\text{-inc})$ denote the $\mathrm{FO}(\mathcal{C})$ sentences in which at most k-ary inclusion atoms (i.e. atoms of the form $\boldsymbol{x} \subseteq \boldsymbol{y}$ where $|\boldsymbol{x}| = |\boldsymbol{y}| \leq k$) may appear. For proving the claim, we define, for each $k \geq 2$, a graph property which is definable in $\mathrm{FO}(\subseteq)(k\text{-inc})$ but not in $\mathrm{FO}(\subseteq)(k-1\text{-inc})$. The non-definability part of the proof will be based on Martin Grohe's work in fixed-point logics in [13] where analogous results were proved for transitive closure logic (TC), least fixed-point logic (LFP), inflationary fixed-point logic (IFP) and partial fixed-point logic (PFP). We will also give a negative answer to the open question presented in [4]; that was, whether the fragments $\mathrm{FO}(\subseteq)(k\forall)$ give rise to an infinite expressivity hierarchy. This will be done by showing that $\mathrm{FO}(\subseteq)(1\forall) = \mathrm{FO}(\subseteq)$. However, if the strict version of team semantics is used, then we obtain that $\mathrm{FO}(\subseteq)(k\forall) < \mathrm{FO}(\subseteq)((k+1)\forall)$ [14].

The article is organized as follows. In Sect. 2 we define inclusion logic and present some of its basic properties. In Sect. 3 we show two results for inclusion logic. First we prove that the universal hierarchy collapses at level 1, and then we show that the arity hierarchy is strict (the full proof of this is in Appendix). In Sect. 4 we relate these results to analogous results in dependence logic and its variants, and conclude the section by presenting open problems.

2 Preliminaries

2.1 Notation

Unless otherwise stated, we use $x_1, x_2, \ldots$ to denote variables and $t_1, t_2, \ldots$ to denote terms. Analogously, bold versions $\boldsymbol{x}_1, \boldsymbol{x}_2, \ldots$ and $\boldsymbol{t}_1, \boldsymbol{t}_2, \ldots$ are used to denote tuples of variables and tuples of terms, respectively. For tuples $\boldsymbol{a}$ and $\boldsymbol{b}$, we write $\boldsymbol{ab}$ for the concatenation of the tuples. If f is a unary function and $(x_1, \ldots, x_n)$ is a sequence listing members of $\mathrm{Dom}(f)$, then we write $f(x_1, \ldots, x_n)$ for $(f(x_1), \ldots, f(x_n))$.

2.2 Inclusion Logic

The syntax of $\mathrm{FO}(\subseteq)$ is obtained by adding inclusion atoms to the syntax of first-order logic.

Definition 1. FO($\subseteq$) *is defined by the following grammar. Note that in an inclusion atom* $\boldsymbol{x}_1 \subseteq \boldsymbol{x}_2$*, the tuples* $\boldsymbol{x}_1$ *and* $\boldsymbol{x}_2$ *must be of the same length.*

$$\phi ::= \boldsymbol{x}_1 \subseteq \boldsymbol{x}_2 \mid t_1 = t_2 \mid \neg t_1 = t_2 \mid R(\boldsymbol{t}) \mid \neg R(\boldsymbol{t}) \mid (\phi \wedge \psi) \mid (\phi \vee \psi) \mid \forall x \phi \mid \exists x \phi.$$

FO(=(...)) and FO($\perp_{\mathrm{c}}$) are obtained from Definition 1 by replacing inclusion atoms $\boldsymbol{x}_1 \subseteq \boldsymbol{x}_2$ with dependence atoms $=(\boldsymbol{x}_1, x_2)$ and conditional independence atoms $\boldsymbol{x}_2 \perp_{\boldsymbol{x}_1} \boldsymbol{x}_3$, respectively. Pure independence logic FO($\perp$) is a fragment of FO($\perp_{\mathrm{c}}$) where only pure independence atoms $\boldsymbol{x}_1 \perp \boldsymbol{x}_2$ (i.e. atoms of the form $\boldsymbol{x}_1 \perp_{\emptyset} \boldsymbol{x}_2$) may appear. Also, for any $\mathcal{C} \subseteq \{\subseteq, =(\ldots), \perp_{\mathrm{c}}, \perp\}$, we use FO($\mathcal{C}$) (omitting the set parentheses of $\mathcal{C}$) to denote the logic obtained from Definition 1 by replacing inclusion atoms with atoms of $\mathcal{C}$.

In order to define semantics for these logics, we need to define the concept of a *team*. Let $\mathfrak{M}$ be a model with domain M. We assume that all our models have at least two elements.[1] An *assignment* over $\mathfrak{M}$ is a finite function that maps variables to elements of M. A *team* X of M with domain $\mathrm{Dom}(X) = \{x_1, \ldots, x_n\}$ is a set of assignments from $\mathrm{Dom}(X)$ into M. If X is a team of M and $F : X \to \mathcal{P}(M) \setminus \{\emptyset\}$, then we use $X[F/x]$ to denote the team $\{s(a/x) \mid s \in X, a \in F(s)\}$ and $X[M/x]$ for $\{s(a/x) \mid s \in X, a \in M\}$. Also one should note that if s is an assignment, then $\mathfrak{M} \models_s \phi$ refers to Tarski semantics and $\mathfrak{M} \models_{\{s\}} \phi$ refers to team semantics.

Definition 2. *For a model* $\mathfrak{M}$*, a team* X *and a formula in* FO($\subseteq, =(\ldots), \perp_{\mathrm{c}}$)*, the satisfaction relation* $\mathfrak{M} \models_X \phi$ *is defined as follows:*

- $\mathfrak{M} \models_X \alpha$ *if* $\forall s \in X (\mathfrak{M} \models_s \alpha)$*, when* α *is a first-order literal,*
- $\mathfrak{M} \models_X \boldsymbol{x}_1 \subseteq \boldsymbol{x}_2$ *if* $\forall s \in X \exists s' \in X \big(s(\boldsymbol{x}_1) = s'(\boldsymbol{x}_2)\big)$,
- $\mathfrak{M} \models_X \boldsymbol{x}_2 \perp_{\boldsymbol{x}_1} \boldsymbol{x}_3$ *if* $\forall s, s' \in X \big(s(\boldsymbol{x}_1) = s'(\boldsymbol{x}_1) \Rightarrow$
  $\exists s'' \in X (s''(\boldsymbol{x}_1) = s(\boldsymbol{x}_1), s''(\boldsymbol{x}_2) = s(\boldsymbol{x}_2), s''(\boldsymbol{x}_3) = s'(\boldsymbol{x}_3))\big)$,
- $\mathfrak{M} \models_X =(\boldsymbol{x}_1, x_2)$ *if* $\forall s, s' \in X \big(s(\boldsymbol{x}_1) = s'(\boldsymbol{x}_1) \Rightarrow s(x_2) = s'(x_2)\big)$,
- $\mathfrak{M} \models_X \phi \wedge \psi$ *if* $\mathfrak{M} \models_X \phi$ *and* $\mathfrak{M} \models_X \psi$,
- $\mathfrak{M} \models_X \phi \vee \psi$ *if* $\mathfrak{M} \models_Y \phi$ *and* $\mathfrak{M} \models_Z \psi$*, for some* $Y \cup Z = X$,
- $\mathfrak{M} \models_X \exists x \phi$ *if* $\mathfrak{M} \models_{X[F/x]} \phi$*, for some* $F : X \to \mathcal{P}(M) \setminus \{\emptyset\}$,
- $\mathfrak{M} \models_X \forall x \phi$ *if* $\mathfrak{M} \models_{X[M/x]} \phi$.

If $\mathfrak{M} \models_X \phi$*, then we say that* X *satisfies* ϕ *in* $\mathfrak{M}$*. If* ϕ *is a sentence and* $\mathfrak{M} \models_{\{\emptyset\}} \phi$[2]*, then we say that* ϕ *is true in* $\mathfrak{M}$*, and write* $\mathfrak{M} \models \phi$.

Note that in Definition 2, we obtain the *lax* version of team semantics. The *strict* version of team semantics is defined as in Definition 2 except that only disjoint subteams are allowed to witness split disjunction and existential quantification ranges over M instead of non-empty subsets of M. (See [1] for more information.)

First-order formulae are *flat* in the following sense (the proof is a straightforward structural induction).

---

[1] This assumption is needed in Theorem 3.

[2] $\{\emptyset\}$ denotes the team that consists of the empty assignment.

Theorem 1 (Flatness). *For a model $\mathfrak{M}$, a first-order formula ϕ and a team X, the following are equivalent:*

- $\mathfrak{M} \models_X \phi$,
- $\mathfrak{M} \models_{\{s\}} \phi$ *for all* $s \in X$,
- $\mathfrak{M} \models_s \phi$ *for all* $s \in X$.

By $\mathrm{Fr}(\phi)$ we denote the set of variables that appear free in ϕ. If X is a team and V a set of variables, then $X \upharpoonright V$ denotes the team $\{s \upharpoonright V \mid s \in X\}$ where $s \upharpoonright V$ is the restriction of the assignment s on V. Now, all formulae satisfy the following *locality* property (note that this is not true under the strict team semantics). The proof is a straightforward structural induction.

Theorem 2 (Locality). *Let $\mathfrak{M}$ be a model, X be a team, $\phi \in \mathrm{FO}(\subseteq, =(\ldots), \perp_c)$ and V a set of variables such that $Fr(\phi) \subseteq V \subseteq Dom(X)$. Then*

$$\mathfrak{M} \models_X \phi \Leftrightarrow \mathfrak{M} \models_{X \upharpoonright V} \phi.$$

We say that formulae $\phi, \psi \in \mathrm{FO}(\subseteq, =(\ldots), \perp_c)$ are *logically equivalent*, written $\phi \equiv \psi$, if for all models $\mathfrak{M}$ and teams X such that $\mathrm{Fr}(\phi) \cup \mathrm{Fr}(\psi) \subseteq \mathrm{Dom}(X)$,

$$\mathfrak{M} \models_X \phi \Leftrightarrow \mathfrak{M} \models_X \psi.$$

We obtain the following normal form theorem.

Theorem 3 ([4]). *Any formula $\phi \in \mathrm{FO}(\subseteq, =(\ldots), \perp_c)$ is logically equivalent to a formula ϕ' such that*

- *ϕ' is of the form $Q^1 x_1 \ldots Q^n x_n \psi$ where ψ is quantifier-free,*
- *any literal or dependency atom which occurs in ϕ' occurred already in ϕ,*
- *the number of universal quantifiers in ϕ' is the same as the number of universal quantifiers in ϕ.*

For logics $\mathcal{L}$ and $\mathcal{L}'$, we write $\mathcal{L} \leq \mathcal{L}'$, if for every signature τ, every $\mathcal{L}[\tau]$-sentence is logically equivalent to some $\mathcal{L}'[\tau]$-sentence. We write $\mathcal{L} \leq_O \mathcal{L}'$ if $\mathcal{L} \leq \mathcal{L}'$ is true in finite linearly ordered models. Equality and inequality relations are obtained from $\leq$ naturally. We end this section with the following list of theorems characterizing the expressive powers of our logics.

Theorem 4 ([2,15,11,5]).

- $\mathrm{FO}(=(\ldots)) = \mathrm{FO}(\perp_c) = \mathrm{FO}(\perp) = \mathrm{ESO}$,
- $\mathrm{FO}(\subseteq) = \mathrm{GFP}$.

3 Hierarchies in Inclusion Logic

In this section we consider universal and arity fragments of inclusion logic. In Subsect. 3.1 we define these fragments and also concepts of strictness and collapse of a hierarchy. In Subsect. 3.2 and 3.3 we prove collapse of the universal hierarchy and strictness of the arity hierarchy, respectively.

3.1 Syntactical Fragments

Definition 3. *Let* $\mathcal{C} \subseteq \{\subseteq, =(\ldots), \perp_{\mathrm{c}}, \perp\}$*. Then universal and arity fragments of* FO($\mathcal{C}$) *are defined as follows:*

- FO($\mathcal{C}$)($k\forall$) *is the class of* FO($\mathcal{C}$) *formulae in which at most* k *universal quantifiers may appear,*
- FO($\mathcal{C}$)(k-inc) *is the class of* FO($\mathcal{C}$) *formulae in which inclusion atoms of the form* $\boldsymbol{x}_1 \subseteq \boldsymbol{x}_2$ *where* $\boldsymbol{x}_1$ *and* $\boldsymbol{x}_2$ *are sequences of length at most* k*, may appear,*
- FO($\mathcal{C}$)(k-dep) *is the class of* FO($\mathcal{C}$) *formulae in which dependence atoms of the form* $=(\boldsymbol{x}_1, x_2)$ *where* $\boldsymbol{x}_1 x_2$ *is a sequence of length at most* $k+1$*, may appear,*
- FO($\mathcal{C}$)(k-ind) *is the class of* FO($\mathcal{C}$) *formulae in which conditional independence atoms of the form* $\boldsymbol{x}_2 \perp_{\boldsymbol{x}_1} \boldsymbol{x}_3$ *where* $\boldsymbol{x}_1\boldsymbol{x}_2\boldsymbol{x}_3$ *is a sequence listing at most* $k+1$ *distinct variables, may appear.*

For an increasing (with respect to $\leq$) sequence of logics $(\mathcal{L}_k)_{k\in\mathbb{N}}$, we say that the $\mathcal{L}_k$-hierarchy collapses at level m if $\mathcal{L}_m = \bigcup_{k\in\mathbb{N}} \mathcal{L}_k$. If the $\mathcal{L}_k$-hierarchy does not collapse at any level, then we say that the hierarchy is infinite. An infinite $\mathcal{L}_k$-hierarchy is called strict if $\mathcal{L}_k < \mathcal{L}_{k+1}$ for all $k \in \mathbb{N}$.

As mentioned before, we show that the FO($\subseteq$)($k\forall$)-hierarchy collapses already at level 1 but FO($\subseteq$)(k-inc) forms a strict hierarchy which holds already in finite models.

3.2 Collapse of the Universal Hierarchy

We first show that the universal hierarchy of inclusion logic collapses. This is done by introducing a translation where all universal quantifiers are removed, and new existential quantifiers, new inclusion atoms and one new universal quantifier are added. The translation holds already at the level of formulae.

Theorem 5. FO($\subseteq$)($1\forall$) = FO($\subseteq$).

Proof. Let $\phi \in$ FO($\subseteq$) be a formula. We define a formula $\phi' \in$ FO($\subseteq$)($1\forall$) such that $\phi \equiv \phi'$. By Theorem 3 we may assume that ϕ is of the form

$$Q^1 x_1 \ldots Q^n x_n \theta$$

where θ is quantifier-free. We let

$$\phi' := \exists x_1 \ldots \exists x_n \forall y(\bigwedge_{\substack{1\leq i\leq n \\ Q^i=\forall}} \boldsymbol{z} x_1 \ldots x_{i-1} y \subseteq \boldsymbol{z} x_1 \ldots x_{i-1} x_i \wedge \theta)$$

where $\boldsymbol{z}$ lists Fr(ϕ). Let now $\mathfrak{M}$ be a model and X a team such that Fr(ϕ) $\subseteq$ Dom(X); we show that $\mathfrak{M} \models_X \phi \Leftrightarrow \mathfrak{M} \models_X \phi'$. By Theorem 2 we may assume without loss of generality that Fr(ϕ) = Dom(X). Assume first that $\mathfrak{M} \models_X \phi$.

Then we find, for $1 \leq i \leq n$, functions $F_i : X[F_1/x_1]\dots[F_{i-1}/x_{i-1}] \to \mathcal{P}(M) \setminus \{\emptyset\}$ such that $F_i(s) = M$ if $Q^i = \forall$, and $\mathfrak{M} \models_{X'} \theta$ where

$$X' := X[F_1/x_1]\dots[F_n/x_n].$$

For $\mathfrak{M} \models_X \phi'$, it suffices to show that

$$\mathfrak{M} \models_{X'[M/y]} \bigwedge_{\substack{1\leq i\leq n\\ Q^i=\forall}} zx_1\dots x_{i-1}y \subseteq zx_1\dots x_{i-1}x_i \wedge \theta. \tag{1}$$

By Theorem 2 $\mathfrak{M} \models_{X'[M/y]} \theta$, so it suffices to consider only the new inclusion atoms of (1). So let $1 \leq i \leq n$ be such that $Q^i = \forall$ and let $s \in X'[M/y]$; we need to find a $s' \in X'[M/y]$ such that $s(zx_1\dots x_{i-1}y) = s'(zx_1\dots x_{i-1}x_i)$. Now, since $Q^i = \forall$, we note that $s(s(y)/x_i) \in X' \upharpoonright (\mathrm{Fr}(\phi) \cup \{x_1,\dots,x_i\})$. Therefore we may choose s' to be any extension of $s(s(y)/x_i)$ in $X'[M/y]$.

For the other direction, assume that $\mathfrak{M} \models_X \phi'$. Then for $1 \leq i \leq n$, there are functions $F_i : X[F_1/x_1]\dots[F_{i-1}/x_{i-1}] \to \mathcal{P}(M) \setminus \{\emptyset\}$ such that (1) holds, for $X' := X[F_1/x_1]\dots[F_n/x_n]$. By Theorem 2 $\mathfrak{M} \models_{X'} \theta$, so it suffices to show that, for all $1 \leq i \leq n$ with $Q^i = \forall$, F_i is the constant function which maps assignments to M. So let i be of the above kind, and let $s \in X[F_1/x_1]\dots[F_{i-1}/x_{i-1}]$ and $a \in M$. We need show that $s(a/x_i) \in X[F_1/x_1]\dots[F_i/x_i]$. First note that since y is universally quantified, $s(a/y)$ has an extension s_0 in $X'[M/y]$. Therefore, by (1), there is $s_1 \in X'[M/y]$ such that $s_0(zx_1\dots x_{i-1}y) = s_1(zx_1\dots x_{i-1}x_i)$. Since now s_1 agrees with s in $\mathrm{Fr}(\phi) \cup \{x_1,\dots,x_{i-1}\}$ and maps x_i to a, we obtain that

$$s(a/x_i) = s_1 \upharpoonright (\mathrm{Fr}(\phi) \cup \{x_1,\dots,x_i\}) \in X[F_1/x_1]\dots[F_i/x_i].$$

□

3.3 Strictness of the Arity Hierarchy

In this section we show that the following strict arity hierarchy holds (already in finite models).

Theorem 6. *For $k \geq 2$, $\mathrm{FO}(\subseteq)(k-1\text{-inc}) < \mathrm{FO}(\subseteq)(k\text{-inc})$.*

For proving this, we use the earlier work of Grohe in [13] where an analogous result was proved for TC, LFP, IFP and PFP. More precisely, it was shown that, for $k \geq 2$,

$$\mathrm{TC}^k \not\leq \mathrm{PFP}^{k-1} \tag{2}$$

where the superscript part gives the maximum arity allowed for the fixed-point operator. Since $\mathrm{TC}^k \leq \mathrm{LFP}^k \leq \mathrm{IFP}^k \leq \mathrm{PFP}^k$, a strict arity hierarchy is obtained for each of these logics.

We start by fixing τ as the signature consisting of one binary relation symbol E and $2k$ constant symbols $b_1,\dots,b_k,c_1,\dots,c_k$. Then the idea is to present a $\mathrm{FO}(\subseteq)(k\text{-inc})[\tau]$-definable graph property, and show that it is not definable in

$\mathrm{FO}(\subseteq)(k-1\text{-inc})[\tau]$. This graph property will actually be a negated version of the one that separates the fragments in (2). For this, we first define a first-order formula indicating that the k-tuples $\boldsymbol{x}$ and $\boldsymbol{y}$ form a $2k$-clique in a graph. Namely, we define $\mathrm{EDGE}_k(\boldsymbol{x},\boldsymbol{y})$ as follows:

$$\mathrm{EDGE}_k(\boldsymbol{x},\boldsymbol{y}) := \bigwedge_{1\le i,j\le k} E(x_i,y_j) \wedge \bigwedge_{1\le i\neq j\le k} (E(x_i,x_j)\wedge E(y_i,y_j)).$$

Using this we define the graph property. Let $\boldsymbol{x},\boldsymbol{y}$ be tuples of disjoint variables and $\boldsymbol{t},\boldsymbol{u}$ tuples of terms, all of the same length. Then for a first-order formula $\psi(\boldsymbol{x},\boldsymbol{y})$, we write $\mathfrak{M}\models\neg[\mathrm{TC}_{\boldsymbol{x},\boldsymbol{y}}\psi](\boldsymbol{t},\boldsymbol{u})$ if $(\boldsymbol{t}^{\mathfrak{M}},\boldsymbol{u}^{\mathfrak{M}})$ is not in the transitive closure of $\{(\boldsymbol{m}_1,\boldsymbol{m}_2)\mid \mathfrak{M}\models\psi(\boldsymbol{m}_1,\boldsymbol{m}_2)\}$. Note that by transitive closure of a binary relation R we denote the smallest transitive relation containing R. Now, given two k-ary tuples of disjoint constant terms $\boldsymbol{b}$ and $\boldsymbol{c}$, the non-trivial part is to show that $\neg[\mathrm{TC}_{\boldsymbol{x},\boldsymbol{y}}\mathrm{EDGE_k}](\boldsymbol{b},\boldsymbol{c})$ is not definable in $\mathrm{FO}(\subseteq)(k-1\text{-inc})[\tau]$. It is definable in $\mathrm{FO}(\subseteq)(k\text{-inc})[\tau]$ by the following theorem.

Theorem 7 **([1]).** *Let $\psi(\boldsymbol{x},\boldsymbol{y})$ be any first-order formula, where $\boldsymbol{x}$ and $\boldsymbol{y}$ are $k-$ary tuples of disjoint variables, for some $k\in\mathbb{N}\setminus\{0\}$. Furthermore, let $\psi'(\boldsymbol{x},\boldsymbol{y})$ be the result of writing $\neg\psi(\boldsymbol{x},\boldsymbol{y})$ in negation normal form. Then for all models $\mathfrak{M}$ containing the signature of ψ, and all pairs $\mathbf{b},\mathbf{c}$ of k-ary constant term tuples of the model,*

$$\mathfrak{M}\models\phi\Leftrightarrow\mathfrak{M}\models\neg[\mathrm{TC}_{\boldsymbol{x},\boldsymbol{y}}\psi](\boldsymbol{b},\boldsymbol{c}),$$

for ϕ defined as

$$\exists\boldsymbol{z}(\boldsymbol{b}\subseteq\boldsymbol{z}\wedge\boldsymbol{z}\neq\boldsymbol{c}\wedge\forall\boldsymbol{w}(\psi'(\boldsymbol{z},\boldsymbol{w})\vee\boldsymbol{w}\subseteq\boldsymbol{z})).$$

Note that ϕ is not yet of the right form since Definition 1 does not allow terms to appear in inclusion atoms. This is however not a problem since all the terms that appear in inclusion atoms can be replaced by using new existentially quantified variables and new identity atoms.

Hence, for Theorem 6, it suffices to prove that $\neg[\mathrm{TC}_{\boldsymbol{x},\boldsymbol{y}}\mathrm{EDGE_k}](\boldsymbol{b},\boldsymbol{c})$ is not definable in $\mathrm{FO}(\subseteq)(k-1\text{-inc})[\tau]$. In this part we follow the work in [13]. For $k,n\geq 1$, we first let $\mathcal{C}_{k,n}$ denote the set of all graphs $\mathfrak{A}$ with universe

$$A:=\{1,\ldots,n\}\times\{-k,\ldots,-1,1,\ldots,k\}.$$

The following theorem generates a graph $\mathfrak{A}\in\mathcal{C}_{k,n}$ which will be used in the non-definability proof. It was originally proved by Grohe using a method of Hrushovski [16] to extend partial isomorphisms of finite graphs.

Theorem 8 **([13]).** *Let $k,n\geq 2$. Then there exists a graph $\mathfrak{A}\in\mathcal{C}_{k,n}$ such that:*

1. *There exists a mapping row: $A\to\{1,\ldots,n\}$ such that*

$$\forall a,b\in A:(E^{\mathfrak{A}}ab\Rightarrow|\mathrm{row}(b)-\mathrm{row}(a)|\leq 1).$$

2. *There exists an automorphism* ε *of* $\mathfrak{A}$ *that is self-inverse and preserves the rows i.e.*
   - $\varepsilon^{-1} = \varepsilon$,
   - $\forall a \in A : \mathrm{row}(\varepsilon(a)) = \mathrm{row}(a)$.
3. *There exist tuples* $\boldsymbol{b}, \boldsymbol{c} \in A^k$ *in the first and last row respectively (i.e.* $\forall i \leq k : (\mathrm{row}(b_i) = 1 \wedge \mathrm{row}(c_i) = n)$*) such that*

$$\mathfrak{A} \models \neg[\mathrm{TC}_{\boldsymbol{x},\boldsymbol{y}} \mathit{EDGE}_{\mathrm{k}}](\boldsymbol{b}, \boldsymbol{c}) \text{ and } \mathfrak{A} \not\models \neg[\mathrm{TC}_{\boldsymbol{x},\boldsymbol{y}} \mathit{EDGE}_{\mathrm{k}}](\boldsymbol{b}, \varepsilon(\boldsymbol{c})).^3$$

4. *For all* $a_1, \ldots, a_{k-1} \in A$ *there exists an automorphism* f *of* $\mathfrak{A}$ *that is self-inverse, preserves the rows, and maps* $a_1, \ldots, a_{k-1}$ *according to* ε*, but leaves all elements in rows of distance* > 1 *from* $\mathrm{row}(a_1), \ldots, \mathrm{row}(a_{k-1})$ *fixed i.e.*
   - $f^{-1} = f$
   - $\forall a \in A : \mathrm{row}(f(a)) = \mathrm{row}(a)$,
   - $\forall i \leq k-1 : f(a_i) = \varepsilon(a_i)$,
   - *for each* $a \in A$ *with* $\forall i \leq k-1 : |\mathrm{row}(a) - \mathrm{row}(a_i)| > 1$ *we have* $f(a) = a$.

Using this theorem we can prove the following lemma.

Lemma 1. *Let* $k \geq 2$ *and let* τ *be a signature consisting of a binary relation symbol* E *and* $2k$ *constant symbols* $b_1, \ldots, b_k, c_1, \ldots, c_k$*. Then* $\neg[\mathrm{TC}_{\boldsymbol{x},\boldsymbol{y}} \mathit{EDGE}_{\mathrm{k}}](\boldsymbol{b}, \boldsymbol{c})$ *is not definable in* $\mathrm{FO}(\subseteq)(k-1\text{-inc})[\tau]$.

The proof of Lemma 1 is in Appendix but the outline of the proof is listed below.

Step 1 First we assume to the contrary that there is a $\phi(\boldsymbol{b}, \boldsymbol{c}) \in \mathrm{FO}(\subseteq)(k-1\text{-inc})[\tau]$ which is equivalent to $\neg[\mathrm{TC}_{\boldsymbol{x},\boldsymbol{y}} \mathrm{EDGE}_{\mathrm{k}}](\boldsymbol{b}, \boldsymbol{c})$. By Theorem 3 we may assume that ϕ is of the form $Q^1 x_1 \ldots Q^m x_m \theta$ where θ is a quantifier free formula from $\mathrm{FO}(\subseteq)(k-1\text{-inc})[\tau]$.

Step 2 We let $n = 2^{m+2}$ and obtain a graph $\mathfrak{A} \in \mathcal{C}_{k,n}$ for which items 1-4 of Theorem 8 hold. In particular, we find two k-ary tuples $\boldsymbol{b}$ and $\boldsymbol{c}$ such that $\mathfrak{A} \models \neg[\mathrm{TC}_{\boldsymbol{x},\boldsymbol{y}} \mathrm{EDGE}_{\mathrm{k}}](\boldsymbol{b}, \boldsymbol{c})$ and $\mathfrak{A} \not\models \neg[\mathrm{TC}_{\boldsymbol{x},\boldsymbol{y}} \mathrm{EDGE}_{\mathrm{k}}](\boldsymbol{b}, \varepsilon(\boldsymbol{c}))$. Then by the assumption $(\mathfrak{A}, \boldsymbol{b}, \boldsymbol{c}) \models \phi$, and hence we find, for $1 \leq i \leq m$, functions

$$F_i : \{\emptyset\}[F_1/x_1] \ldots [F_{i-1}/x_{i-1}] \to \mathcal{P}(A) \setminus \{\emptyset\}$$

such that $F_i(s) = A$ if $Q^i = \forall$, and

$$(\mathfrak{A}, \boldsymbol{b}, \boldsymbol{c}) \models_X \theta \tag{3}$$

where $X := \{\emptyset\}[F_1/x_1] \ldots [F_m/x_m]$.

Step 3 From X we will construct a team X^* such that

$$(\mathfrak{A}, \boldsymbol{b}, \varepsilon(\boldsymbol{c})) \models_{X^*} \theta. \tag{4}$$

[3] In [13], $\boldsymbol{c}$ and $\varepsilon(\boldsymbol{c})$ are here placed the other way round. This is however not a problem since ε is self-inverse and preserves the rows.

Step 4 At last, we will show that X^* can be constructed by quantifying the prefix $Q^1x_1 \dots Q^m x_m$ in $\mathfrak{A}$ over $\{\emptyset\}$. Hence we will obtain that $\mathfrak{A} \models \phi(\boldsymbol{b}, \varepsilon(\boldsymbol{c}))$. But now, since

$$\mathfrak{A} \not\models \neg[\mathrm{TC}_{\boldsymbol{x},\boldsymbol{y}}\mathrm{EDGE_k}](\boldsymbol{b}, \varepsilon(\boldsymbol{c})),$$

this contradicts the assumption that $\phi(\boldsymbol{b}, \boldsymbol{c})$ defines $\neg[\mathrm{TC}_{\boldsymbol{x},\boldsymbol{y}}\mathrm{EDGE_k}](\boldsymbol{b}, \boldsymbol{c})$. Hence the assumption is false and we obtain the result of Lemma 1.

The most difficult part of the proof is Step 3, and we will sketch this next. The idea is that we define two operations auto and swap for teams Y of A with $\mathrm{Dom}(Y) = \{x_1, \dots, x_m\}$. Then we let $X^* := \mathrm{swap}(\mathrm{auto}(X))$ and show by induction on the complexity of θ that (4) follows from (3). In the induction proof, the only non-trivial case is the base case where θ is either a first-order literal or an (at most $k-1$-ary) inclusion atom. Let us now explore this base case in more detail and examine how the two operations arise from it.

Consider first the case where θ is a first-order literal. For this, we define the operation swap. For a team Y of A with $\mathrm{Dom}(Y) = \{x_1, \dots, x_m\}$, $\mathrm{swap}(Y)$ will be defined as a team $\{s' \mid s \in Y\}$ where s' maps each variable x_i either to $s(x_i)$ or $\varepsilon \circ s(x_i)$. Here ε is the automorphism introduced in item 2 of Theorem 8. The idea is to define swap so that for all teams Y of A with $\mathrm{Dom}(Y) = \{x_1, \dots, x_m\}$ and all literals $\theta \in \mathrm{FO}[\tau]$ we obtain that

$$(\mathfrak{A}, \boldsymbol{b}, \boldsymbol{c}) \models_Y \theta \Rightarrow (\mathfrak{A}, \boldsymbol{b}, \varepsilon(\boldsymbol{c})) \models_{\mathrm{swap}(Y)} \theta.^4 \tag{5}$$

Consider then the case where $\theta = y_1 \dots y_l \subseteq z_1 \dots z_l$ for $l \le k-1$. First note that in this case (5) does not necessarily hold. Since, assuming the left-hand side of (5) and given any $s_0 \in Y$, we first notice that there is an $s_1 \in Y$ such that $s_0(y_i) = s_1(z_i)$ for all i. Now, given s'_0 of s_0, a natural choice from $\mathrm{swap}(Y)$ for the inclusion atom would be s'_1. However, this might not work since possibly $s'_0(y_i) = \varepsilon \circ s'_1(z_i)$ for some i. To overcome this, we again apply Theorem 8. This time we compose assignments of teams with automorphisms obtained by item 4 of Theorem 8. Namely, we first let I consist of all i for which $s'_0(y_i) = \varepsilon \circ s'_1(z_i)$. Then by item 4 of Theorem 8 we find an automorphism f which maps $s'_1(z_i)$ to $\varepsilon \circ s'_1(z_i)$ for $i \in I$, but leaves all elements in rows of distance > 1 from $(s'_1(z_i))_{i \in I}$ fixed. Then for $i \in I$, we obtain that

$$s'_0(y_i) = \varepsilon \circ s'_1(z_i) = f \circ s'_1(z_i),$$

and for $i \notin I$, we obtain that $s'_0(y_i) = s'_1(z_i) = f \circ s'_1(z_i)$.

[4] A detailed definition of swap is in Appendix but it can be illustrated by using Ehrenfeucht-Fraïssé games [17,18]. The point is that we have chosen $n := 2^{m+2}$ so that Duplicator has a winning strategy in the m-round Ehrenfeucht-Fraïssé game $\mathrm{EF}_m((\mathfrak{A}, \boldsymbol{b}, \boldsymbol{c}), (\mathfrak{A}, \boldsymbol{b}, \varepsilon(\boldsymbol{c}))$. Then for each $s \in Y$, we let s' correspond to Duplicator's choices in a single play of $\mathrm{EF}_m((\mathfrak{A}, \boldsymbol{b}, \boldsymbol{c}), (\mathfrak{A}, \boldsymbol{b}, \varepsilon(\boldsymbol{c}))$ where Spoiler picks members of $(\mathfrak{A}, \boldsymbol{b}, \boldsymbol{c})$ according to s and Duplicator picks members of $(\mathfrak{A}, \boldsymbol{b}, \varepsilon(\boldsymbol{c}))$ according to her winning strategy (see Fig. 1 in Appendix). Then we obtain that (5) holds.

Using the above idea, we now define the operation auto. First we let $\mathcal{F} = (F, \circ)$ be the group generated by the automorphisms $f_{\boldsymbol{a}}$ where $f_{\boldsymbol{a}}$ is obtained by item 4 of Theorem 8 for a sequence $\boldsymbol{a}$ listing $a_1, \ldots, a_{k-1} \in A$ where no a_i is located in the first or the last row. For a team Y of A, we then let

$$\mathrm{auto}(Y) := \{f \circ s \mid f \in \mathcal{F}, s \in Y\}.$$

Finally, we will show that these two approaches can be combined. We let $X^* := \mathrm{swap}(\mathrm{auto}(X))$ and prove by structural induction that (4) follows from (3). This concludes the sketch of Step 3.

By Theorem 7 and Lemma 1 we now conclude that Theorem 6 holds.

4 Conclusion

We have shown that the arity fragments of inclusion logic give rise to a strict expressivity hierarchy. Earlier, analogous results have been proved for dependence logic and independence logic. We also observed that the $\mathrm{FO}(\subseteq)(k\forall)$-hierarchy collapses at a very low level as it is the case with the $\mathrm{FO}(\perp_c)(k\forall)$-hierarchy. However, the $\mathrm{FO}(=(\ldots))(k\forall)$-hierarchy is infinite since it can be related to the strict $\mathrm{ESO}_f(k\forall)$-hierarchy. From the results of [10,4] and this article, we obtain the following classification for syntactical hierarchies of dependence, independence and inclusion logic under the lax semantics.

	Arity of Dependency Atom	Number of $\forall$
$\mathrm{FO}(=(\ldots))$	strict $\mathrm{FO}(=(\ldots))(k\text{-dep}) <$ $\mathrm{FO}(=(\ldots))(k+1\text{-dep})$	infinite $\mathrm{FO}(=(\ldots))(k\forall) <$ $\mathrm{FO}(=(\ldots))(2k+2\forall)$
$\mathrm{FO}(\perp_c)$	strict $\mathrm{FO}(\perp_c)(k\text{-ind}) <$ $\mathrm{FO}(\perp_c)(k+1\text{-ind})$	collapse at 2 $\mathrm{FO}(\perp_c)(2\forall) = \mathrm{FO}(\perp_c)$
$\mathrm{FO}(\subseteq)$	**strict** $\mathrm{FO}(\subseteq)(k\text{-inc}) <$ $\mathrm{FO}(\subseteq)(k+1\text{-inc})$	**collapse at 1** $\mathrm{FO}(\subseteq)(1\forall) = \mathrm{FO}(\subseteq)$

Since $\mathrm{FO}(\subseteq)$ captures PTIME in finite ordered models, it would be interesting to investigate syntactical fragments of inclusion logic in that setting. It appears that then the techniques used in this article would be of no use. Namely, we cannot hope to construct two ordered models in the style of Theorem 8. In fixed-point logics, this same question has been studied in the 90s. Imhof showed in [19] that the arity hierarchy of PFP remains strict in ordered models ($\mathrm{PFP}^k <_{\mathcal{O}} \mathrm{PFP}^{k+1}$) by relating the PFP^k-fragments to the degree hierarchy within PSPACE. For LFP and IFP, the same question appears to be more difficult, since both collapse and its negation have strong complexity theoretical consequences. That is, for both IFP and LFP in ordered models, collapse of arity hierarchy implies PTIME < PSPACE, and infinite arity hierarchy implies

LOGSPACE < PTIME. It might be possible to prove similar results for inclusion logic by relating the fragments FO($\subseteq$)(k-inc) to arity fragments of fixed-point logics. However, the translations between FO($\subseteq$) and GFP provided in [5] do not respect arities. It remains open whether collapse of the FO($\subseteq$)(k-inc)-hierarchy or its negation have such strong consequences or whether it is possible to relate the FO($\subseteq$)(k-inc)-fragments in ordered models to the degree hierarchy within PTIME? Another line would be to find some other syntactical parameter that would fit for this purpose.

Acknowledgements. The author was supported by grant 264917 of the Academy of Finland.

References

1. Galliani, P.: Inclusion and exclusion dependencies in team semantics: On some logics of imperfect information. Annals of Pure and Applied Logic 163(1), 68 (2012)
2. Väänänen, J.: Dependence Logic. Cambridge University Press (2007)
3. Hodges, W.: Compositional Semantics for a Language of Imperfect Information. Journal of the Interest Group in Pure and Applied Logics 5(4), 539–563 (1997)
4. Galliani, P., Hannula, M., Kontinen, J.: Hierarchies in independence logic. In: Rocca, S.R.D. (ed.) Computer Science Logic 2013 (CSL 2013). Leibniz International Proceedings in Informatics (LIPIcs), vol. 23, pp. 263–280. Schloss Dagstuhl–Leibniz-Zentrum fuer Informatik, Dagstuhl (2013)
5. Galliani, P., Hella, L.: Inclusion Logic and Fixed Point Logic. In: Rocca, S.R.D. (ed.) Computer Science Logic 2013 (CSL 2013). Leibniz International Proceedings in Informatics (LIPIcs), vol. 23, pp. 281–295. Schloss Dagstuhl–Leibniz-Zentrum fuer Informatik, Dagstuhl (2013)
6. Abramsky, S., Väänänen, J.: From IF to BI. Synthese 167, 207–230 (2009), doi:10.1007/s11229-008-9415-6
7. Yang, F.: Expressing Second-order Sentences in Intuitionistic Dependence Logic. Studia Logica 101(2), 323–342 (2013)
8. Grädel, E.: Model-checking games for logics of imperfect information. Theor. Comput. Sci. 493, 2–14 (2013)
9. Kontinen, J.: Coherence and computational complexity of quantifier-free dependence logic formulas. In: Kontinen, J., Väänänen, J. (eds.) Proceedings of Dependence and Independence in Logic. ESSLLI, pp. 58–77 (2010)
10. Durand, A., Kontinen, J.: Hierarchies in dependence logic. ACM Transactions on Computational Logic (TOCL) 13(4), 31 (2012)
11. Galliani, P., Väänänen, J.A.: On dependence logic. In: Baltag, A., Smets, S. (eds.) Johan van Benthem on Logical and Informational Dynamics, pp. 101–119. Springer (2014)
12. Ajtai, M.: Σ^1_1-formulae on finite structures. Ann. Pure Appl. Logic 4(1), 1–48 (1983)
13. Grohe, M.: Arity hierarchies. Ann. Pure Appl. Logic 82(2), 103–163 (1996)
14. Hannula, M., Kontinen, J.: Hierarchies in independence and inclusion logic with strict semantics. Journal of Logic and Computation (2014), doi: 10.1093/logcom/exu057

15. Grädel, E., Väänänen, J.: Dependence and independence. Studia Logica 101(2), 399–410 (2013)
16. Hrushovski, E.: Extending partial isomorphisms of graphs. Combinatorica 12(4), 411–416 (1992)
17. Fraïssé, R.: Sur une nouvelle classification des systmes de relations. Comptes Rendus 230, 1022–1024 (1950)
18. Ehrenfeucht, A.: An application of games to the completeness problem for formalized theories. Fundamenta Mathematicae 49, 129–141 (1961)
19. Imhof, H.: Computational aspects of arity hierarchies. In: van Dalen, D., Bezem, M. (eds.) CSL 1996. LNCS, vol. 1258, pp. 211–225. Springer, Heidelberg (1997)

Appendix

The proof of Lemma 1 is presented in the following.

Proof (Lemma 1). We may start from Step 3 of the outline of the proof presented after Lemma 1. Hence we have

$$(\mathfrak{A}, \boldsymbol{b}, \boldsymbol{c}) \models_X \theta, \tag{6}$$

for $X := \{\emptyset\}[F_1/x_1]\dots[F_m/x_m]$, and the first step is to construct a team X^* such that

$$(\mathfrak{A}, \boldsymbol{b}, \varepsilon(\boldsymbol{c})) \models_{X^*} \theta. \tag{7}$$

For this, we first define the operation auto. By item 4 of Theorem 8, for all $\boldsymbol{a}$ listing $a_1, \dots, a_{k-1} \in A$ there exists an automorphism $f_{\boldsymbol{a}}$ which maps $\boldsymbol{a}$ pointwise to $\varepsilon(\boldsymbol{a})$, but leaves all elements in rows of distance > 1 from $\text{row}(a_1), \dots, \text{row}(a_{k-1})$ fixed. Let $\mathcal{F} = (F, \circ)$ be the group generated by the automorphisms $f_{\boldsymbol{a}}$ where $f_{\boldsymbol{a}}$ is obtained from item 4 of Theorem 8 and $\boldsymbol{a}$ is a sequence listing $a_1, \dots, a_{k-1} \in A$ such that $2 < \text{row}(a_i) < n-1$, for $1 \le i \le k-1$. For a team Y of A, we then let

$$\text{auto}(Y) := \{f \circ s \mid f \in \mathcal{F}, s \in Y\}.$$

Next we will define the operation swap. For this, we will first define mappings mid and h. We let mid map m-sequences of $\{1, \dots, n\}$ into $\{1, \dots, n\}$ so that, for any $\boldsymbol{p} := (p_1, \dots, p_m)$ and $\boldsymbol{q} := (q_1, \dots, q_m)$ in $\{1, \dots, n\}^m$,

1. $1 < \text{mid}(\boldsymbol{p}) < n$,
2. $\forall i \le m : \text{mid}(\boldsymbol{p}) \ne p_i$,
3. $\forall l \le m$: if $\boldsymbol{p} \restriction \{1, \dots, l\} = \boldsymbol{q} \restriction \{1, \dots, l\}$, then $\forall i \le l : p_i < \text{mid}(\boldsymbol{p})$ iff $q_i < \text{mid}(\boldsymbol{q})$.

This can be done by following the strategy illustrated in Fig. 1.[5] We shall explain this in detail in the following. Let $\boldsymbol{p} := (p_1, \ldots, p_m)$ be a sequence listing natural numbers of $\{1, \ldots, n\}$. For mid($\boldsymbol{p}$), we first show how to choose M and N, for each $0 \le i \le m$, so that

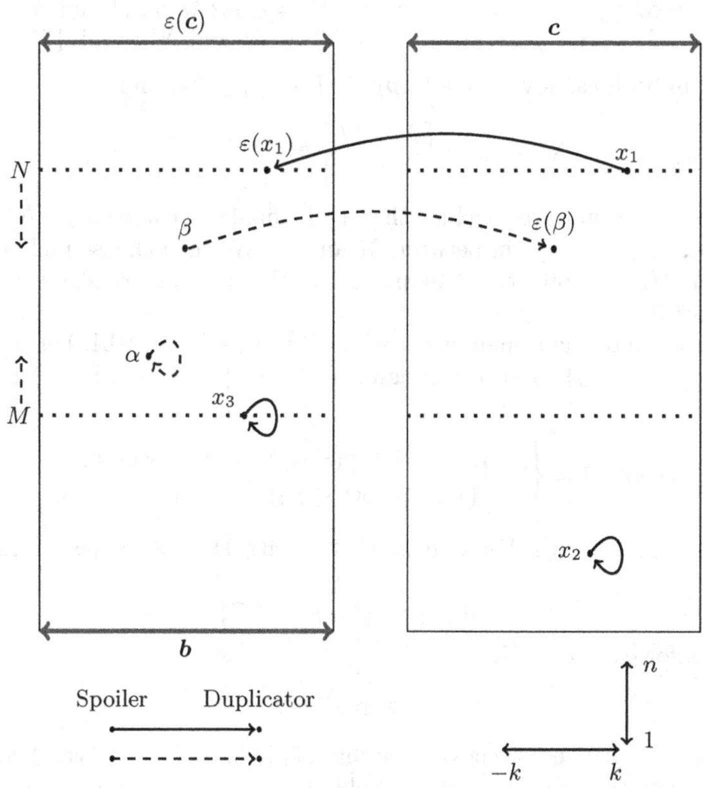

Fig. 1

[5] A play of $\mathrm{EF}_m((\mathfrak{A}, \boldsymbol{b}, \boldsymbol{c}), (\mathfrak{A}, \boldsymbol{b}, \varepsilon(\boldsymbol{c}))$ where Spoiler picks members of $(\mathfrak{A}, \boldsymbol{b}, \boldsymbol{c})$ and Duplicator picks members of $(\mathfrak{A}, \boldsymbol{b}, \varepsilon(\boldsymbol{c}))$ according to her winning strategy is illustrated in Fig. 1. The idea is that after each round $i \le m$, M and N are placed so that $N - M \ge 2^{m+1-i}$. Also for each $j \le i$, $y_j = x_j$ if $\mathrm{row}(x_j) \le M$, and $y_j = \varepsilon(x_j)$ if $\mathrm{row}(x_j) \ge N$, where y_j and x_j represent Duplicator's and Spoiler's choices, respectively. In the picture, α and β represent two alternative choices Spoiler can make at the fourth round. If Spoiler chooses $x_4 := \alpha$, then Duplicator chooses $y_4 := \alpha$, and M is moved to $\mathrm{row}(\alpha)$. If Spoiler chooses $x_4 := \beta$, then Duplicator chooses $y_4 := \varepsilon(\beta)$, and N is moved to $\mathrm{row}(\beta)$. Proceeding in this way we obtain that at the final stage m, $(\mathfrak{A}, \boldsymbol{b}, \boldsymbol{c}, x_1, \ldots, x_m)$ and $(\mathfrak{A}, \boldsymbol{b}, \varepsilon(\boldsymbol{c}), y_1, \ldots, y_m)$ agree on all atomic $\mathrm{FO}[\tau]$ formulae.

- $N - M \geq 2^{m+1-i}$,
- $\forall j \leq i : p_j \leq M$ or $p_j \geq N$.

We do this inductively as follows. We let $M := 1$ and $N := n$, for $i = 0$. Since $n = 2^{m+2}$, clearly the conditions above hold. Assume that M and N are defined for i so that the conditions above hold; we define M' and N' for $i+1$ as follows:

1. If $p_{i+1} - M \leq N - p_{i+1}$, then we let $M' := \max\{M, p_{i+1}\}$ and $N' := N$.
2. If $p_{i+1} - M > N - p_{i+1}$, then we let $M' := M$ and $N' := \min\{N, p_{i+1}\}$.

Note that in both cases $\forall j \leq i+1 : p_j \leq M'$ or $p_j \geq N'$, and

$$N' - M' \geq \left\lceil \frac{N-M}{2} \right\rceil \geq 2^{m+1-(i+1)}.$$

Proceeding in this way we conclude that at the final stage m we have $N - M \geq 2$ with no $p_1, \ldots, p_m$ stricly in between M and N. We then choose $\mathrm{mid}(\boldsymbol{p})$ as any number in $]M, N[$. Note that defining mid in this way we are able to meet the conditions 1-3.

After this we define a mapping $h : {}^{\{x_1,\ldots,x_m\}}A \to {}^{\{x_1,\ldots,x_m\}}A$. For an assignment $s : \{x_1, \ldots, x_m\} \to A$, the assignment $h(s) : \{x_1, \ldots, x_m\} \to A$ is defined as follows:

$$h(s)(x_i) = \begin{cases} s(x_i) & \text{if } \mathrm{row}(s(x_i)) < \mathrm{mid}(\mathrm{row}(s(\boldsymbol{x}))), \\ \varepsilon \circ s(x_i) & \text{if } \mathrm{row}(s(x_i)) > \mathrm{mid}(\mathrm{row}(s(\boldsymbol{x}))), \end{cases}$$

where $\boldsymbol{x} := (x_1, \ldots, x_m)$. For a team Z of A with $\mathrm{Dom}(Z) = \{x_1, \ldots, x_m\}$, we now let

$$\mathrm{swap}(Z) := \{h(s) \mid s \in Z\},$$

and define, for each $Y \subseteq X$,

$$Y^* := \mathrm{swap}(\mathrm{auto}(Y)).$$

With X^* now defined, we next show that (7) holds. Without loss of generality we may assume that if a constant symbol b_j (or c_j) appears in an atomic subformula α of θ, then α is of the form $x_i = b_j$ (or $x_i = c_j$) where x_i is an existentially quantified variable of the quantifier prefix. Hence and by (6), it now suffices to show that for all $Y \subseteq X$ and all quantifier-free $\psi \in \mathrm{FO}(\subseteq)(k-1\text{-inc})[\tau]$ with the above restriction for constants,

$$(\mathfrak{A}, \boldsymbol{b}, \boldsymbol{c}) \models_Y \psi \Rightarrow (\mathfrak{A}, \boldsymbol{b}, \varepsilon(\boldsymbol{c})) \models_{Y^*} \psi.$$

This can be done by induction on the complexity of the quantifier-free ψ. Since $Y^* \cup Z^* = (Y \cup Z)^*$, for $Y, Z \subseteq X$, it suffices to consider only the case where ψ is an atomic or negated atomic formula. For this, assume that $(\mathfrak{A}, \boldsymbol{b}, \boldsymbol{c}) \models_Y \psi$; we show that

$$(\mathfrak{A}, \boldsymbol{b}, \varepsilon(\boldsymbol{c})) \models_{Y^*} \psi. \tag{8}$$

Now ψ is either of the form $x_i = b_j$, $x_i = c_j$, $x_i = x_j$, $\neg x_i = x_j$, $E(x_i, x_j)$, $\neg E(x_i, x_j)$ or $\boldsymbol{y} \subseteq \boldsymbol{z}$ where b_j, c_j are constant symbols and $\boldsymbol{y}, \boldsymbol{z}$ are sequences of variables from $\{x_1, \ldots, x_m\}$ with $|\boldsymbol{y}| = |\boldsymbol{z}| \leq k-1$.

- Assume first that ψ is of the form $x_i = b_j$ or $x_i = c_j$, and let $s \in Y^*$ be arbitrary. For (8), it suffices to show by Theorem 1 that
$$(\mathfrak{A}, \boldsymbol{b}, \varepsilon(\boldsymbol{c})) \models_s \psi. \tag{9}$$
First note that $s = h(f \circ t)$ for some automorphism $f \in \mathcal{F}$ and assignment $t \in Y$ for which, by the assumption and Theorem 1, $(\mathfrak{A}, \boldsymbol{b}, \boldsymbol{c}) \models_t \psi$. Hence for (9), we only need to show that $s(x_i) = t(x_i)$ in case $t(x_i)$ is listed in $\boldsymbol{b}$, and $s(x_i) = \varepsilon \circ t(x_i)$ in case $t(x_i)$ is listed in $\boldsymbol{c}$. For this, first recall that $\mathcal{F}$ is the group generated by automorphisms $f_{\boldsymbol{a}}$ where $f_{\boldsymbol{a}}$ is obtained from item 4 of Theorem 8 and $\boldsymbol{a}$ is a sequence listing $a_1, \ldots, a_{k-1} \in A$ such that $2 < \mathrm{row}(a_i) < n-1$, for $1 \leq i \leq k-1$. Therefore f leaves all elements in the first and the last row fixed when $f(\boldsymbol{b}) = \boldsymbol{b}$ and $f(\boldsymbol{c}) = \boldsymbol{c}$. On the other hand, by the definition of mid, $1 < \mathrm{mid}(\mathrm{row}(f \circ t(\boldsymbol{x}))) < n$, and hence $h(f \circ t)(x_i) = f \circ t(x_i)$ if $f \circ t(x_i)$ is in the first row, and $h(f \circ t)(x_i) = \varepsilon \circ f \circ t(x_i)$ if $f \circ t(x_i)$ is in the last row. Since tuples $\boldsymbol{b}$ and $\boldsymbol{c}$ are in the first and the last row, respectively, we conclude that the claim holds. The case where ψ is of the form $x_i = x_j$ or $\neg x_i = x_j$ is straightforward.
- Assume that ψ is of the form $E(x_i, x_j)$ or $\neg E(x_i, x_j)$. Again, let $s \in Y^*$ when $s = h(f \circ t)$ for some $f \in \mathcal{F}$ and $t \in Y$. For (9), consider first the cases where
$$\mathrm{row}(t(x_i)), \mathrm{row}(t(x_j)) < \mathrm{mid}(\mathrm{row}(t(\boldsymbol{x}))), \text{ or} \tag{10}$$
$$\mathrm{row}(t(x_i)), \mathrm{row}(t(x_j)) > \mathrm{mid}(\mathrm{row}(t(\boldsymbol{x}))). \tag{11}$$
Since f is a row-preserving automorphism, we conclude by the definition of h that s maps both x_i and x_j either according to $f \circ t$ or according to $\varepsilon \circ f \circ t$. Since ε is also an automorphism, we obtain (9) in both cases. Assume then that (10) and (11) both fail. Then by symmetry suppose we have
$$\mathrm{row}(t(x_i)) < \mathrm{mid}(\mathrm{row}(t(\boldsymbol{x}))) < \mathrm{row}(t(x_j)).$$
Since $(\mathfrak{A}, \boldsymbol{b}, \boldsymbol{c}) \models_t \psi$, we have by item 1 of Theorem 8 that ψ is $\neg E(x_i, x_j)$. Since f and ε preserve the rows, we have
$$\mathrm{row}(s(x_i)) < \mathrm{mid}(\mathrm{row}(s(\boldsymbol{x}))) < \mathrm{row}(s(x_j)).$$
Therefore we obtain $(\mathfrak{A}, \boldsymbol{b}, \boldsymbol{c}) \models_s \neg E(x_i, x_j)$ which concludes this case.
- Assume that ϕ is $\boldsymbol{y} \subseteq \boldsymbol{z}$, for some $\boldsymbol{y} = y_1 \ldots y_l$ and $\boldsymbol{z} = z_1 \ldots z_l$ where $l \leq k-1$. Let $s \in Y^*$ be arbitrary. For (8), we show that there exists a $s' \in Y^*$ such that $s(\boldsymbol{y}) = s'(\boldsymbol{z})$. Now $s = h(f \circ t)$ for some $f \in \mathcal{F}$ and $t \in Y$, and $(\mathfrak{A}, \boldsymbol{b}, \boldsymbol{c}) \models_Y \psi$ by the assumption. Hence there exists a $t' \in Y$ such that $t(\boldsymbol{y}) = t'(\boldsymbol{z})$. Let now I list the indices $1 \leq i \leq l$ for which (i) or (ii) hold:[6]

[6] An example where $\boldsymbol{y} := y_1y_2y_3$ and $\boldsymbol{z} := z_1z_2z_3$ is illustrated in Fig. 2. Note that in the example, $I = \{2\}$ since the index number 2 satisfies (ii). Then letting $s_0 := h(f \circ t')$, we obtain $s(y_1y_3) = s_0(z_1z_3)$ but only $s(y_2) = \varepsilon \circ s_0(z_2)$. Fig. 3 shows that choosing $s' := h(f_a \circ f \circ t')$, for $a := f \circ t'(z_2)$, we obtain $s(\boldsymbol{y}) = s'(\boldsymbol{z})$.

(i) $\mathrm{row}(t(y_i)) < \mathrm{mid}(\mathrm{row}(t(\boldsymbol{x})))$ and $\mathrm{row}(t'(z_i)) > \mathrm{mid}(\mathrm{row}(t'(\boldsymbol{x})))$,
(ii) $\mathrm{row}(t(y_i)) > \mathrm{mid}(\mathrm{row}(t(\boldsymbol{x})))$ and $\mathrm{row}(t'(z_i)) < \mathrm{mid}(\mathrm{row}(t'(\boldsymbol{x})))$.

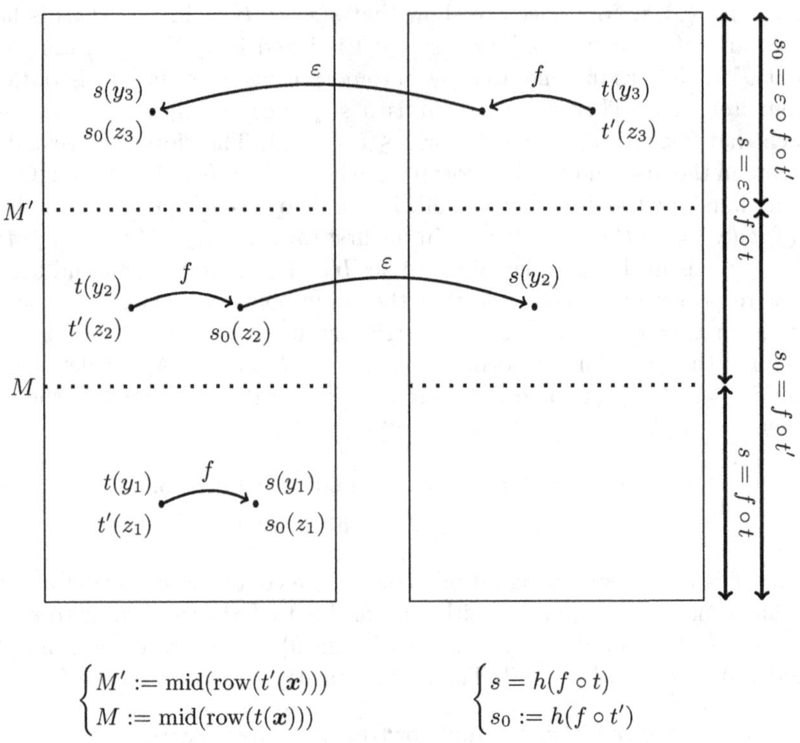

Fig. 2

Since $|I| \leq k-1$, choosing $\boldsymbol{a} := (f \circ t'(z_i))_{i \in I}$ we find by item 4 of Theorem 8 an automorphism $f_{\boldsymbol{a}}$ that swaps $f \circ t'(z_i)$ to $\varepsilon \circ f \circ t'(z_i)$, for each $i \in I$, but leaves all elements in rows of distance > 1 from $(\mathrm{row}(f \circ t'(z_i)))_{i \in I}$ fixed. We now let $s' := h(f_{\boldsymbol{a}} \circ f \circ t')$. Since

$$1 < \mathrm{mid}(\mathrm{row}(t(\boldsymbol{x}))), \mathrm{mid}(\mathrm{row}(t'(\boldsymbol{x}))) < n$$

by the definition, we have $2 < \mathrm{row}(t'(z_i)) < n-1$, for $i \in I$. Hence $f_{\boldsymbol{a}} \in \mathcal{F}$ and $s' \in Y^*$. Moreover, for $i \in I$, we obtain that

(i) $s(y_i) = f \circ t(y_i) = f \circ t'(z_i) = \varepsilon \circ f_{\boldsymbol{a}} \circ f \circ t'(z_i) = s'(z_i)$, or
(ii) $s(y_i) = \varepsilon \circ f \circ t(y_i) = \varepsilon \circ f \circ t'(z_i) = f_{\boldsymbol{a}} \circ f \circ t'(z_i) = s'(z_i)$.

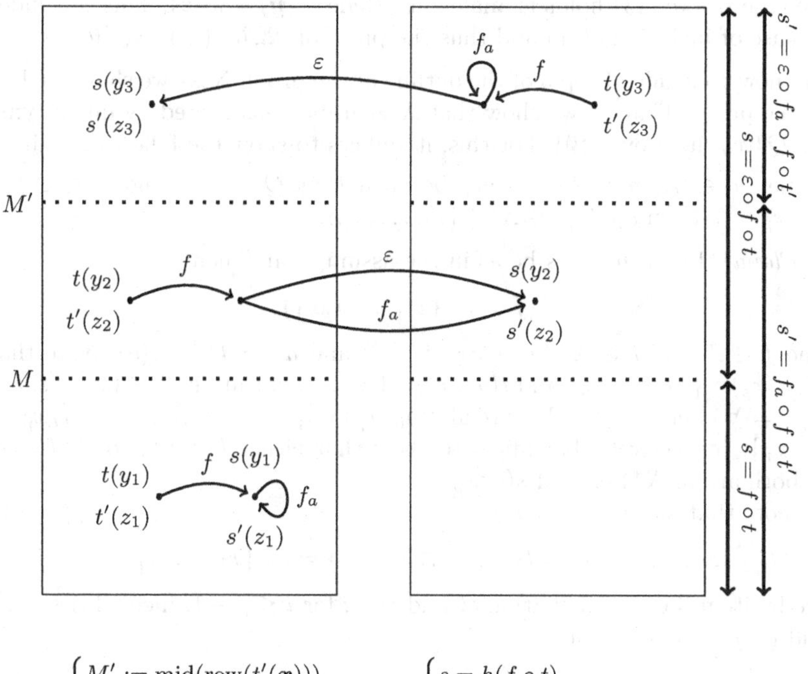

$$\begin{cases} M' := \mathrm{mid}(\mathrm{row}(t'(\boldsymbol{x}))) \\ M := \mathrm{mid}(\mathrm{row}(t(\boldsymbol{x}))) \end{cases} \qquad \begin{cases} s = h(f \circ t) \\ s' := h(f_a \circ f \circ t'), \text{ for } a := f \circ t'(z_2) \end{cases}$$

Fig. 3

For the first and last equalities note that f_a and f preserve the rows. For (i) recall also that ε is self-inverse.
Let then $1 \leq j \leq l$ be such that $j \notin I$ when both (i) and (ii) and fail for j. Then we obtain

$$\mathrm{row}(t(y_j)) > \mathrm{mid}(\mathrm{row}(t(\boldsymbol{x}))) \text{ and } \mathrm{row}(t'(z_j)) > \mathrm{mid}(\mathrm{row}(t'(\boldsymbol{x}))), \text{ or} \quad (12)$$
$$\mathrm{row}(t(y_j)) < \mathrm{mid}(\mathrm{row}(t(\boldsymbol{x}))) \text{ and } \mathrm{row}(t'(z_j)) < \mathrm{mid}(\mathrm{row}(t'(\boldsymbol{x}))). \quad (13)$$

Assume first that (12) holds and let $i \in I$. Then either

(i) $\mathrm{row}(t(y_i)) < \mathrm{mid}(\mathrm{row}(t(\boldsymbol{x}))) < \mathrm{row}(t(y_j))$, or
(ii) $\mathrm{row}(t'(z_i)) < \mathrm{mid}(\mathrm{row}(t'(\boldsymbol{x}))) < \mathrm{row}(t'(z_j))$.

Since $t(y_j) = t'(z_j)$, $t(y_i) = t'(z_i)$, and f preserves the rows, in both cases we conclude that

$$|\mathrm{row}(f \circ t'(z_j)) - \mathrm{row}(f \circ t'(z_i))| > 1.$$

Therefore f_a leaves $f \circ t'(z_j)$ fixed. By (12) we now have

$$s(y_j) = \varepsilon \circ f \circ t(y_j) = \varepsilon \circ f \circ t'(z_j) = \varepsilon \circ f_a \circ f \circ t'(z_j) = s'(z_j).$$

The case where (13) holds is analogous. Hence $s(\boldsymbol{y}) = s'(\boldsymbol{z})$. This concludes the case of inclusion atom and thus the proof of $(\mathfrak{A}, \boldsymbol{b}, \varepsilon(\boldsymbol{c})) \models_{X^*} \theta$.

We have now concluded Step 3 of the outline of the proof. Next we show the last part of the proof. That is, we show that X^* can be constructed by quantifying $Q^1x_1 \dots Q^m x_m$ in $\mathfrak{A}$ over $\{\emptyset\}$. For this, it suffices to show the following claim.

Claim. Let $a \in A$, $p \in \{1, \dots, m\}$ be such that $Q^p = \forall$, and $s \in X^* \upharpoonright \{x_1, \dots, x_{p-1}\}$. Then $s(a/x_p) \in X^* \upharpoonright \{x_1, \dots, x_p\}$.

Proof (Claim). Let a, p and s be as in the assumption. Then

$$s = h(f \circ t) \upharpoonright \{x_1, \dots, x_{p-1}\},$$

for some $f \in \mathcal{F}$ and $t \in X$. Let $a_0 = f^{-1}(a)$ and $a_1 = f^{-1} \circ \varepsilon(a)$. Note that both $t(a_0/x_p) \upharpoonright \{x_1, \dots, x_p\}$ and $t(a_1/x_p) \upharpoonright \{x_1, \dots, x_p\}$ are in $X \upharpoonright \{x_1, \dots, x_p\}$ since $Q^p = \forall$. Let $t_0, t_1 \in X$ extend $t(a_0/x_p) \upharpoonright \{x_1, \dots, x_p\}$ and $t(a_1/x_p) \upharpoonright \{x_1, \dots, x_p\}$, respectively. It suffices to show that either $h(f \circ t_0)$ or $h(f \circ t_1)$ (which both are in X^*) extend $s(a/x_p)$.

First note that since

$$t_0 \upharpoonright \{x_1, \dots, x_{p-1}\} = t_1 \upharpoonright \{x_1, \dots, x_{p-1}\} = t \upharpoonright \{x_1, \dots, x_{p-1}\}$$

we have by item 3 of the definition of mid that, for $i \leq p-1$, inequalities (14), (15) and (16) are equivalent:

$$\mathrm{row}(t_0(x_i)) < \mathrm{mid}(\mathrm{row}(t_0(\boldsymbol{x}))), \tag{14}$$

$$\mathrm{row}(t_1(x_i)) < \mathrm{mid}(\mathrm{row}(t_1(\boldsymbol{x}))), \tag{15}$$

$$\mathrm{row}(t(x_i)) < \mathrm{mid}(\mathrm{row}(t(\boldsymbol{x}))). \tag{16}$$

Since also f preserves the rows, we have by the definition of h that $h(f \circ t_0)$, $h(f \circ t_1)$ and $h(f \circ t)$ all agree in variables $x_1, \dots, x_{p-1}$. Note that also ε preserves the rows, so have $\mathrm{row}(a_0) = \mathrm{row}(a_1)$. Since then $\mathrm{row}(t_0(x_i)) = \mathrm{row}(t_1(x_i))$, for $i \leq p$, we have by item 3 of the definition of mid that

$$\mathrm{row}(t_0(x_p)) < \mathrm{mid}(\mathrm{row}(t_0(\boldsymbol{x}))) \text{ iff } \mathrm{row}(t_1(x_p)) < \mathrm{mid}(\mathrm{row}(t_1(\boldsymbol{x}))).$$

Therefore, either

$$\mathrm{row}(t_0(x_p)) < \mathrm{mid}(\mathrm{row}(t_0(\boldsymbol{x}))) \text{ or } \mathrm{row}(t_1(x_p)) > \mathrm{mid}(\mathrm{row}(t_1(\boldsymbol{x}))).$$

Then in the first case $h(f \circ t_0)(x_p) = f \circ t_0(x_p) = a$, and in the second case $h(f \circ t_1)(x_p) = \varepsilon \circ f \circ t_1(x_p) = \varepsilon \circ \varepsilon(a) = a$. Hence $s(a/x_p) \in X^* \upharpoonright \{x_1, \dots, x_p\}$. This concludes the proof of Claim. □

We have now shown that X^* can be constructed by quantifying $Q^1x_1 \dots Q^m x_m$ in $\mathfrak{A}$ over $\{\emptyset\}$. Also previously we showed that $(\mathfrak{A}, \boldsymbol{b}, \varepsilon(\boldsymbol{c})) \models_{X^*} \theta$. Therefore, since $\phi = Q^1x_1 \dots Q^m x_m \theta$, we obtain that $(\mathfrak{A}, \boldsymbol{b}, \varepsilon(\boldsymbol{c})) \models \phi$. Hence the assumption that $\phi(\boldsymbol{b}, \boldsymbol{c})$ defines $\neg[\mathrm{TC}_{\boldsymbol{x},\boldsymbol{y}}\mathrm{EDGE_k}](\boldsymbol{b}, \boldsymbol{c})$ is false. Otherwise $\mathfrak{A} \models \neg[\mathrm{TC}_{\boldsymbol{x},\boldsymbol{y}}\mathrm{EDGE_k}](\boldsymbol{b}, \boldsymbol{c})$ would yield $(\mathfrak{A}, \boldsymbol{b}, \boldsymbol{c}) \models \phi$ from which $(\mathfrak{A}, \boldsymbol{b}, \varepsilon(\boldsymbol{c})) \models \phi$ follows. Therefore we would obtain $\mathfrak{A} \models \neg[\mathrm{TC}_{\boldsymbol{x},\boldsymbol{y}}\mathrm{EDGE_k}](\boldsymbol{b}, \varepsilon(\boldsymbol{c}))$ which contradicts the fact that $\mathfrak{A} \not\models \neg[\mathrm{TC}_{\boldsymbol{x},\boldsymbol{y}}\mathrm{EDGE_k}](\boldsymbol{b}, \varepsilon(\boldsymbol{c}))$ by Theorem 8. This concludes the proof of Lemma 1. □

A Modal Logic for Non-deterministic Information Systems

Md. Aquil Khan

Discipline of Mathematics,
Indian Institute of Technology Indore,
Indore 452017, India
aquilk@iiti.ac.in

Abstract. In this article, we propose a modal logic for non-deterministic information systems. A deductive system for the logic is presented and corresponding soundness and completeness theorems are proved. The logic is also shown to be decidable.

1 Introduction

Rough set theory, introduced by Pawlak in the early 1980s [17] offers an approach to deal with uncertainty inherent in real-life problems, more specifically that stemming from inconsistency or vagueness in data. The notion of an approximation space, viz. a tuple (U, R), where U is a non-empty set and R an equivalence relation, plays a crucial role in Pawlak's rough set theory. A useful natural generalization is where the relation R is not necessarily an equivalence (cf. e.g. [20,23,12]). Any concept represented as a subset (say) X of the domain U is approximated from within and outside, by its lower and upper approximations, denoted as $\underline{X}_R$ and $\overline{X}_R$ respectively, and are defined as follows:

$$\underline{X}_R := \{x \in U : R(x) \subseteq X\}, \quad \overline{X}_R := \{x \in U : R(x) \cap X \neq \emptyset\},$$

where $R(x) := \{y \in U : (x, y) \in R\}$.

A practical realization of approximation space is a *non-deterministic information system* [16], formally defined as follows.

Definition 1. *A non-deterministic information system* $\mathcal{S} := (U, \mathcal{A}, \bigcup_{a \in \mathcal{A}} \mathcal{V}_a, F)$, *written in brief as NIS, comprises a non-empty set* U *of objects, a non-empty finite set* $\mathcal{A}$ *of attributes, a non-empty finite set* $\mathcal{V}_a$ *of attribute-values for each* $a \in \mathcal{A}$, *and* $F : U \times \mathcal{A} \to 2^{\bigcup_{a \in \mathcal{A}} \mathcal{V}_a}$ *such that* $F(x, a) \subseteq \mathcal{V}_a$.

In the special case when $F(x, a)$ is singleton for each $(x, a) \in U \times \mathcal{A}$, $\mathcal{S}$ is called a *deterministic information system.*

One may attach different interpretations with '$F(x, a) = V$'. For instance, as exemplified in [4,5], if a is the attribute "speaking a language", then $F(x, a) =$ {German, English} can be interpreted as (i) x speaks German and English and no other languages, (ii) x speaks German and English and possibly other languages, (iii) x speaks German or English but not both, or (iv) x speaks German

M. Banerjee and S.N. Krishna (eds.): ICLA 2015, LNCS 8923, pp. 119–131, 2015.

or English or both. Motivated by these interpretations, several relations are defined on NISs (e.g. [16,21,22,5]). We list a few of them below. Consider a NIS $\mathcal{S} := (U, A, \bigcup_{a\in A} \mathcal{V}_a, F)$ and an attribute set B.

Indiscernibility. $(x,y) \in Ind_B^{\mathcal{S}}$ if and only if $F(x,a) = F(y,a)$ for all $a \in B$.
Similarity. $(x,y) \in Sim_B^{\mathcal{S}}$ if and only if $F(x,a) \cap F(y,a) \neq \emptyset$ for all $a \in B$.
Inclusion. $(x,y) \in In_B^{\mathcal{S}}$ if and only if $F(x,a) \subseteq F(y,a)$ for all $a \in B$.
Negative similarity. $(x,y) \in NSim_B^{\mathcal{S}}$ if and only if $\sim F(x,a) \cap \sim F(y,a) \neq \emptyset$ for all $a \in B$, where $\sim$ is the complementation relative to $\mathcal{V}_a$.
Complementarity. $(x,y) \in Com_B^{\mathcal{S}}$ if and only if $F(x,a) = \sim F(y,a)$ for all $a \in B$.
Weak indiscernibility. $(x,y) \in wInd_B^{\mathcal{S}}$ if and only if $F(x,a) = F(y,a)$ for some $a \in B$.
Weak similarity. $(x,y) \in wSim_B^{\mathcal{S}}$ if and only if $F(x,a) \cap F(y,a) \neq \emptyset$ for some $a \in B$.
Weak inclusion. $(x,y) \in wIn_B^{\mathcal{S}}$ if and only if $F(x,a) \subseteq F(y,a)$ for some $a \in B$.
Weak negative similarity. $(x,y) \in wNSim_B^{\mathcal{S}}$ if and only if $\sim F(x,a) \cap \sim F(y,a) \neq \emptyset$ for some $a \in B$.
Weak complementarity. $(x,y) \in wCom_B^{\mathcal{S}}$ if and only if $F(x,a) = \sim F(y,a)$ for some $a \in B$.

Each of the relations defined above gives rise to a generalized approximation space, where the relation may not be an equivalence. Thus, one can approximate any subset of the domain using the lower and upper approximations defined on these generalized approximation spaces.

Search for a logic which can be used to reason about the approximations of concepts remains an important area of research in rough set theory. For a comprehensive survey on this direction of research, we refer to [6,2]. It is not difficult to observe that in a logic for information systems, one would like to have the following two features. (i) The logic should be able to describe aspects of information systems such as attribute, and attribute-values. (ii) It should also be able to capture concept approximations induced by different sets of attributes. In literature, one can find logics with the first feature (e.g. [14,15,21,1]) as well as logics with the second feature (e.g. [13,16,14,15,19,24,7,8]). But proposal for a sound and complete modal logic of NIS having both feature is not known to us. In [9], a sound and complete logic for *deterministic* information systems with both of the above features was proposed. But recall that in the case of NISs, an object takes a set of attribute-values instead of just one as in the case of deterministic information system. Moreover, unlike deterministic information system, many relations other than indiscernibility are relevant and studied in the context of NISs. Therefore, in this article, our aim is to present a sound and complete modal logic with semantics directly based on NIS, having both of the above mentioned properties.

The remainder of this article is organized as follows. In Section 2, we introduced the syntax and semantics of the logic LNIS of NISs. This logic can capture the approximations relative to indiscernibilitry, similarity and inclusion relations. An axiomatic system of the logic LNIS is presented in Section 3, and corresponding soundness and completeness theorems are proved. The decidability of the logic is proved in Section 4. In Section 5, it is shown that the current work can be extended to other types of relations defined on NISs as well. Section 6 concludes the article.

2 Logic for Non-deterministic Information Systems

In this section, we shall propose a logic LNIS (logic for non-deterministic information systems) which can be used to reason about the rough set approximations with respect to indiscernibility, similarity and inclusion relations induced from NISs. Moreover, it can also talk about the attributes, attribute-values of the objects.

2.1 Syntax

The language $\mathbb{L}$ of LNIS contains (i) a non-empty finite set $\mathcal{A}$ of attribute constants, (ii) for each $a \in \mathcal{A}$, a non-empty finite set $\mathcal{V}_a$ of attribute-value constants, (iii) a non-empty countable set PV of propositional variables, and (iv) the propositional constants $\top$, $\bot$.

The atomic well-formed formulae (wffs) are the propositional variables p from PV, and the tuples of the form (a, v) where $a \in \mathcal{A}$ and $v \in \mathcal{V}_a$. The tuples (a, v) are called *descriptors* [18]. The set of all descriptors will be denoted by $\mathcal{D}$.

Using the atomic wffs, Boolean connectives $\neg$ (negation), $\wedge$ (conjunction), and the modal operators $\Box^1_B$, $\Box^2_B$ and $\Box^3_B$ for each $B \subseteq \mathcal{A}$, the wffs of $\mathbb{L}$ are then defined recursively as:

$$\alpha ::= p \mid (a, v) \mid \neg\alpha \mid \alpha \wedge \beta \mid \Box^1_B\alpha \mid \Box^2_B\alpha \mid \Box^3_B\alpha.$$

For $a \in \mathcal{A}$, we shall simply write $\Box^k_{\{a\}}$, $k \in \{1, 2, 3\}$, as $\Box^k_a$. Derived connectives are the usual ones: $\vee$ (disjunction), $\rightarrow$ (implication), $\Diamond^1_B$, $\Diamond^2_B$, $\Diamond^3_B$ (diamonds). We shall use the same symbol $\mathbb{L}$ to denote the set of all wffs of the language $\mathbb{L}$.

2.2 Semantics

As we desired, the semantics of $\mathbb{L}$ will be based on NISs. Thus we have the following definition of models. Recall that for an equivalence relation R on U, and for $x \in U$, $[x]_R$ denotes the equivalence class of x with respect to R.

Definition 2. *A model of $\mathbb{L}$ is defined as a tuple $\mathfrak{M} := (\mathcal{S}, V)$, where $\mathcal{S} := (W, \mathcal{A}, \bigcup_{a\in\mathcal{A}} \mathcal{V}_a, F)$ is a non-deterministic information system, and $V : PV \rightarrow 2^W$ is a valuation function.*

The satisfiability of a wff α in a model $\mathfrak{M} := (\mathcal{S}, V)$, where $\mathcal{S} := (W, \mathcal{A}, \bigcup_{a\in\mathcal{A}} \mathcal{V}_a, F)$, at an object $w \in W$, denoted as $\mathfrak{M}, w \models \alpha$, is defined as follows. We omit the cases of propositional constants and Boolean connectives.

Definition 3

$\mathfrak{M}, w \models (a, v)$, *if and only if* $v \in F(w, a)$.
$\mathfrak{M}, w \models \alpha$, *if and only if* $w \in V(\alpha)$, *for* $\alpha \in PV$.
$\mathfrak{M}, w \models \Box^1_B\alpha$, *if and only if for all* $w' \in W$ *with* $(w, w') \in Ind^{\mathcal{S}}_B$, $\mathfrak{M}, w' \models \alpha$.
$\mathfrak{M}, w \models \Box^2_B\alpha$, *if and only if for all* $w' \in W$ *with* $(w, w') \in Sim^{\mathcal{S}}_B$, $\mathfrak{M}, w' \models \alpha$.
$\mathfrak{M}, w \models \Box^3_B\alpha$, *if and only if for all* $w' \in W$ *with* $(w, w') \in In^{\mathcal{S}}_B$, $\mathfrak{M}, w' \models \alpha$.

Note that, unlike the logic in [9], the semantics of LNIS is directly based on information systems. The extension of a wff α relative to a model $\mathfrak{M}$, denoted as $[\![\alpha]\!]_{\mathfrak{M}}$, is given by the set $\{w : \mathfrak{M}, w \models \alpha\}$. A wff α is said to be valid in a model $\mathfrak{M}$ with domain U if $[\![\alpha]\!]_{\mathfrak{M}} = U$. A wff α is called valid, denoted as $\models \alpha$, if α is valid in all models.

The following proposition shows that the operators $\Box^1_B$, $\Box^2_B$ and $\Box^3_B$ capture the lower approximations relative to the attribute set B with respect to indiscernibility, similarity and inclusion relations respectively, while $\Diamond^1_B, \Diamond^2_B$ and $\Diamond^3_B$ respectively capture the corresponding upper approximations.

Proposition 1. *Let* $\mathfrak{M} := (\mathcal{S}, V)$ *be a model, where* $\mathcal{S} := (U, \mathcal{A}, \bigcup_{a\in\mathcal{A}} \mathcal{V}_a, F)$. *Then the following hold.*

1. $[\![\Box^1_B\alpha]\!]_{\mathfrak{M}} = \underline{[\![\alpha]\!]_{\mathfrak{M}}}_{Ind^{\mathcal{S}}_B}$, $[\![\Diamond^1_B\alpha]\!]_{\mathfrak{M}} = \overline{[\![\alpha]\!]_{\mathfrak{M}}}_{Ind^{\mathcal{S}}_B}$.
2. $[\![\Box^2_B\alpha]\!]_{\mathfrak{M}} = \underline{[\![\alpha]\!]_{\mathfrak{M}}}_{Sim^{\mathcal{S}}_B}$, $[\![\Diamond^2_B\alpha]\!]_{\mathfrak{M}} = \overline{[\![\alpha]\!]_{\mathfrak{M}}}_{Sim^{\mathcal{S}}_B}$.
3. $[\![\Box^3_B\alpha]\!]_{\mathfrak{M}} = \underline{[\![\alpha]\!]_{\mathfrak{M}}}_{In^{\mathcal{S}}_B}$, $[\![\Diamond^3_B\alpha]\!]_{\mathfrak{M}} = \overline{[\![\alpha]\!]_{\mathfrak{M}}}_{In^{\mathcal{S}}_B}$.

The presence of descriptors in the language helps LNIS to talk about the attributes, attribute-values of the objects, and its effect on the approximation operators. For instance, the wff $(a, v) \wedge \bigwedge_{u\in\mathcal{V}_a\setminus\{v\}} \neg(a, u) \rightarrow \Box^2_B p$ represents a decision rule according to which if an object x takes the value v for the attribute a, then x is in the lower approximation of the set represented by p with respect to similarity relation corresponding to the attribute set B. The valid wff $(b, v) \wedge \Box^2_{B\cup\{b\}}\alpha \rightarrow \Box^2_B((b, v) \rightarrow \alpha)$ corresponds to the fact that $Y \cap \underline{X}_{Sim^{\mathcal{S}}_{B\cup\{b\}}} \subseteq \underline{X \cup Y^c}_{Sim^{\mathcal{S}}_B}$, where $Y = \{y : v \in F(y, b)\}$, and Y^c is the set-theoretic complement of Y.

3 Axiomatic System

We now present an axiomatic system for LNIS, and prove the corresponding soundness and completeness theorems.

We recall that $\mathcal{D}$ denotes the set of all descriptors. Consider the wffs which are a conjunction of literals from the set $\bigcup_{\alpha\in\mathcal{D}}\{\alpha, \neg\alpha\}$, and which contain precisely one of (a, v), or $\neg(a, v)$ for each $(a, v) \in \mathcal{D}$. Let Θ be the set of all such wffs.

Let $\mathfrak{M} := (\mathcal{S}, V)$ be a model, where $\mathcal{S} := (W, \mathcal{A}, \bigcup_{a\in\mathcal{A}} \mathcal{V}_a, F)$. Then it is not difficult to obtain the following.

Proposition 2. **(V1)** $W/Ind^{\mathcal{S}}_{\mathcal{A}} \subseteq \{[\![i]\!]_{\mathfrak{M}} : i \in \Theta\}$.
(V2) $\{[\![i]\!]_{\mathfrak{M}} : i \in \Theta\} \subseteq W/Ind^{\mathcal{S}}_{\mathcal{A}} \cup \{\emptyset\}$.
(V3) $[\![i]\!]_{\mathfrak{M}} \cap [\![j]\!]_{\mathfrak{M}} = \emptyset$, *for* $i, j \in \Theta$, *and* $i \neq j$.
(V4) *For* $i, j \in \Theta$, *if* $(x, y) \notin Ind^{\mathcal{S}}_{\mathcal{A}}$, $[x]_{Ind^{\mathcal{S}}_{\mathcal{A}}} = [\![i]\!]_{\mathfrak{M}}$ *and* $[y]_{Ind^{\mathcal{S}}_{\mathcal{A}}} = [\![j]\!]_{\mathfrak{M}}$, *then* $i \neq j$.

From conditions (V1)-V(4), it is evident that the element of Θ are used as nominals to name the equivalence classes of $Ind^{\mathcal{S}}_{\mathcal{A}}$ such that different equivalence classes are provided with different names. We use these nominals to give the following deductive system. Let $B \subseteq \mathcal{A}$, $\Box_B \in \{\Box^1_B, \Box^2_B, \Box^3_B\}$, and $i \in \Theta$.

Axiom schema:

1. All axioms of classical propositional logic.
2. $\Box_B(\alpha \to \beta) \to (\Box_B\alpha \to \Box_B\beta)$.
3. $\Box^1_\emptyset\alpha \to \alpha$.
4. $\alpha \to \Box^1_\emptyset\Diamond^1_\emptyset\alpha$.
5. $\Diamond^1_\emptyset\Diamond^1_\emptyset\alpha \to \Diamond^1_\emptyset\alpha$.
6. $\Box^m_\emptyset\alpha \leftrightarrow \Box^n_\emptyset\alpha$, where $m, n \in \{1, 2, 3\}$.
7. $\Box_C\alpha \to \Box_B\alpha$ for $C \subseteq B$.
8. $(a, v) \to \Box^k_a(a, v)$ for $k \in \{1, 3\}$.
9. $\neg(a, v) \to \Box^1_a\neg(a, v)$.
10. $i \to \Box^2_a\left(\bigvee_{v\in\mathcal{V}_a} ((a, v) \wedge \Box^1_\emptyset(i \to (a, v)))\right)$.
11. $i \wedge \Box^1_{B\cup\{b\}}\alpha \to \Box^1_B\left(\bigwedge_{v\in\mathcal{V}_b}\left((b, v) \leftrightarrow \Box^1_\emptyset(i \to (b, v))\right) \to \alpha\right)$.
12. $(b, v) \wedge \Box^2_{B\cup\{b\}}\alpha \to \Box^2_B((b, v) \to \alpha)$.
13. $i \wedge \Box^3_{B\cup\{b\}}\alpha \to \Box^3_B\left(\bigwedge_{v\in\mathcal{V}_b}\left(\Box^1_\emptyset(i \to (b, v)) \to (b, v)\right) \to \alpha\right)$.
14. $i \wedge (a, v) \to \Box^1_\emptyset(i \to (a, v))$.
15. $i \wedge \neg(a, v) \to \Box^1_\emptyset(i \to \neg(a, v))$.
16. $\bigvee_{i\in\Theta} i$.
17. $\neg i \vee \neg j$ for distinct elements i and j of Θ.
18. $i \to \Box^1_{\mathcal{A}} i$.

Rules of inference:

$$N.\ \frac{\alpha}{\Box_B\alpha} \qquad MP.\ \frac{\alpha \quad \alpha \to \beta}{\beta}$$

The notion of theoremhood is defined in the usual way, and we write $\vdash \alpha$ to indicate that α is a theorem of the above deductive system.

Axioms 7-13 relate attribute, attribute-values of the objects with the relations corresponding to the modal operators $\Box^k_B$, $k \in \{1, 2, 3\}$. For instance, let R^2_B be the relation corresponding to the modal operator $\Box^2_B$, and let us see how the axioms 7, 10 and 12 relate attribute, attribute-values of the objects with R^2_B. Axiom 7 for $\Box^2_B$ corresponds to the condition $R^2_B \subseteq R^2_C$ for $C \subseteq B$. Axiom 10 gives the condition that if $(x, y) \in R^2_{\{a\}}$, then there exists a $v \in \mathcal{V}_a$ such that $v \in F(x, a) \cap F(y, a)$. Thus axiom 7 for $\Box^2_B$ and axiom 10 correspond to the condition that if $(x, y) \in R^2_B$, then $F(x, a) \cap F(y, a) \neq \emptyset$ for all $a \in B$. On the other hand, axiom 12 corresponds to the condition that if $(x, y) \in R^2_B$, and $v \in F(x, b) \cap F(y, b)$, then $(x, y) \in R^2_{B \cup \{b\}}$. In particular, axiom 12 with $B = \emptyset$ gives the condition that if $v \in F(x, b) \cap F(y, b)$, then $(x, y) \in R^2_{\{b\}}$. Thus, axiom 12 inductively imposes the condition that if $F(x, a) \cap F(y, a) \neq \emptyset$ for all $a \in B$, then $(x, y) \in R^2_B$. Hence axioms 10, 12 and axiom 7 for $\Box^2_B$ correspond to the condition that $(x, y) \in R^2_B$ if and only if $F(x, a) \cap F(y, a) \neq \emptyset$ for all $a \in B$. Note that this is the defining condition of similarity relation.

Axiom 6 says that relations corresponding to modal operators $\Box^1_\emptyset$, $\Box^2_\emptyset$ and $\Box^3_\emptyset$ are all same. Axioms 14-18 capture the conditions $(V1) - (V4)$ mentioned in Proposition 2.

Observe that the wffs from the set Θ appear in the axioms 10, 11, 13-18 acting as nominals. Such a use of nominals was not required for the axiomatic system presented in [9] for deterministic information systems. We also note that unlike hybrid logic (cf. [3]), nominals are not required to append to the language of LNIS, and wffs from Θ are used for the purpose.

It is not difficult to obtain the following soundness theorem.

Theorem 1 (Soundness). *If* $\vdash \alpha$, *then* $\models \alpha$.

3.1 Completeness Theorem

The completeness theorem is proved following the standard modal logic technique [3]. As in normal modal logic, we have the following result.

Proposition 3. *Every consistent set of wffs has a maximally consistent extension.*

Let $W := \{\Gamma : \Gamma \text{ is maximal consistent}\}$. Consider the equivalence relation $R^1_\emptyset$ defined on W as follows.

$$(\Gamma, \Delta) \in R^1_\emptyset \text{ if and only if for all wff } \Box^1_\emptyset \alpha \in \Gamma,\ \alpha \in \Delta.$$

Let Σ be a given element of W and consider the equivalence class W^Σ of Σ with respect to relation $R^1_\emptyset$. We now describe the canonical model $\mathfrak{M}^\Sigma$ for LNIS corresponding to the given Σ.

Definition 4 (Canonical Model). $\mathfrak{M}^\Sigma := (\mathcal{S}^\Sigma, V^\Sigma)$, *where*

- $\mathcal{S}^\Sigma := (W^\Sigma, \mathcal{A}, \cup_{a \in \mathcal{A}} \mathcal{V}_a, F^\Sigma)$,
- $F^\Sigma(\Gamma, a) := \{v \in \mathcal{V}_a : (a, v) \in \Gamma\}$,
- $V^\Sigma(p) := \{\Gamma \in W^\Sigma : p \in \Gamma\}$ *for* $p \in PV$.

Note that, unlike the cases of standard modal logics and the logic presented in [9], the canonical model of LNIS is based on a NIS $\mathcal{S}^\Sigma$, and therefore a natural question would be about the connections between the canonical relations

obtained corresponding to the modal operators $\Box^k_B$ and the indiscernibility, similarity and inclusion relations obtained from the information system $\mathcal{S}^\Sigma$. Next, we capture this aspect, which will also lead us to the Truth Lemma.
We recall the following definition. For, $B \subseteq \mathcal{A}$, and $k \in \{1,2,3\}$, we define $R^k_B \subseteq W^\Sigma \times W^\Sigma$ such that

$$(\Gamma, \Delta) \in R^k_B \text{ if and only if } \Box^k_B \alpha \in \Gamma \text{ implies } \alpha \in \Delta. \tag{1}$$

For $a \in \mathcal{A}$, we again simply write $R^k_{\{a\}}$ as R^k_a. For a $\Gamma \in W^\Sigma$, and $a \in \mathcal{A}$, let Γ_a denote the set $\{(a,v) : (a,v) \in \Gamma\}$. Then we have the following.

Proposition 4. *1. $R^k_B \subseteq R^k_C$ for $C \subseteq B$, $k \in \{1,2,3\}$.*
2. $\Gamma_a = \Delta_a$ if and only if $(\Gamma, \Delta) \in R^1_a$.
3. $\Gamma_a \cap \Delta_a \neq \emptyset$ if and only if $(\Gamma, \Delta) \in R^2_a$.
4. $\Gamma_a \subseteq \Delta_a$ if and only if $(\Gamma, \Delta) \in R^3_a$.
5. If $(\Gamma, \Delta) \in R^1_B$ and $\Gamma_a = \Delta_a$, then $(\Gamma, \Delta) \in R^1_{B\cup\{a\}}$.
6. If $(\Gamma, \Delta) \in R^2_B$ and $\Gamma_a \cap \Delta_a \neq \emptyset$, then $(\Gamma, \Delta) \in R^2_{B\cup\{a\}}$.
7. If $(\Gamma, \Delta) \in R^3_B$ and $\Gamma_a \subseteq \Delta_a$, then $(\Gamma, \Delta) \in R^3_{B\cup\{a\}}$.
8. $R^k_B = \bigcap_{a\in B} R^k_a$, $k \in \{1,2,3\}$.

Proof. We provide the proofs of Items 2 and 3.
(2): First suppose $\Gamma_a = \Delta_a$, and let $\Box^1_a \alpha \in \Gamma$. We need to show $\alpha \in \Delta$. From axiom 16, we obtain $i \in \Gamma$ for some $i \in \Theta$. Therefore, $i \wedge \Box^1_a \alpha \in \Gamma$. Now using axiom 11 for $B = \emptyset$, we obtain

$$\Box^1_\emptyset \Bigg(\bigwedge_{v\in\mathcal{V}_a} \Big((a,v) \leftrightarrow \Box^1_\emptyset (i \rightarrow (a,v)) \Big) \rightarrow \alpha \Bigg) \in \Gamma.$$

Since $(\Gamma, \Delta) \in R^1_\emptyset$, we obtain

$$\bigwedge_{v\in\mathcal{V}_a} \Big((a,v) \leftrightarrow \Box^1_\emptyset (i \rightarrow (a,v)) \Big) \rightarrow \alpha \in \Delta.$$

If possible, let $\alpha \notin \Delta$. Then there exists a $v \in \mathcal{V}_a$ such that either $(a,v) \rightarrow \Box^1_\emptyset(i \rightarrow (a,v)) \notin \Delta$, or $\Box^1_\emptyset(i \rightarrow (a,v)) \rightarrow (a,v) \notin \Delta$. First suppose, $(a,v) \rightarrow \Box^1_\emptyset(i \rightarrow (a,v)) \notin \Delta$. Then $(a,v) \in \Delta$, and $\Box^1_\emptyset(i \rightarrow (a,v)) \notin \Delta$. Now using the fact that $\Gamma_a = \Delta_a$, we obtain $(a,v) \in \Gamma$, and hence $i \wedge (a,v) \in \Gamma$. Therefore, from axiom 14, we obtain $\Box^1_\emptyset(i \rightarrow (a,v)) \in \Gamma$. Again using the fact that $(\Gamma, \Delta) \in R^1_\emptyset$, we obtain $i \rightarrow (a,v) \in \Delta$. Hence by axiom 14, we have $\Box^1_\emptyset(i \rightarrow (a,v)) \in \Delta$, a contradiction. Similarly, $\Box^1_\emptyset(i \rightarrow (a,v)) \rightarrow (a,v) \notin \Delta$ will also lead us to a contradiction.

Conversely suppose $(\Gamma, \Delta) \in R^1_a$. We need to show $(a,v) \in \Gamma$ if and only if $(a,v) \in \Delta$. First let $(a,v) \in \Gamma$. Then from axiom 8, we obtain $\Box^1_a(a,v) \in \Gamma$, and hence $(a,v) \in \Delta$. Now suppose $(a,v) \in \Delta$. If $(a,v) \notin \Gamma$, then using axiom 9, we obtain $\Box^1_a(\neg(a,v)) \in \Gamma$, and hence $\neg(a,v) \in \Delta$, a contradiction.

(3): First suppose $\Gamma_a \cap \Delta_a \neq \emptyset$ and we show $(\Gamma, \Delta) \in R^2_a$. Let $(a,v) \in \Gamma_a \cap \Delta_a$. Let $\Box^2_a \alpha \in \Gamma$. We need to show $\alpha \in \Delta$. We have $(a,v) \wedge \Box^2_a \alpha \in \Gamma$, and hence by axiom 12, $\square^2_\emptyset((a,v) \to \alpha) \in \Gamma$. This gives $\square^1_\emptyset((a,v) \to \alpha) \in \Gamma$ (due to axiom 6), and therefore we obtain $(a,v) \to \alpha \in \Delta$ as $(\Gamma, \Delta) \in R^1_\emptyset$. This gives us $\alpha \in \Delta$ as $(a,v) \in \Delta$.

Conversely, suppose $(\Gamma, \Delta) \in R^2_a$, and we prove $\Gamma_a \cap \Delta_a \neq \emptyset$. Axiom 16 guarantees the existence of an $i \in \Theta$ such that $i \in \Gamma$. Therefore, by axiom 10, we obtain $\square^2_a \left(\bigvee_{v \in \mathcal{V}_a}((a,v) \wedge \square^1_\emptyset(i \to (a,v)))\right) \in \Gamma$ and hence $\bigvee_{v \in \mathcal{V}_a}((a,v) \wedge \square^1_\emptyset(i \to (a,v))) \in \Delta$. Therefore, for some $v \in \mathcal{V}_a$, $(a,v) \wedge \square^1_\emptyset(i \to (a,v)) \in \Delta$. Since $i \in \Gamma$, we obtain $(a,v) \in \Gamma \cap \Delta$ and hence $\Gamma_a \cap \Delta_a \neq \emptyset$. □

The following proposition relates the canonical relations corresponding to the modal operators $\square^k_B$ and the indiscernibility, similarity and inclusion relations obtained from the information system $\mathcal{S}^\Sigma$. Let us use the notations Ind^Σ_B, Sim^Σ_B and In^Σ_B to denote the indiscernibility, similarity and inclusion relations relative to the attribute set B induced from the NIS $\mathcal{S}^\Sigma$.

Proposition 5. *1.* $R^1_B = Ind^\Sigma_B$.
2. $R^2_B = Sim^\Sigma_B$.
3. $R^3_B = In^\Sigma_B$.

Proof. We provide the proof for the indiscernibility relation. Since $Ind^\Sigma_B = \bigcap_{a \in B} Ind^\Sigma_a$ and $R^1_B = \bigcap_{a \in B} R^1_a$ (by Item 8 of Proposition 4), it is enough to prove the result for singleton B. So let $B = \{b\}$. Suppose $(\Gamma, \Delta) \in Ind^\Sigma_b$. Then, we obtain $F^\Sigma(\Gamma, b) = F^\Sigma(\Delta, b)$. This implies $\Gamma_b = \Delta_b$, and hence by Item (2) of Proposition 4, we obtain $(\Gamma, \Delta) \in R^1_b$, as desired. Conversely, let $(\Gamma, \Delta) \in R^1_b$. Then by Item (2) of Proposition 4, we obtain $\Gamma_b = \Delta_b$. This gives $F^\Sigma(\Gamma, b) = F^\Sigma(\Delta, b)$ and hence $(\Gamma, \Delta) \in Ind^\Sigma_b$. □

Once we have the Proposition 5, giving the same argument as in normal modal logic, we obtain

Proposition 6 (Truth Lemma). *For any wff* β *and* $\Gamma \in W^\Sigma$,

$$\beta \in \Gamma \text{ if and only if } \mathfrak{M}^\Sigma, \Gamma \models \beta.$$

Using the Truth Lemma and Proposition 3 we have

Proposition 7. *If* α *is consistent then there exists a maximal consistent set* Σ *containing* α *such that* $\mathfrak{M}^\Sigma, \Sigma \models \alpha$.

This gives the desired completeness theorem.

Theorem 2 (Completeness). *For any wff* α, *if* $\models \alpha$, *then* $\vdash \alpha$.

4 Decidability

In this section we shall prove the following decidability result.

Theorem 3. *We can decide for a given wff* α, *whether* $\models \alpha$.

For this, we prove that LNIS has the finite model property (Proposition 11 below). We will employ the standard filtration technique [3] with necessary modifications to prove this result. Let Σ denote a finite sub-formula closed set of wffs.

Consider a model $\mathfrak{M} := (\mathcal{S}, V)$, where $\mathcal{S} := (U, \mathcal{A}, \bigcup_{a\in\mathcal{A}} \mathcal{V}_a, F)$. We define a binary relation $\equiv_\Sigma$ on U as follows:

$w \equiv_\Sigma w'$, if and only if for all $\beta \in \Sigma \cup \mathcal{D}$, $\mathfrak{M}, w \models \beta$ if and only if $\mathfrak{M}, w' \models \beta$.

Definition 5 (Filtration model). *Given a model $\mathfrak{M} := (\mathcal{S}, V)$, where $\mathcal{S} := (U, \mathcal{A}, \bigcup_{a\in\mathcal{A}} \mathcal{V}_a, F)$ and Σ as above, we define a model $\mathfrak{M}^f = (\mathcal{S}^f, V^f)$, where*

- $\mathcal{S}^f := (U^f, \mathcal{A}, \bigcup_{a\in\mathcal{A}} \mathcal{V}_a, F^f)$
- $U^f := \{[w] : w \in U\}$, *$[w]$ is the equivalence class of w with respect to the equivalence relation $\equiv_\Sigma$;*
- $F^f([w], a) = F(w, a)$;
- $V^f(p) := \{[w] \in U^f : w \in V(p)\}$

$\mathfrak{M}^f$ is the filtration of $\mathfrak{M}$ through the sub-formula closed set Σ.

First note that the definition of F^f is well defined as for all $w' \in [w]$, $F(w, a) = F(w', a)$. Moreover,

Proposition 8. *For any model $\mathfrak{M}$, if $\mathfrak{M}^f$ is a filtration of $\mathfrak{M}$ through Σ, then the domain U^f of $\mathfrak{M}^f$ contains at most 2^n elements, where $|\Sigma \cup \mathcal{D}| = n$.*

Proof. Define the map $g : U^f \to 2^{\Sigma\cup\mathcal{D}}$ where $g([w]) := \{\beta \in \Sigma\cup\mathcal{D} : \mathfrak{M}, w \models \beta\}$. Since g is injective, U^f contains at most 2^n elements. □

We also note the following fact.

Proposition 9. *1. $([w], [u]) \in Ind_B^{\mathcal{S}^f}$ if and only if $(w, u) \in Ind_B^{\mathcal{S}}$.*
2. $([w], [u]) \in Sim_B^{\mathcal{S}^f}$ if and only if $(w, u) \in Sim_B^{\mathcal{S}}$.
3. $([w], [u]) \in In_B^{\mathcal{S}^f}$ if and only if $(w, u) \in In_B^{\mathcal{S}}$.

Proof. $([w], [u]) \in Ind_B^{\mathcal{S}^f} \Leftrightarrow F^f([w], a) = F^f([u], a)$ for all $a \in B \Leftrightarrow F(w, a) = F(u, a)$ for all $a \in B \Leftrightarrow (w, u) \in Ind_B^{\mathcal{S}}$.

For the similarity and inclusion relations, it can be done in the same way. □

Once we have Proposition 9, it is not difficult to obtain the following.

Proposition 10 (Filtration Theorem). *Let Σ be a finite sub-formula closed set of wffs. For all wffs $\beta \in \Sigma \cup \mathcal{D}$, all models $\mathfrak{M}$, and all objects $w \in W$,*

$$\mathfrak{M}, w \models \beta \text{ if and only if } \mathfrak{M}^f, [w] \models \beta.$$

Finally, from Propositions 8 and 10 we have

Proposition 11 (Finite Model Property). *Let α be a wff and Σ be the set of all sub-wffs of α. If α is satisfiable, then it is satisfiable in a finite model with at most $2^{|\Sigma\cup\mathcal{D}|}$ elements.*

5 Logic for Other Types of Relations

Although the logic LNIS is confined to capture the approximations of concepts relative to indiscernibility, similarity and inclusion relations only, but it can easily be extended to capture other types of relations as well. In this section, we briefly sketch the extensions to capture the indistinguishability relations defined in Section 1. The syntax and semantics can be modified in a natural way to accommodate these relations. As far as axiomatization is concerned, main task is to come up with the axioms relating the relations with the attributes and attribute-values of the objects. We list below the axioms for the relations defined in Section 1. These axioms along with the axioms 1-6, 14-18 will give us the desired sound and complete deductive system.

Negative Similarity Relation

- $\Box_C \alpha \rightarrow \Box_B \alpha$ for $C \subseteq B \subseteq \mathcal{A}$.
- $i \rightarrow \Box_a(\bigvee_{v \in \mathcal{V}_a}(\neg(a,v) \wedge \Box_\emptyset(i \rightarrow \neg(a,v))))$.
- $\neg(b,v) \wedge \Box_{B\cup\{b\}}\alpha \rightarrow \Box_B(\neg(b,v) \rightarrow \alpha)$.

Complementarity Relation

- $\Box_C \alpha \rightarrow \Box_B \alpha$ for $C \subseteq B \subseteq \mathcal{A}$.
- $(a,v) \rightarrow \Box_a(\neg(a,v))$.
- $\neg(a,v) \rightarrow \Box_a((a,v))$.
- $i \wedge \Box_{B\cup\{b\}}\alpha \rightarrow \Box_B\left(\bigwedge_{v \in \mathcal{V}_b}\left(\neg(b,v) \leftrightarrow \Box_\emptyset(i \rightarrow (b,v))\right) \rightarrow \alpha\right)$.

Weak Indiscernibility Relation

- $\Box_B \alpha \rightarrow \Box_C \alpha$ for $\emptyset \neq C \subseteq B \subseteq \mathcal{A}$.
- $i \rightarrow \Box_B\left(\bigvee_{b \in B}\bigwedge_{v \in \mathcal{V}_b}\left((b,v) \leftrightarrow \Box^1_\emptyset(i \rightarrow (b,v))\right)\right)$, $B \neq \emptyset$.
- $\Box_\emptyset \bot$.

Weak Similarity Relation

- $\Box_B \alpha \rightarrow \Box_C \alpha$ for $\emptyset \neq C \subseteq B \subseteq \mathcal{A}$.
- $i \rightarrow \Box_B\left(\bigvee_{b \in B}\bigvee_{v \in \mathcal{V}_b}\left((b,v) \wedge \Box^1_\emptyset(i \rightarrow (b,v))\right)\right)$, $B \neq \emptyset$.
- $\Box_\emptyset \bot$.

Weak Inclusion Relation

- $\Box_B \alpha \rightarrow \Box_C \alpha$ for $\emptyset \neq C \subseteq B \subseteq \mathcal{A}$.
- $i \rightarrow \Box_B\left(\bigvee_{b \in B}\bigwedge_{v \in \mathcal{V}_b}\left(\Box^1_\emptyset(i \rightarrow (b,v)) \rightarrow d(b,v)\right)\right)$, $B \neq \emptyset$.
- $\Box_\emptyset \bot$.

Weak Negative Similarity Relation

- $\Box_B \alpha \to \Box_C \alpha$ for $\emptyset \neq C \subseteq B \subseteq \mathcal{A}$.
- $i \to \Box_B \Big(\bigvee_{b \in B} \bigvee_{v \in \mathcal{V}_b} \Big(\neg(b,v) \wedge \Box^1_\emptyset (i \to \neg(b,v)) \Big) \Big)$, $B \neq \emptyset$.
- $\Box_\emptyset \bot$.

Weak Complementarity Relation

- $\Box_B \alpha \to \Box_C \alpha$ for $\emptyset \neq C \subseteq B \subseteq \mathcal{A}$.
- $i \to \Box_B \Big(\bigvee_{b \in B} \bigwedge_{v \in \mathcal{V}_b} \Big(\neg(b,v) \leftrightarrow \Box^1_\emptyset (i \to (b,v)) \Big) \Big)$, $B \neq \emptyset$.
- $\Box_\emptyset \bot$.

We sketch briefly how the three axioms of negative similarity relation given above will give the corresponding completeness theorem. One can proceed similarly for the other relations. Let R_B be the canonical relation corresponding to the modal operator $\Box_B$ for the negative similarity relation (cf. (1) of page 125). Then corresponding to the Proposition 4, we will obtain the following:

1. $R_B \subseteq R_C$ for $C \subseteq B$ (using first axiom).
2. $\Gamma^c_a \cap \Delta^c_a \neq \emptyset$ if and only if $(\Gamma, \Delta) \in R_a$ (using second axiom and third axiom for $B = \emptyset$).
3. If $(\Gamma, \Delta) \in R_B$ and $\Gamma^c_a \cap \Delta^c_a \neq \emptyset$, then $(\Gamma, \Delta) \in R_{B \cup \{a\}}$ (using third axiom).

As a consequence of these facts, we will obtain R_B to be the negative similarity relation induced from the canonical information system (cf. Definition 4) and consequently we get the Truth Lemma, and hence completeness theorem.

6 Conclusions

In NISs, object-attribute pair is mapped to a set of attribute values. This represents uncertainty, in the sense that we know some possible attribute values that an object can take for an attribute, but we do not know exactly which ones. In this kind of situation, one could get information which reduces or removes this uncertainty. In the line of work in [10,11], a natural question would be about the proposal of an update logic for NISs, which can capture such an information and its effect on the approximations of concepts. The current work has completed the first step in this direction and given the static part of such an update logic. It remains to extend the language of LNIS to obtain the update logic for NISs.

References

1. Balbiani, P., Orłowska, E.: A hierarchy of modal logics with relative accessibility relations. Journal of Applied Non-Classical Logics 9(2-3), 303–328 (1999)
2. Banerjee, M., Khan, M.A.: Propositional logics from rough set theory. In: Peters, J.F., Skowron, A., Düntsch, I., Grzymała-Busse, J.W., Orłowska, E., Polkowski, L. (eds.) Transactions on Rough Sets VI. LNCS, vol. 4374, pp. 1–25. Springer, Heidelberg (2007)
3. Blackburn, P., de Rijke, M., Venema, Y.: Modal Logic. Cambridge University Press (2001)
4. Düntsch, I., Gediga, G., Orłowska, E.: Relational attribute systems. International Journal of Human Computer Studies 55(3), 293–309 (2001)
5. Düntsch, I., Gediga, G., Orłowska, E.: Relational attribute systems II: Reasoning with relations in information structures. In: Peters, J.F., Skowron, A., Marek, V.W., Orłowska, E., Słowiński, R., Ziarko, W.P. (eds.) Transactions on Rough Sets VII. LNCS, vol. 4400, pp. 16–35. Springer, Heidelberg (2007)
6. Demri, S., Orłowska, E.: Incomplete Information: Structure, Inference, Complexity. Springer (2002)
7. Fan, T.F., Hu, W.C., Liau, C.J.: Decision logics for knowledge representation in data mining. In: COMPSAC 2001, pp. 626–631. IEEE Computer Society, Washington, DC (2001)
8. Fan, T.F., Liu, D.R., Tzeng, G.H.: Arrow decision logic. In: Ślęzak, D., Wang, G., Szczuka, M.S., Düntsch, I., Yao, Y. (eds.) RSFDGrC 2005. LNCS (LNAI), vol. 3641, pp. 651–659. Springer, Heidelberg (2005)
9. Khan, M.A., Banerjee, M.: A logic for complete information systems. In: Sossai, C., Chemello, G. (eds.) ECSQARU 2009. LNCS (LNAI), vol. 5590, pp. 829–840. Springer, Heidelberg (2009)
10. Khan, M.A., Banerjee, M.: Logics for information systems and their dynamic extensions. ACM Transactions on Computational Logic 12(4), 29 (2011)
11. Khan, M.A., Banerjee, M., Rieke, R.: An update logic for information systems. International Journal of Approximate Reasoning 55(1), 436–456 (2014)
12. Komorowski, J., Pawlak, Z., Polkowski, L., Skowron, A.: Rough sets: a tutorial. In: Pal, S.K., Skowron, A. (eds.) Rough Fuzzy Hybridization: A New Trend in Decision-Making, pp. 3–98. Springer Verlag, Singapore (1999)
13. Orłowska, E.: Dynamic information system. Fundamenta Informaticae 5, 101–118 (1982)
14. Orłowska, E.: A logic of indiscernibility relations. In: Goos, G., Hartmanis, J. (eds.) Proceedings of the Symposium on Computation Theory, Zabrów 1984. LNCS, vol. 208, pp. 177–186. Springer, Heidelberg (1985)
15. Orłowska, E.: Logic of nondeterministic information. Studia Logica 1, 91–100 (1985)
16. Orłowska, E., Pawlak, Z.: Representation of nondeterministic information. Theoretical Computer Science 29, 27–39 (1984)
17. Pawlak, Z.: Rough sets. International Journal of Computer and Information Science 11(5), 341–356 (1982)
18. Pawlak, Z.: Rough Sets. Theoretical Aspects of Reasoning about Data. Kluwer Academic Publishers, Dordrecht (1991)
19. Rauszer, C.M.: Rough logic for multiagent systems. In: Masuch, M., Polos, L. (eds.) Logic at Work 1992. LNCS, vol. 808, pp. 161–181. Springer, Heidelberg (1994)
20. Skowron, A., Stepaniuk, J.: Tolerance approximation spaces. Fundamenta Informaticae 27, 245–253 (1996)

21. Vakarelov, D.: Modal logics for knowledge representation systems. Theoretical Computer Science 90, 433–456 (1991)
22. Vakarelov, D.: Information systems, similarity relations and modal logics. In: Orłowska, E. (ed.) Incomplete Information: Rough Set Analysis. STUDFUZZ, vol. 13, pp. 492–550. Springer, Heidelberg (1998)
23. Yao, Y.Y.: Generalized rough set models. In: Polkowski, L., Skowron, A. (eds.) Rough Sets in Knowledge Discovery, pp. 286–318. Physica-Verlag, Heidelberg (1998)
24. Yao, Y.Y., Liu, Q.: A generalized decision logic in interval-set-valued information tables. In: Zhong, N., Skowron, A., Ohsuga, S. (eds.) RSFDGrC 1999. LNCS (LNAI), vol. 1711, pp. 285–293. Springer, Heidelberg (1999)

Tableaux for Non-normal Public Announcement Logic

Minghui Ma[1], Katsuhiko Sano[2], François Schwarzentruber[3], and Fernando R. Velázquez-Quesada[4]

[1] Center for the Study of Logic and Intelligence, Southwest University, China
[2] School of Information Science, Japan Advanced Institute of Science and Technology, Japan
[3] ENS Rennes, Campus de Ker Lann 35170 Bruz, France
[4] Grupo de Lógica, Lenguaje e Informaci6n, Universidad de Sevilla, Spain

Abstract. This paper presents a tableau calculus for two semantic interpretations of public announcements over monotone neighbourhood models: the intersection and the subset semantics, developed by Ma and Sano. We show that both calculi are sound and complete with respect to their corresponding semantic interpretations and, moreover, we establish that the satisfiability problem of this public announcement extensions is NP-complete in both cases. The tableau calculi has been implemented in Lotrecscheme.

1 Introduction

Public announcement logic (**PAL**; [7,21]) studies the effect of the most basic communicative action on the knowledge of epistemic logic agents (**EL**; [12,6]), and it has served as the basis for the study of more complex announcements [3] and other forms of epistemic changes [27,25]. Under the standard **EL** semantic model, relational models, **PAL** relies on a natural interpretation of what the public announcement of a formula φ does: it eliminates those epistemic possibilities that do not satisfy φ. Despite its simplicity, **PAL** has proved to be a fruitful field for interesting research, as the characterisation of successful formulas (those that are still true after being truthfully announced: [28,14]), the characterisation of schematic validities [13] and many others [24].

However, relational models are not the unique structures for interpreting **EL** formulas, and recently there have been approaches that, using the so called *minimal* or *neighborhood* models [23,18,19,4], have studied not only epistemic phenomena but also their dynamics [31,26,17,30]. The set of **EL** validities under neighborhood models is smaller than that under relational models, so the agent's knowledge has less 'built-in' properties, which allows a finer representation of epistemic notions and their dynamics without resorting to 'syntactic' awareness models [5].

In [17], the authors presented two ways of updating (monotone) neighborhood models and thus of representing public announcements: one intersecting the current neighborhoods with the new information ($\cap$-*semantics*, already proposed in [31]), and another preserving only those neighborhoods which are subsets of the new information ($\subseteq$-*semantics*). The two updates behave differently, as their provided sound and complete axiom systems show. The present work continues the study of such updates, first, by extending the tableau system for monotone neighborhood models of [15] with rules for dealing with its public announcement extensions, and second, by showing how the satisfiability problem is NP-complete for both the intersection and the subset semantics.

M. Banerjee and S.N. Krishna (eds.): ICLA 2015, LNCS 8923, pp. 132–145, 2015.

2 Preliminaries

This section recalls some basic concepts from [17]. We work on the single agent case, but the results obtained can be easily extended to multi-agent scenarios.

Throughout this paper, let Prop be a countable set of atomic propositions. The language $\mathcal{L}_{\mathbf{EL}}$ extends the classical propositional language with formulas of the form $\Box\varphi$, read as "the agent knows that φ". Formally,

$$\varphi \ ::= \ p \mid \neg\varphi \mid \varphi \wedge \psi \mid \Box\varphi$$

with $p \in$ Prop. Other propositional connectives ($\vee$, $\rightarrow$ and $\leftrightarrow$) are defined as usual. The dual of $\Box$ is defined as $\Diamond\varphi := \neg\Box\neg\varphi$.

A *monotone neighborhood frame* is a pair $\mathcal{F} = (W, \tau)$ where $W \neq \emptyset$ is the domain, a set of possible worlds, and $\tau : W \rightarrow \wp(\wp(W))$ is a neighborhood function satisfying the following monotonicity condition: for all $w \in W$ and all $X, Y \subseteq W$, $X \in \tau(w)$ and $X \subseteq Y$ implies $Y \in \tau(w)$. A *monotone neighborhood model (MNM)* $\mathcal{M} = (\mathcal{F}, V)$ is a monotone neighborhood frame $\mathcal{F}$ together with a valuation function V : Prop $\rightarrow \wp(W)$. Given a $\mathcal{M} = (W, \tau, V)$ and a $\mathcal{L}_{\mathbf{EL}}$-formula φ, the notion of φ being true at a state w in the model $\mathcal{M}$ (written $\mathcal{M}, w \models \varphi$) is defined inductively as follows:

$\mathcal{M}, w \models p$ iff $w \in V(p)$, $\qquad$ $\mathcal{M}, w \models \varphi \wedge \psi$ iff $\mathcal{M}, w \models \varphi$ and $\mathcal{M}, w \models \psi$,
$\mathcal{M}, w \models \neg\varphi$ iff $\mathcal{M}, w \not\models \varphi$, $\qquad$ $\mathcal{M}, w \models \Box\varphi$ iff $[\![\varphi]\!]_{\mathcal{M}} \in \tau(w)$.

where $[\![\varphi]\!]_{\mathcal{M}} := \{u \in W \mid \mathcal{M}, u \models \varphi\}$ is the truth set of φ in $\mathcal{M}$. Since $\mathcal{M}$ is a *MNM*, the satisfaction clause for $\Box$ can be equivalently rewritten as follows:

$$\mathcal{M}, w \models \Box\varphi \text{ iff } X \subseteq [\![\varphi]\!]_{\mathcal{M}} \text{ for some } X \in \tau(w).$$

The language $\mathcal{L}_{\mathbf{PAL}}$ extends $\mathcal{L}_{\mathbf{EL}}$ with the public announcement operator $[\varphi]$, allowing the construction of formulas of the form $[\varphi]\psi$, read as "ψ is true after the public announcement of φ". (Define $\langle\varphi\rangle\psi := \neg[\varphi]\neg\psi$.) For the semantic interpretation, we recall the intersection and subset semantics of [17].

Definition 1. *Let $\mathcal{M} = (W, \tau, V)$ be a MNM. For any non-empty $U \subseteq W$, define the function V^U : Prop $\rightarrow U$ by $V^U(p) := V(p) \cap U$ for each $p \in$ Prop.*

- *The* intersection submodel *of $\mathcal{M}$ induced by U, $\mathcal{M}^{\cap U} = (U, \tau^{\cap U}, V^U)$, is given by $\tau^{\cap U}(u) := \{P \cap U \mid P \in \tau(u)\}$, for every $u \in U$.*
- *The* subset submodel *of $\mathcal{M}$ induced by U, $\mathcal{M}^{\subseteq U} = (U, \tau^{\subseteq U}, V^U)$, is given by $\tau^{\subseteq U}(u) := \{P \in \tau(u) \mid P \subseteq U\}$, for every $u \in U$.*

If $\mathcal{M}$ is monotone, then so are $\mathcal{M}^{\cap U}$ and $\mathcal{M}^{\subseteq U}$, as shown in [17].

Given a *MNM* $\mathcal{M} = (W, \tau, V)$, formulas φ, ψ in $\mathcal{L}_{\mathbf{PAL}}$, the notion of a formula being true at a state of a model extends that for formulas in $\mathcal{L}_{\mathbf{EL}}$ with the following clauses:

- $\mathcal{M}, w \models_{\cap} [\varphi]\psi$ iff $\mathcal{M}, w \models_{\cap} \varphi$ implies $\mathcal{M}^{\cap\varphi}, w \models_{\cap} \psi$,
- $\mathcal{M}, w \models_{\subseteq} [\varphi]\psi$ iff $\mathcal{M}, w \models_{\subseteq} \varphi$ implies $\mathcal{M}^{\subseteq\varphi}, w \models_{\subseteq} \psi$;

where $\mathcal{M}^{\cap\varphi}$ abbreviates $\mathcal{M}^{\cap[\![\varphi]\!]_{\mathcal{M}}}$ and $\mathcal{M}^{\subseteq\varphi}$ abbreviates $\mathcal{M}^{\subseteq[\![\varphi]\!]_{\mathcal{M}}}$. If we use the symbol $* \in \{\cap, \subseteq\}$ to denote either semantics then, from $\langle\varphi\rangle\psi$'s definition,

- $\mathcal{M}, w \models_* \langle\varphi\rangle\psi$ iff $\mathcal{M}, w \models_* \varphi$ and $\mathcal{M}^{*\varphi}, w \models_* \psi$.

The subscript $* \in \{\cap, \subseteq\}$ will be dropped from $\models_*$ when its meaining is clear from the context. A sound and complete axiomatization for $\mathcal{L}_{\mathbf{PAL}}$ w.r.t. the provided semantics under *MNM*s can be found in [17]. The purpose of this paper is to develop tableau systems for both logics. The following proposition is a generalization of the monotonicity of $\Box$ under *MNM*s ($[\![\varphi]\!]_{\mathcal{M}} \subseteq [\![\psi]\!]_{\mathcal{M}}$ implies $[\![\Box\varphi]\!]_{\mathcal{M}} \subseteq [\![\Box\psi]\!]_{\mathcal{M}}$) to the public announcements extensions and it will be key for providing $\Box$'s rules for both intersection and subset semantics.

Proposition 1. *Let ρ_i $(1 \leq i \leq n)$, θ_j $(1 \leq j \leq m)$ and φ be $\mathcal{L}_{\mathbf{PAL}}$-formulas and $\mathcal{M} = (W, \tau, V)$ be a MNM.*

(i) $[\![[\rho_1]\cdots[\rho_n]\varphi]\!]_{\mathcal{M}} \subseteq [\![[\theta_1]\cdots[\theta_m]\psi]\!]_{\mathcal{M}}$ *implies* $[\![\Box\varphi]\!]_{\mathcal{M}^{\cap\rho_1;\cdots;\cap\rho_n}} \subseteq [\![\Box\psi]\!]_{\mathcal{M}^{\cap\theta_1;\cdots;\cap\theta_m}}$
(ii) $[\![\langle\rho_1\rangle\cdots\langle\rho_n\rangle\varphi]\!]_{\mathcal{M}} \subseteq [\![\langle\theta_1\rangle\cdots\langle\theta_m\rangle\psi]\!]_{\mathcal{M}}$ *implies* $[\![\Box\varphi]\!]_{\mathcal{M}^{\subseteq\rho_1;\cdots;\subseteq\rho_n}} \subseteq [\![\Box\psi]\!]_{\mathcal{M}^{\subseteq\theta_1;\cdots;\subseteq\theta_m}}$

Proof. For (i), assume $[\![[\rho_1]\cdots[\rho_n]\varphi]\!]_{\mathcal{M}} \subseteq [\![[\theta_1]\cdots[\theta_m]\psi]\!]_{\mathcal{M}}$. Now fix any $w \in W$ with $\mathcal{M}^{\cap\rho_1;\cdots;\cap\rho_n}, w \models_\cap \Box\varphi$. By semantic interpretation, there is $X \in \tau^{\cap\rho_1;\cdots;\cap\rho_n}(w)$ s.t. $X \subseteq [\![\varphi]\!]_{\mathcal{M}^{\cap\rho_1;\cdots;\cap\rho_n}}$; then, by the definition of $\tau^{\cap\rho_1;\cdots;\cap\rho_n}(w)$, there is $Y \in \tau(w)$ s.t. $(Y \cap [\![\rho_1]\!]_{\mathcal{M}} \cap \cdots \cap [\![\rho_n]\!]_{\mathcal{M}^{\cap\rho_1;\cdots;\cap\rho_{n-1}}}) \subseteq [\![\varphi]\!]_{\mathcal{M}^{\cap\rho_1;\cdots;\cap\rho_n}}$, i.e., $Y \subseteq [\![[\rho_1]\cdots[\rho_n]\varphi]\!]_{\mathcal{M}}$ and hence, by assumption, $Y \subseteq [\![[\theta_1]\cdots[\theta_m]\psi]\!]_{\mathcal{M}}$. Thus, $Y \subseteq [\![\psi]\!]_{\mathcal{M}^{\cap\theta_1;\cdots;\cap\theta_m}}$ for $Y \in \tau^{\cap\theta_1;\cdots;\cap\theta_m}(w)$ so $\mathcal{M}^{\cap\theta_1;\cdots;\cap\theta_m}, w \models_\cap \Box\psi$, as needed.

For (ii), assume $[\![\langle\rho_1\rangle\cdots\langle\rho_n\rangle\varphi]\!]_{\mathcal{M}} \subseteq [\![\langle\theta_1\rangle\cdots\langle\theta_m\rangle\psi]\!]_{\mathcal{M}}$. Now fix any $w \in W$ with $\mathcal{M}^{\subseteq\rho_1;\cdots;\subseteq\rho_n}, w \models_\subseteq \Box\varphi$. Then there is $X \in \tau^{\subseteq\rho_1;\cdots;\subseteq\rho_n}(w)$ s.t. $X \subseteq [\![\varphi]\!]_{\mathcal{M}^{\subseteq\rho_1;\cdots;\subseteq\rho_n}}$ and, by definition of $\tau^{\subseteq\rho_1;\cdots;\subseteq\rho_n}(w)$, both $X \in \tau(w)$ and $X \subseteq ([\![\rho_1]\!]_{\mathcal{M}} \cap \cdots \cap [\![\rho_n]\!]_{\mathcal{M}^{\subseteq\rho_1;\cdots;\cap\rho_{n-1}}} \cap [\![\varphi]\!]_{\mathcal{M}^{\subseteq\rho_1;\cdots;\cap\rho_n}})$, i.e., $X \subseteq [\![\langle\rho_1\rangle\cdots\langle\rho_n\rangle\varphi]\!]_{\mathcal{M}}$ and hence, by assumption, $X \subseteq [\![\langle\theta_1\rangle\cdots\langle\theta_m\rangle\psi]\!]_{\mathcal{M}}$. Thus, $X \in \tau^{\subseteq\theta_1;\cdots;\subseteq\theta_m}(w)$ and $X \subseteq [\![\psi]\!]_{\mathcal{M}^{\subseteq\theta_1;\cdots;\subseteq\theta_m}}$ so $\mathcal{M}^{\subseteq\theta_1;\cdots;\subseteq\theta_m}, w \models_\subseteq \Box\psi$, as needed.

3 Tableaux for Non-normal Monotone (Static) Epistemic Logic

There are several works on tableau calculus of non-normal modal logic. Kripke [16] proposed a calculus based on Kripke semantics which allow the notion of *normal world*, and [8] constructed a uniform framework for tableau calculi for neighborhood semantics employing labels for both a states and set of states. More recently, Indrzejczak [15] avoided the label for set of states while presenting tableau calculi for several non-normal logics over neighborhood semantics.

As a prelude to our contribution, here we recall the tableau method for non-normal monotone modal logic of Indrzejczak [15], of which our proposal is an extension, as well as the argument for soundness and completeness. Then we recall why the satisfiability problem for non-normal monotone modal logic is NP-complete [29].

$$\frac{(\sigma:\varphi\wedge\psi)}{(\sigma:\varphi)(\sigma:\psi)}\,(\wedge) \qquad \frac{(\sigma:\neg(\varphi\wedge\psi))}{(\sigma:\neg\varphi)\mid(\sigma:\neg\psi)}\,(\neg\wedge) \qquad \frac{(\sigma:\neg\neg\varphi)}{(\sigma:\varphi)}\,(\neg\neg) \qquad \frac{(\sigma:\Box\varphi)(\sigma:\neg\Box\psi)}{(\sigma_{new}:\varphi)(\sigma_{new}:\neg\psi)}\,(\Box)$$

Fig. 1. Tableau rules for non-normal monotone logic [15]

The terms in the tableau rules (Figure 1), of the form $(\sigma : \varphi)$, indicate that formula φ is true in state (prefix) σ. Rules $(\wedge)$, $(\neg\wedge)$ and $(\neg\neg)$ correspond to propositional reasoning, and rule $(\Box)$ is the *prefix generating rule*. There are two general constraints on the construction of tableaus: (1) The prefix generating rule is never applied twice to the same premise on the same branch; (2) A formula is never added to a tableau branch where it already occurs.

As usual, a tableau is *saturated* when no more rules that satisfy the constraints can be applied. A branch is *saturated* if it belongs to a saturated tableau, and it is *closed* if it contains formulas $(\sigma : \varphi)$ and $(\sigma : \neg\varphi)$ for some σ and φ (otherwise, the branch is *open*). A tableau is *closed* if all its branches are closed, and it is *open* if at least one of its branches is open.

Rule $(\Box)$ might surprise readers familiar with tableaux for normal modal logic, but it states a straightforward fact: if both $\Box\varphi$ and $\neg\Box\psi$ hold in a world σ, then while $\Box\varphi$ imposes the existence of a neighborhood in $\tau(\sigma)$ containing only φ-worlds, $\neg\Box\psi$ imposes a $\neg\psi$-world in every neighborhood in $\tau(\sigma)$. The world σ_{new} denotes exactly that.

3.1 Soundness and Completeness

Definition 2. *Given a branch* Θ, Prefix(Θ) *is the set of all its prefixes. We say that* Θ *is* faithful *to a MNM* $\mathcal{M} = (W, \tau, V)$ *if there is a mapping* f : Prefix(Θ) $\to W$ *such that* $(\sigma : \varphi) \in \Theta$ *implies* $\mathcal{M}, f(\sigma) \models \varphi$ *for all* $\sigma \in$ Prefix(Θ).

Lemma 1. *Let* Θ *be any branch of a tableau and* $\mathcal{M} = (W, \tau, V)$ *a MNM. If* Θ *is faithful to* $\mathcal{M}$*, and a tableau rule is applied to it, then it produces at least one extension* Θ' *such that* Θ' *is faithful to* $\mathcal{M}$*.*

For the proof, see Appendix A.1.

Theorem 1 (Soundness). *Given any formula* φ*, if there is a closed tableau for* $(\sigma_{initial} : \neg\varphi)$*, then* φ *is valid in the class of all MNMs.*

Proof. We show the contrapositive. Suppose that $\neg\varphi$ is satisfiable, i.e., there is a *MNM* $\mathcal{M} = (W, \tau, V)$ and a $w \in W$ s.t. $\mathcal{M}, w \not\models \varphi$. Then the initial tableau $\Theta = \{(\sigma_{initial} : \neg\varphi)\}$ is faithful to $\mathcal{M}$ and hence, by Lemma 1, only faithful tableau to *MNM* will be produced. A faithful branch cannot be closed. Hence $(\sigma_{initial} : \neg\varphi)$ can have no closed tableau.

Lemma 2. *Given an open saturated branch* Θ*, define the model* $\mathcal{M}^\Theta = (W^\Theta, \tau^\Theta, V^\Theta)$ *as* $W^\Theta :=$ Prefix(Θ), $V^\Theta(p) := \{\sigma \in W^\Theta \mid (\sigma : p) \in \Theta\}$ *and, for every* $\sigma \in W^\Theta$*,*

$$X \in \tau^\Theta(\sigma) \quad \textit{iff} \quad \textit{there is } \varphi \textit{ s.t.} (\sigma : \Box\varphi) \in \Theta \textit{ and } \{\sigma' \in W^\Theta \mid (\sigma' : \varphi) \in \Theta\} \subseteq X$$

Then, for all formulas φ *and all prefix* σ*,* (i) $(\sigma : \varphi) \in \Theta$ *implies* $\mathcal{M}^\Theta, \sigma \models \varphi$ *and* (ii) $(\sigma : \neg\varphi) \in \Theta$ *implies* $\mathcal{M}^\Theta, \sigma \not\models \varphi$*.*

Note that τ^Θ is clearly monotone and thus, if Θ is non-empty, $\mathcal{M}^\Theta$ is a *MNM*. For the proof, see Appendix A.2.

Theorem 2 (Completeness). *Given any formula* φ*, if there is an open saturated tableau for* $(\sigma_{initial} : \varphi)$*, then* φ *is satisfiable in a MNM.*

Proof. If there is an open saturated branch Θ containing $(\sigma_{initial} : \varphi)$, Lemma 2 yields $\mathcal{M}^\Theta, \sigma_{initial} \models \varphi$ so φ is satisfiable in a *MNM*.

3.2 Complexity

Normal modal logics as **K** and **KT** are PSPACE-complete, and negative introspection $\neg\Box p \rightarrow \Box\neg\Box p$ makes any modal logics between **K** and **S4** NP-complete [11]. Tableau systems for such logics have been given in [10].

The satisfiability problem (deciding whether a given φ is satisfiable) for non-normal monotone modal logic is NP-complete [29]. A known method is to build a tableau from $\{(\sigma_{initial} : \varphi)\}$; at each step, the process adds non-deterministically a term of the form $(\sigma : \psi)$ with σ is a symbol and ψ is a subformula or a negation of a subformula of φ.

Proposition 2. *When executing the tableau method from* $\{(\sigma_{initial} : \varphi)\}$*, the number of terms* $(\sigma : \psi)$ *that can be added is polynomial in the length of* φ.

Proof. As ψ is a subformula or a negation of a subformula of φ, the number of possible ψ is linear in the size of φ. The number of possible world symbols σ is polynomial in the size of φ, as they are created only for pairs of the form $\Box\psi_1$, $\neg\Box\psi_2$. Thus, the number of such σ is bounded by $|\varphi|^2$, and hence the number of possible terms $(\sigma : \psi)$ is bounded by $|\varphi|^3$.

Corollary 1. *The satisfiability problem in non-normal monotone modal logic is NP-complete.*

Proof. NP-hardness comes from the fact that the satisfiability problem for classical propositional logic is polynomially reducible to the satisfiability problem for non-normal monotone modal logic. Now let us figure out why it is in NP. In the non-deterministic algorithm shown below, the size of Θ is polynomial in the length of φ (Proposition 2). Testing that Θ is saturated or non-deterministically applying a rule can be implemented in polynomial time in the size of Θ; then, these operations are polynomial in the length of φ. As we add a term to Θ at each iteration of the **while** loop, there are at most a polynomial number of iterations. Therefore, the tableau method can be implemented in polynomial time on a non-deterministic machine.

```
procedure sat(φ)
    Θ := {(σ_initial : φ)}
    while Θ is not saturated
        Θ := result of the (non-deterministic) application of a rule on Θ
        if Θ is closed then reject
    accept
```

4 Tableaux for Non-normal Public Annoucement Logics

Tableaux for public announcements for normal modal logic already appeared in [2], where the tableau formalism needed to represent the information of accessibility relation. Since we are concerned with non-normal modal logic characterized by neighborhood models, our tableau calculus will not introduce any formalism for accessibility relation. In this sense, our work is not a trivial generalization of [2]. For non-normal monotone modal logic, this section adapts the tableau method of Section 3 to deal with public announcements under both the ∩- and the ⊆-semantics. Here, terms in the tableau rules can be either

- of the form $(\sigma :_L \varphi)$ with σ a world symbol, L a list of announced formulas (ϵ is the empty list) and φ a formula, indicating that σ survives the successive announcements of the elements of L and afterwards it satisfies φ, or
- of the form $(\sigma :_L \times)$, indicating that σ does not survive successive announcements of the elements of L.

Figure 2 shows the tableau rules for non-normal public annoucement logics. We define the rule set for the $\cap$-semantics as all the common rules plus $(\Box^{\cap})$, while the rule set for the $\subseteq$-semantics as all the common rules plus $(\Box^{\subseteq})$.

Common rules: $\dfrac{(\sigma :_L p)}{(\sigma :_\epsilon p)}\ (\downarrow \epsilon)$ $\quad \dfrac{(\sigma :_L \neg p)}{(\sigma :_\epsilon \neg p)}\ (\downarrow \epsilon\neg)$ $\quad \dfrac{(\sigma :_{L;\varphi} \times)}{(\sigma :_L \neg\varphi) \mid (\sigma :_L \times)}\ (\times Back)$

$$\frac{(\sigma :_L \varphi \wedge \psi)}{(\sigma :_L \varphi)(\sigma :_L \psi)}\ (\wedge) \qquad \frac{(\sigma :_L \neg(\varphi \wedge \psi))}{(\sigma :_L \neg\varphi) \mid (\sigma :_L \neg\psi)}\ (\neg\wedge) \qquad \frac{(\sigma :_L \neg\neg\varphi)}{(\sigma :_L \varphi)}\ (\neg\neg)$$

$$\frac{(\sigma :_{L;\varphi} \psi)}{(\sigma :_L \varphi)}\ (Back) \qquad \frac{(\sigma :_L [\varphi]\psi)}{(\sigma :_L \neg\varphi) \mid (\sigma :_{L;\varphi} \psi)}\ ([\cdot]) \qquad \frac{(\sigma :_L \neg[\varphi]\psi)}{(\sigma :_{L;\varphi} \neg\psi)}\ (\neg[\cdot])$$

For $\cap$-semantics: $\dfrac{(\sigma :_L \Box\varphi)(\sigma :_{L'} \neg\Box\psi)}{(\sigma_{new} :_L \varphi)(\sigma_{new} :_{L'} \neg\psi) \mid (\sigma_{new} :_L \times)(\sigma_{new} :_{L'} \neg\psi)}\ (\Box^{\cap})$

For $\subseteq$-semantics: $\dfrac{(\sigma :_L \Box\varphi)(\sigma :_{L'} \neg\Box\psi)}{(\sigma_{new} :_{L'} \neg\psi)(\sigma_{new} :_L \varphi) \mid (\sigma_{new} :_{L'} \times)(\sigma_{new} :_L \varphi)}\ (\Box^{\subseteq})$

Fig. 2. Tableau rules for handling public announcements

Rules $(\wedge)$, $(\neg\wedge)$ and $(\neg\neg)$ deal with propositional reasoning. Rules $(\downarrow \epsilon)$, $(\downarrow \epsilon\neg)$ indicate that valuations do not change after a sequence of announcements. Rule $(\neg[\cdot])$ states that if $\neg[\varphi]\psi$ holds in σ after a sequence of announcements L then $\neg\psi$ must hold in σ after the sequence of announcements $L;\varphi$. Rule $([\cdot])$ states that if $[\varphi]\psi$ holds in σ after a sequence of announcements L, then either φ fails in σ after a sequence of announcements L or else ψ holds in σ after the sequence of announcements $L;\varphi$. Rule $(Back)$ deals with a world surviving a sequence of announcements, and rule $(\times Back)$ deals with a world not surviving it.

The rule of $(\Box^{\cap})$ is a rewriting of the first item of Proposition 1 into the rule of tableau calculus. For simplicity, let us assume that $L \equiv \rho;\rho'$ and $L' \equiv \theta$. By taking the contrapositive implication of Proposition 1.(i), we obtain the following rule:

$$\frac{(\sigma :_{\rho;\rho'} \Box\varphi)(\sigma :_\theta \neg\Box\psi)}{(\sigma_{new} :_\epsilon [\rho][\rho']\varphi)(\sigma_{new} :_\epsilon \neg[\theta]\psi)}$$

While $(\sigma_{new} :_\epsilon \neg[\theta]\psi)$ generates $(\sigma_{new} :_\theta \neg\psi)$ by the rule $(\neg[\cdot])$, we have two cases for expanding $(\sigma_{new} :_\epsilon [\rho][\rho']\varphi)$. First, assume that σ_{new} survives after the successive updates of ρ and ρ'. Then, we may add $(\sigma_{new} :_{\rho:\rho'} \varphi)$ to the branch. Second, suppose that σ_{new} does not survive after the successive updates of ρ and ρ'. Then, we add $(\sigma_{new} :_{\rho;\rho'} \times)$ to the branch. This also explains the soundness of $(\Box^{\cap})$ for $\cap$-semantics.

Rule ($\Box^{\subseteq}$) can also be explained in terms of the second item of Proposition 1. Let L and L' as above. By taking the contrapositive implication of Proposition 1.(ii) and rewriting the diamond $\langle\gamma\rangle$ in terms of the dual $[\gamma]$, we obtain the following:

$$\frac{(\sigma :_{\rho;\rho'} \Box\varphi)(\sigma :_{\theta} \neg\Box\psi)}{(\sigma_{new} :_{\epsilon} \neg[\rho][\rho']\neg\varphi)(\sigma_{new} :_{\epsilon} [\theta]\neg\psi)}$$

By a procedure similar to the used for ($\Box^{\cap}$) we can justify the rule ($\Box^{\subseteq}$).

As before, there are two constraints on the construction of tableaus: A prefix generating rule is never applied twice to the same premise on the same branch; A formula is never added to a tableau branch where it already occurs. The notions of *saturated tableau* and *saturated branch* are as before. In order to deal with terms of the form $(\sigma :_L \times)$, the notion of closed branch is extended as follows: a branch of a tableau is *closed* when (1) it contains terms $(\sigma :_L \varphi)$ and $(\sigma :_L \neg\varphi)$ for some σ, L and φ, or (2) it contains $(\sigma :_{\epsilon} \times)$ for some σ; otherwise, the branch is called *open*. The notions of *closed* and *open tableau* are defined as before.

4.1 Soundness

We start with the $\cap$-semantics. As before, given a branch Θ, Prefix(Θ) denotes the set of all prefixes in Θ.

Definition 3. *Given a branch Θ and a MNM $\mathcal{M} = (W, \tau, V)$, Θ is* faithful *to $\mathcal{M}$ if there is a mapping f : Prefix(Θ) $\to W$ such that, for all $\sigma \in$ Prefix(Θ),*

- $(\sigma :_{\psi_1;\cdots;\psi_n} \varphi) \in \Theta$ *implies* $\mathcal{M}^{\cap\psi_1;\cdots;\cap\psi_n}, f(\sigma) \models \varphi$, *and*
- $(\sigma :_{\psi_1;\cdots;\psi_n} \times) \in \Theta$ *implies that* $f(\sigma)$ *is not in* $\mathcal{M}^{\cap\psi_1;\cdots;\cap\psi_n}$*'s domain.*

Lemma 3. *Let Θ be any branch of a tableau and $\mathcal{M} = (W, \tau, V)$ a MNM. If Θ is faithful to $\mathcal{M}$, and a tableau rule is applied to it, then it produces at least one extension Θ' such that Θ' is faithful to $\mathcal{M}$.*

Proof. We only show the case for rule ($\Box^{\cap}$). For the cases of rules ($\downarrow \epsilon$), ($[\cdot]$), ($\times$*Back*), (*Back*), see Appendix A.3. Throughout this proof, let $L \equiv \rho_1;\cdots;\rho_n$. Let $L' \equiv \theta_1;\cdots;\theta_m$ in the rule ($\Box^{\cap}$) of Table 2. Since $(\sigma :_L \Box\varphi), (\sigma :_{L'} \neg\Box\psi) \in \Theta$, there is an f s.t. $f(\sigma) \in [\![\Box\varphi]\!]_{\mathcal{M}^{\cap\rho_1;\cdots;\cap\rho_n}}$ and $f(\sigma) \notin [\![\Box\psi]\!]_{\mathcal{M}^{\cap\theta_1;\cdots;\cap\theta_m}}$. Thus, $[\![\Box\varphi]\!]_{\mathcal{M}^{\cap\rho_1;\cdots;\cap\rho_n}} \not\subseteq [\![\Box\psi]\!]_{\mathcal{M}^{\cap\theta_1;\cdots;\cap\theta_m}}$ and hence, by Proposition 1, $[\![[\rho_1]\cdots[\rho_n]\varphi]\!]_{\mathcal{M}} \not\subseteq [\![[\theta_1]\cdots[\theta_m]\psi]\!]_{\mathcal{M}}$: there is u in $\mathcal{M}$ such that $u \in [\![[\rho_1]\cdots[\rho_n]\varphi]\!]_{\mathcal{M}}$ but $u \notin [\![[\theta_1]\cdots[\theta_m]\psi]\!]_{\mathcal{M}}$. From the latter it follows that u survives the successive intersection updates of $\theta_1, \ldots, \theta_n$ but $\mathcal{M}^{\cap\theta_1;\cdots;\cap\theta_m}, u \not\models \psi$. From the former, suppose (1) u is in the domain of $\mathcal{M}^{\cap\rho_1;\cdots;\cap\rho_n}$; then $\mathcal{M}^{\cap\rho_1;\cdots;\cap\rho_n}, u \models \varphi$ and we can take $\Theta' := \Theta \cup \{(\sigma_{new} :_L \varphi), (\sigma_{new} :_{L'} \neg\psi)\}$ and extend the original f into f' : Prefix(Θ') $\to W$ by defining $f'(\sigma_{new}) := u$. It follows that $\mathcal{M}^{\cap\rho_1;\cdots;\cap\rho_n}, f(\sigma_{new}) \models \varphi$ and $\mathcal{M}^{\cap\theta_1;\cdots;\cap\theta_m}, f(\sigma_{new}) \not\models \psi$, and so Θ' is faithful to $\mathcal{M}$. Otherwise, (2) u is not in the domain of $\mathcal{M}^{\cap\rho_1;\cdots;\cap\rho_n}$, an a similar argument shows that $\Theta' = \Theta \cup \{(\sigma_{new} :_L \times), (\sigma_{new} :_{L'} \neg\psi)\}$ is faithful to $\mathcal{M}$.

Theorem 3. *Given any formula φ and any list $L \equiv \rho_1;\cdots;\rho_n$, if there is a closed tableau for $(\sigma_{\text{initial}} :_L \neg\varphi)$, then φ is valid in $\mathcal{M}^{\cap\rho_1;\cdots;\cap\rho_n}$ for all MNMs $\mathcal{M}$.*

Proof. We show the contrapositive. Suppose that there is a *MNM* $\mathcal{M} = (W, \tau, V)$ and a $w \in W$ such that $\mathcal{M}^{\cap\rho_1;\cdots;\cap\rho_n}, w \not\models \varphi$. Then the initial tableau $\Theta = \{(\sigma_{initial} :_L \neg\varphi)\}$ is faithful to $\mathcal{M}$ and hence, by Lemma 3, only faithful tableau to *MNM* will be produced. A faithful branch cannot be closed. Hence $(\sigma_{initial} :_L \neg\varphi)$ can have no closed tableau.

Now, for the $\subseteq$-semantics, we have the following.

Lemma 4. *Let Θ be any branch of a tableau and $\mathcal{M} = (W, \tau, V)$ a MNM. If Θ is faithful to $\mathcal{M}$, and a tableau rule is applied to it, then it produces at least one extension Θ' such that Θ' is faithful to $\mathcal{M}$.*

Proof. We only show the case for the rule $(\Box^{\subseteq})$. Let $L \equiv \rho_1; \cdots ; \rho_n$ and $L' \equiv \theta_1; \cdots ; \theta_m$ in the rule $(\Box^{\subseteq})$ of Table 2. Since $(\sigma :_L \Box\varphi), (\sigma :_{L'} \neg\Box\psi) \in \Theta$, there is an f s.t. $f(\sigma) \in [\![\Box\varphi]\!]_{\mathcal{M}^{\subseteq\rho_1;\cdots;\subseteq\rho_n}}$ and $f(\sigma) \notin [\![\Box\psi]\!]_{\mathcal{M}^{\subseteq\theta_1;\cdots;\subseteq\theta_m}}$. Thus, $[\![\Box\varphi]\!]_{\mathcal{M}^{\subseteq\rho_1;\cdots;\subseteq\rho_n}} \nsubseteq [\![\Box\psi]\!]_{\mathcal{M}^{\subseteq\theta_1;\cdots;\subseteq\theta_m}}$ and hence, by Proposition 1, $[\![\langle\rho_1\rangle \cdots \langle\rho_n\rangle\varphi]\!]_{\mathcal{M}} \nsubseteq [\![\langle\theta_1\rangle \cdots \langle\theta_m\rangle\psi]\!]_{\mathcal{M}}$. Then, there is u in $\mathcal{M}$ such that $u \in [\![\langle\rho_1\rangle \cdots \langle\rho_n\rangle\varphi]\!]_{\mathcal{M}}$ but $u \notin [\![\langle\theta_1\rangle \cdots \langle\theta_m\rangle\psi]\!]_{\mathcal{M}}$. From the former it follows that u survives the successive subset updates of $\rho_1, \ldots, \rho_n$ and $\mathcal{M}^{\subseteq\rho_1;\cdots;\subseteq\rho_n}, u \models \varphi$. From the latter, suppose (1) u is in the domain of $\mathcal{M}^{\subseteq\theta_1;\cdots;\subseteq\theta_m}$; then $\mathcal{M}^{\subseteq\theta_1;\cdots;\subseteq\theta_m}, u \models \neg\psi$ and we can take $\Theta' := \Theta \cup \{(\sigma_{new} :_{L'} \neg\psi), (\sigma_{new} :_L \varphi)\}$ and extend the original f into $f' : \text{Prefix}(\Theta') \to W$ by defining $f'(\sigma_{new}) := u$. It follows that $\mathcal{M}^{\subseteq\theta_1;\cdots;\subseteq\theta_m}, f(\sigma_{new}) \not\models \psi$ and $\mathcal{M}^{\subseteq\rho_1;\cdots;\subseteq\rho_n}, f(\sigma_{new}) \models \varphi$, and so Θ' is faithful to $\mathcal{M}$. Otherwise, (2) u is not in the domain of $\mathcal{M}^{\subseteq\theta_1;\cdots;\subseteq\theta_m}$, and a similar argument shows that $\Theta' := \Theta \cup \{(\sigma_{new} :_{L'} \times), (\sigma_{new} :_L \varphi)\}$ is faithful to $\mathcal{M}$.

Theorem 4. *Given any formula φ and any list $L \equiv \rho_1; \cdots ; \rho_n$, if there is a closed tableau for $(\sigma_{initial} :_L \varphi)$, then φ is valid in $\mathcal{M}^{\subseteq\rho_1;\cdots;\subseteq\rho_n}$ for all MNMs $\mathcal{M}$.*

4.2 Completeness

We start with the $\cap$-semantics. Define the function len : $\mathcal{L}_{\mathbf{PAL}} \cup \{\times, L\} \to \mathbb{N}$ as

$$\begin{array}{lll} \text{len}(\times) := 1, & \text{len}(\neg\varphi) := \text{len}(\varphi) + 1, & \text{len}(\varphi \wedge \psi) := \text{len}(\varphi) + \text{len}(\psi) + 1, \\ \text{len}(p) := 1, & \text{len}(\Box\varphi) := \text{len}(\varphi) + 1, & \text{len}([\varphi]\psi) := \text{len}(\varphi) + \text{len}(\psi) + 1, \\ & \text{len}(L) := \text{len}(\varphi_1) + \cdots + \text{len}(\varphi_n) & \text{for } L \equiv \varphi_1; \cdots ; \varphi_n. \end{array}$$

Lemma 5. *Given an open saturated branch Θ, define the model $\mathcal{M}^\Theta = (W^\Theta, \tau^\Theta, V^\Theta)$ as $W^\Theta :=$ Prefix(Θ), $V^\Theta(p) := \{\sigma \in W^\Theta \mid (\sigma :_\epsilon p) \in \Theta\}$ and, for every $\sigma \in W^\Theta$, $X \in \tau^\Theta(\sigma)$ iff there are φ and L such that*

$$(\sigma :_L \Box\varphi) \in \Theta \textit{ and } \{\sigma' \in W^\Theta \mid (\sigma' :_L \times) \in \Theta \textit{ or } (\sigma' :_L \varphi) \in \Theta\} \subseteq X.$$

Then, for all lists $L = \rho_1; \cdots ; \rho_n$ and all formulas φ,

(i) $(\sigma :_L \varphi) \in \Theta$ *implies* $(\mathcal{M}^\Theta)^{\cap\rho_1;\cdots;\cap\rho_n}, \sigma \models \varphi$
(ii) $(\sigma :_L \neg\varphi) \in \Theta$ *implies* $(\mathcal{M}^\Theta)^{\cap\rho_1;\cdots;\cap\rho_n}, \sigma \not\models \varphi$
(iii) $(\sigma :_L \times) \in \Theta$ *implies σ is not in the domain of* $(\mathcal{M}^\Theta)^{\cap\rho_1;\cdots;\cap\rho_n}$.

Proof. All of (i), (ii) and (iii) are proved by simultaneous induction on len($*$) + len(L), where $*$ is a formula φ or $\times$. We show the cases (i) and (ii) for $\Box\gamma$. In Appendix A.4, the reader can find arguments for case (iii) fully and the cases for φ of the form p, $[\psi]\gamma$.

Let $\varphi \equiv \Box\gamma$. For (i), assume $(\sigma :_{\rho_1;\cdots;\rho_n} \Box\gamma) \in \Theta$; we show that $[\![\gamma]\!]_{(\mathcal{M}^\Theta)^{\cap\rho_1;\cdots;\cap\rho_n}} \in (\tau^\Theta)^{\cap\rho_1;\cdots;\cap\rho_n}(\sigma)$ or, equivalently, $[\![[\rho_1]\cdots[\rho_n]\gamma]\!]_{\mathcal{M}^\Theta} \in \tau^\Theta(\sigma)$. It suffices to show both

- $(\sigma :_L \Box\gamma) \in \Theta$,
- $\{\sigma' \in W^\Theta \mid (\sigma' :_L \times) \in \Theta$ or $(\sigma' :_L \gamma) \in \Theta\} \subseteq [\![[\rho_1]\cdots[\rho_n]\gamma]\!]_{\mathcal{M}^\Theta}$.

The first is the assumption; the second holds by induction hypothesis. For (ii), assume $(\sigma :_L \neg\Box\gamma) \in \Theta$; we show that $[\![\gamma]\!]_{(\mathcal{M}^\Theta)^{\cap\rho_1;\cdots;\cap\rho_n}} \notin (\tau^\Theta)^{\cap\rho_1;\cdots;\cap\rho_n}(\sigma)$ or, equivalently, $[\![[\rho_1]\cdots[\rho_n]\gamma]\!]_{\mathcal{M}^\Theta} \notin \tau^\Theta(\sigma)$, i.e., for all φ and L',

$$(\sigma :_{L'} \Box\varphi) \in \Theta \text{ implies } \{\sigma' {\in} W^\Theta \mid (\sigma' :_{L'} \times) {\in} \Theta \text{ or } (\sigma' :_{L'} \varphi) \in \Theta\} \nsubseteq [\![[\rho_1]\cdots[\rho_n]\gamma]\!]_{\mathcal{M}^\Theta}$$

Thus, take any φ and L' such that $(\sigma :_{L'} \Box\varphi) \in \Theta$. By the saturatedness of Θ and rule $(\Box^\cap)$ we obtain, for some fresh σ_{new}, either

$$(\sigma_{new} :_{L'} \varphi), (\sigma_{new} :_L \neg\gamma) \in \Theta \text{ or } (\sigma_{new} :_{L'} \times), (\sigma_{new} :_L \neg\gamma) \in \Theta.$$

In either case, it follows from $(\sigma_{new} :_L \neg\gamma) \in \Theta$ and induction hypothesis that γ is false at σ_{new} in $(\mathcal{M}^\Theta)^{\cap\rho_1;\cdots;\cap\rho_n}$, which is equivalent to $\mathcal{M}^\Theta, \sigma_{new} \not\models [\rho_1]\cdots[\rho_n]\gamma$. This finishes establishing our goal; $\{\sigma' \in W^\Theta \mid (\sigma' :_{L'} \times) \in \Theta$ or $(\sigma' :_{L'} \varphi) \in \Theta\} \nsubseteq [\![[\rho_1]\cdots[\rho_n]\gamma]\!]_{\mathcal{M}^\Theta}$.

Theorem 5. *Given any formula φ, if there is an open saturated tableau for $(\sigma_{initial} :_\epsilon \varphi)$, then φ is satisfiable in the class of all MNMs for intersection semantics.*

Proof. By assumption, there is an open saturated branch Θ containing $(\sigma_{initial} :_\epsilon \varphi)$. By Lemma 5, $\mathcal{M}^\Theta, \sigma_{initial} \models \varphi$, which implies the satisfiability of φ in the class of all *MNM*s for intersection semantics.

Now, let us move to the $\subseteq$-semantics.

Lemma 6. *Given an open saturated branch Θ, define the model $\mathcal{M}^\Theta = (W^\Theta, \tau^\Theta, V^\Theta)$ as in Lemma 5 except that, for every $\sigma \in W^\Theta$, $X \in \tau^\Theta(\sigma)$ iff*

$$(\sigma :_L \Box\varphi) \in \Theta \text{ and } \{\sigma' \in W^\Theta \mid (\sigma' :_L \varphi) \in \Theta\} \subseteq X \quad \text{for some } \varphi \text{ and } L.$$

Then, for all lists $L = \rho_1; \cdots; \rho_n$ and all formulas φ,

(i) $(\sigma :_L \varphi) \in \Theta$ *implies* $(\mathcal{M}^\Theta)^{\subseteq\rho_1;\cdots;\subseteq\rho_n}, \sigma \models \varphi$
(ii) $(\sigma :_L \neg\varphi) \in \Theta$ *implies* $(\mathcal{M}^\Theta)^{\subseteq\rho_1;\cdots;\subseteq\rho_n}, \sigma \not\models \varphi$
(iii) $(\sigma :_L \times) \in \Theta$ *implies σ is not in the domain of* $(\mathcal{M}^\Theta)^{\subseteq\rho_1;\cdots;\subseteq\rho_n}$.

Proof. All of (i), (ii) and (iii) are proved by simultaneous induction on len($*$) + len(L), where $*$ is a formula φ or $\times$. We show cases (i) and (ii) for φ of the form $\Box\gamma$.

For (i), assume $(\sigma :_L \Box\gamma) \in \Theta$; we show that $(\mathcal{M}^\Theta)^{\subseteq\rho_1;\cdots;\subseteq\rho_n}, \sigma \models \Box\gamma$, i.e., $[\![\gamma]\!]_{(\mathcal{M}^\Theta)^{\subseteq\rho_1;\cdots;\subseteq\rho_n}} \in (\tau^\Theta)^{\subseteq\rho_1;\cdots;\subseteq\rho_n}(\sigma)$ or, equivalently, $[\![\langle\rho_1\rangle\cdots\langle\rho_n\rangle\gamma]\!]_{\mathcal{M}} \in \tau^\Theta(\sigma)$. It suffices to show

$$(\sigma :_L \Box\gamma) \in \Theta \text{ and } \{\sigma' \in W^\Theta \mid (\sigma' :_L \gamma) \in \Theta\} \subseteq [\![\langle\rho_1\rangle\cdots\langle\rho_n\rangle\gamma]\!]_{\mathcal{M}^\Theta}$$

The first conjunct is the assumption. For the second conjunct, suppose $(\sigma' :_L \gamma) \in \Theta$; we show that $\mathcal{M}^\Theta, \sigma' \models \langle\rho_1\rangle \cdots \langle\rho_n\rangle\gamma$, i.e.,

$$\mathcal{M}^\Theta, \sigma' \models \rho_1, \ \ldots, \ (\mathcal{M}^\Theta)^{\subseteq\rho_1;\cdots;\subseteq\rho_{n-1}}, \sigma' \models \rho_n \ \text{ and } \ (\mathcal{M}^\Theta)^{\subseteq\rho_1;\cdots;\subseteq\rho_n}, \sigma' \models \gamma.$$

These can be derived from $(\sigma' :_L \gamma) \in \Theta$, the rule (*Back*) and induction hypothesis. Therefore, $\mathcal{M}^\Theta, \sigma' \models \langle\rho_1\rangle \cdots \langle\rho_n\rangle\gamma$, as required.

For (ii), assume $(\sigma :_L \neg\Box\gamma) \in \Theta$; we show that $(\mathcal{M}^\Theta)^{\subseteq\rho_1;\cdots;\subseteq\rho_n}, \sigma \not\models \Box\gamma$, i.e., $[\![\langle\rho_1\rangle \cdots \langle\rho_n\rangle\gamma]\!]_{\mathcal{M}} \notin \tau^\Theta(\sigma)$. It suffices to show that, for all L' and all φ,

$$(\sigma :_{L'} \Box\varphi) \in \Theta \quad \text{implies} \quad \{\sigma' \in W^\Theta \mid (\sigma' :_{L'} \varphi) \in \Theta\} \nsubseteq [\![\langle\rho_1\rangle \cdots \langle\rho_n\rangle\gamma]\!]_{\mathcal{M}^\Theta}.$$

Thus, take any L' and φ such that $(\sigma :_{L'} \Box\varphi) \in \Theta$. Since $(\sigma :_{L'} \Box\varphi), (\sigma :_L \neg\Box\gamma) \in \Theta$, Θ's saturatedness and rule $(\Box^{\subseteq})$ imply, for some fresh σ_{new}, either

$$(\sigma_{new} :_L \neg\gamma), (\sigma_{new} :_{L'} \varphi) \in \Theta \text{ or } (\sigma_{new} :_L \times), (\sigma_{new} :_{L'} \varphi) \in \Theta$$

In either case, it follows from induction hypothesis that either σ_{new} is not in $(\mathcal{M}^\Theta)^{\subseteq\rho_1;\cdots;\subseteq\rho_n}$, or else $(\mathcal{M}^\Theta)^{\subseteq\rho_1;\cdots;\subseteq\rho_n}, \sigma_{new} \not\models \gamma$, which is equivalent with $\mathcal{M}^\Theta, \sigma_{new} \not\models \langle\rho_1\rangle \cdots \langle\rho_n\rangle\gamma$. This finishes establishing our goal; $\{\sigma' \in W^\Theta \mid (\sigma' :_{L'} \varphi) \in \Theta\} \nsubseteq [\![\langle\rho_1\rangle \cdots \langle\rho_n\rangle\gamma]\!]_{\mathcal{M}^\Theta}$.

Theorem 6. *Given any formula φ, if there is an open saturated tableau for $(\sigma_{initial} :_\epsilon \varphi)$, then φ is satisfiable in the class of all MNMs for subset semantics.*

Proof. Similar to Theorem 5, using Lemma 6 instead.

4.3 Termination and Complexity

The same argument works for both semantics. In order to check φ's satisfiability, start the tableau method from the set of terms $\{(\sigma_{initial} :_\epsilon \varphi)\}$ where $\sigma_{initial}$ is the initial symbol. At each step, add non-deterministically at least one term of the form $(\sigma :_L *)$ where σ is a symbol, L is a list of subformulas or negation subformulas of φ and $*$ is a subformula or a negation of a subformula of φ or the symbol $\times$.

Proposition 3. *When executing the tableau method from $\{(\sigma_{initial} :_\epsilon \varphi)\}$, the number of terms $\{(\sigma :_L *)\}$ that can be added is polynomial in the length of φ.*

Proof. As $*$ is a subformula or a negation of a subformula of φ or the symbol $\times$, the number of possible $*$ is linear in the size of φ. The number of possible L is linear in the size of φ since each entry corresponds to an occurrence of an operator $[\psi]$ in φ. The number of possible σ is polynomial in the size of φ since new world symbols σ are created for 4-tuple of subformulas of the form $\Box\psi_1, \neg\Box\psi_2$. Thus, the number of possible terms $(\sigma : \psi)$ is bounded by a polynomial in $|\varphi|$.

Corollary 2. *The satisfiability problem in non-normal monotone public announcement logic is NP-complete.*

Proof. The proof is similar to the proof of Corollary 1 except that we use Proposition 3 instead of Proposition 2 and that we start with $\Theta := \{(\sigma_{initial} :_\epsilon \varphi)\}$ instead of $\Theta := \{(\sigma_{initial} : \varphi)\}$.

4.4 Implementation

We implemented the tableau method for both $\cap$-semantics and $\subseteq$-semantics in Lotrecscheme [22]. The tool and the files for logics are available, respectively, at:

```
http://people.irisa.fr/Francois.Schwarzentruber/lotrecscheme/
http://people.irisa.fr/Francois.Schwarzentruber/publications/ICLA2015/
```

Appendix A.5 shows an output of Lotrecscheme.

5 Conclusion

We develop tableau system for both intersection and subset **PAL** based on monotone modal logic. Here we present some problems for future work.

- We may generalize our tableau systems to the general dynamic epistemic logic setting. Intersection **DEL** is already proposed in [31] and subset **DEL** is also proposed in [17]. Our idea for developing tableau system for **PAL**s is to take finite sequences of public announcements into consider. In the **DEL** setting, we may consider histories of actions in the action model. Thus we may develop the tableau rules for operations as it is done in [1] for the **DEL** extension of modal logic **K**.
- It is well-known that modal formulas corresponds to conditions on neighborhood frames ([19]). Thus we may consider how tableau systems can be developed for extensions of monotone modal logic with additional modal axioms, and then consider their dynamics extensions. The problem is to take those special frame conditions into account in the tableau rules for modal operations.
- As the satisfiability problems for both intersection and subset **PAL** are in NP, they are reducible to the satisfiability problem for classical propositional logic [20]. We aim at finding elegant reductions for obtaining efficient solvers for both intersection and subset **PAL**.

References

1. Aucher, G., Schwarzentruber, F.: On the complexity of dynamic epistemic logic. CoRR, abs/1310.6406 (2013)
2. Balbiani, P., van Ditmarsch, H.P., Herzig, A., DeLima, T.: A tableau method for public announcement logics. In: Olivetti, N. (ed.) TABLEAUX 2007. LNCS (LNAI), vol. 4548, pp. 43–59. Springer, Heidelberg (2007)
3. Baltag, A., Moss, L., Solecki, S.: The logic of public announcements, common knowledge and private suspicions. In: Proceedings of TARK, pp. 43–56. Morgan Kaufmann Publishers (1998)
4. Chellas, B.F.: Modal Logic: An Introduction. Cambridge University Press, Cambridge (1980)
5. Fagin, R., Halpern, J.Y.: Belief, awareness, and limited reasoning. Artificial Intelligence 34(1), 39–76 (1988)
6. Fagin, R., Halpern, J.Y., Moses, Y., Vardi, M.Y.: Reasoning about knowledge. The MIT Press, Cambridge (1995)
7. Gerbrandy, J., Groeneveld, W.: Reasoning about information change. Journal of Logic, Language, and Information 6(2), 147–196 (1997)

8. Governatori, G., Luppi, A.: Labelled tableaux for non-normal modal logics. In: Lamma, E., Mello, P. (eds.) AI*IA 1999. LNCS (LNAI), vol. 1792, pp. 119–130. Springer, Heidelberg (2000)
9. Grossi, D., Roy, O., Huang, H. (eds.): LORI. LNCS, vol. 8196. Springer, Heidelberg (2013)
10. Halpern, J.Y., Moses, Y.: A guide to completeness and complexity for modal logics of knowledge and belief. Artif. Intell. 54(2), 319–379 (1992)
11. Halpern, J.Y., Rêgo, L.C.: Characterizing the np-pspace gap in the satisfiability problem for modal logic. J. Log. Comput. 17(4), 795–806 (2007)
12. Hintikka, J.: Knowledge and Belief. Cornell University Press, Ithaca (1962)
13. Holliday, W.H., Hoshi, T., Icard III, T.F.: Schematic validity in dynamic epistemic logic: Decidability. In: van Ditmarsch, H., Lang, J., Ju, S. (eds.) LORI 2011. LNCS, vol. 6953, pp. 87–96. Springer, Heidelberg (2011)
14. Holliday, W.H., Icard, T.F.: Moorean phenomena in epistemic logic. In: Beklemishev, L., Goranko, V., Shehtman, V. (eds.) Advances in Modal Logic, pp. 178–199. College Publications (2010)
15. Indrzejczak, A.: Labelled tableau calculi for weak modal logics. Bulletin of the Section of Logic 36(3/4), 159–171 (2007)
16. Kripke, S.A.: Semantical analysis of modal logic II. non-normal modal propositional calculi. In: Symposium on the Theory of Models. North-Holland Publ. Co., Amsterdam (1965)
17. Ma, M., Sano, K.: How to update neighborhood models. In Grossi et al. [9], pp. 204–217
18. Montague, R.: Universal grammar. Theoria 36(3), 373–398 (1970)
19. Pacuit, E.: Neighborhood Semantics for Modal Logic. An Introduction, Lecture notes for the ESSLLI course A Course on Neighborhood Structures for Modal Logic (2007)
20. Papadimitriou, C.H.: Computational complexity. Addison-Wesley (1994)
21. Plaza, J.: Logics of public communications. Synthese 158(2), 165–179 (2007)
22. Schwarzentruber, F.: Lotrecscheme. Electr. Notes Theor. Comput. Sci. 278, 187–199 (2011)
23. Scott, D.: Advice in modal logic. In: Lambert, K. (ed.) Philosophical Problems in Logic, pp. 143–173. Reidel, Dordrecht (1970)
24. van Benthem, J.: Open problems in logical dynamics. In: Gabbay, D., Goncharov, S.S., Zakharyaschev, M. (eds.) Mathematical Problems from Applied Logic I. International Mathematical Series, vol. 4, pp. 137–192. Springer, New York (2006)
25. van Benthem, J.: Logical Dynamics of Information and Interaction. Cambridge University Press (2011)
26. van Benthem, J., Pacuit, E.: Dynamic logics of evidence-based beliefs. Studia Logica 99(1), 61–92 (2011)
27. van Ditmarsch, H., Kooi, B., van der Hoek, W.: Dynamic Epistemic Logic. Springer (2007)
28. van Ditmarsch, H., Kooi, B.P.: The secret of my success. Synthese 151(2), 201–232 (2006)
29. Vardi, M.Y.: On the complexity of epistemic reasoning. In: Proceedings of the Fourth Annual IEEE, Symposium on Logic in Computer Science (LICS 1989), pp. 243–252 (1989)
30. Velázquez-Quesada, F.R.: Explicit and implicit knowledge in neighbourhood models. In: Grossi et al. [9], pp. 239–252
31. Zvesper, J.A.: Playing with Information. PhD thesis, Institute for Logic, Language and Computation, Universiteit van Amsterdam (2010)

A Appendix

A.1 Proof of Lemma 1

Proof. We work only with ($\Box$). Assume Θ is faithful to $\mathcal{M}$; applying ($\Box$) to $(\sigma : \Box\varphi)$ and $(\sigma : \neg\Box\psi)$ in Θ yields $\Theta' := \Theta \cup \{(\sigma_{new} : \varphi), (\sigma_{new} : \neg\psi)\}$. Since $\{(\sigma : \Box\varphi), (\sigma : \neg\Box\psi)\} \subseteq \Theta$, the assumption implies both $\mathcal{M}, f(\sigma) \models \Box\varphi$ and $\mathcal{M}, f(\sigma) \models \neg\Box\psi$; then $f(\sigma) \notin [\![\Box\varphi \to \Box\psi]\!]_{\mathcal{M}} \neq W$ and hence $[\![\varphi \to \psi]\!]_{\mathcal{M}} \neq W$, so there is $v \in W$ s.t. $v \in [\![\varphi]\!]_{\mathcal{M}}$ and $v \notin [\![\psi]\!]_{\mathcal{M}}$. Now, since Θ is faithful to $\mathcal{M}$, there is f s.t. $\mathcal{M}, f(\sigma) \models \gamma$ for all $(\sigma : \gamma) \in \Theta$. The function $f' : \text{Prefix}(\Theta') \to W$, extending f by defining $f'(\sigma_{new}) := v$ (and thus yielding $\mathcal{M}, f'(\sigma_{new}) \models \varphi$, $\mathcal{M}, f'(\sigma_{new}) \models \neg\psi$), is a witness showing that Θ' is faithful to $\mathcal{M}$.

A.2 Proof of Lemma 2

Proof. Both (i) and (ii) are proved by simultaneous induction on φ. We only check the cases where φ is atomic and of the form $\Box\psi$. First, if φ is an atom p, (i) is immediate from the definition of V^{Θ}. For (ii), assume $(\sigma : \neg p) \in \Theta$; since Θ is open, $(\sigma : p) \notin \Theta$, and hence it follows from V^{Θ}'s definition that $\mathcal{M}^{\Theta}, \sigma \not\models p$.

Second, suppose φ is $\Box\psi$. For (i), assume $(\sigma : \Box\psi) \in \Theta$. In order to show $[\![\psi]\!]_{\mathcal{M}^{\Theta}} \in \tau^{\Theta}(\sigma)$, our candidate for a witness of $[\![\psi]\!]_{\mathcal{M}^{\Theta}} \in \tau^{\Theta}(\sigma)$ is, of course, ψ. Thus, it suffices to show that $\{\sigma' \in W^{\Theta} \mid (\sigma' : \psi) \in \Theta\} \subseteq [\![\psi]\!]_{\mathcal{M}^{\Theta}}$, so suppose $(\sigma' : \psi) \in \Theta$; by induction hypothesis, we obtain $\sigma' \in [\![\psi]\!]_{\mathcal{M}^{\Theta}}$.

For (ii), suppose $(\sigma : \neg\Box\psi) \in \Theta$; we show that $[\![\psi]\!]_{\mathcal{M}^{\Theta}} \notin \tau^{\Theta}(\sigma)$, i.e., for all formulas γ, $(\sigma : \Box\gamma) \in \Theta$ implies $\{\sigma' \in W^{\Theta} \mid (\sigma' : \gamma) \in \Theta\} \nsubseteq [\![\psi]\!]_{\mathcal{M}^{\Theta}}$. So take any γ such that $(\sigma : \Box\gamma) \in \Theta$. Since Θ is saturated, it follows from the rule ($\Box$) that there is a prefix $\sigma_{new} \in W^{\Theta}$ such that $(\sigma_{new} : \gamma), (\sigma_{new} : \neg\psi) \in \Theta$. Then $\sigma_{new} \in \{\sigma' \in W^{\Theta} \mid (\sigma' : \gamma) \in \Theta\}$ but, by induction hypothesis, $\sigma_{new} \notin [\![\psi]\!]_{\mathcal{M}^{\Theta}}$.

A.3 Proof of Lemma 3

Here we provide arguments for the remaining cases in the proof of Lemma 3.

($\downarrow \epsilon$): Since $(\sigma :_L p) \in \Theta$, we obtain $\mathcal{M}^{\cap\rho_1;\cdots;\cap\rho_n}, f(\sigma) \models p$ so $\mathcal{M}, f(\sigma) \models p$. Hence, $\Theta \cup \{(\sigma :_\epsilon p)\}$ is faithful to $\mathcal{M}$.

([·]): Since $(\sigma :_L [\varphi]\psi) \in \Theta$, we obtain $\mathcal{M}^{\cap\rho_1;\cdots;\cap\rho_n}, f(\sigma) \models [\varphi]\psi$. Thus, either $\mathcal{M}^{\cap\rho_1;\cdots;\cap\rho_n}, f(\sigma) \not\models \varphi$ or else $\mathcal{M}^{\cap\rho_1;\cdots;\cap\rho_n;\cap\varphi}, f(\sigma) \models \psi$, so either $\Theta \cup \{(\sigma :_L \neg\varphi)\}$ or else $\Theta \cup \{(\sigma :_{L;\varphi} \psi)\}$ is faithful to $\mathcal{M}$.

($\times$*Back*): Since $(\sigma :_{L;\varphi} \times) \in \Theta$, $f(\sigma)$ is not in the domain of $\mathcal{M}^{\cap\rho_1;\cdots;\cap\rho_n;\cap\varphi}$. If $f(\sigma)$ is in the domain of $\mathcal{M}^{\cap\rho_1;\cdots;\cap\rho_n}$, then φ fails at $f(\sigma)$ in $\mathcal{M}^{\cap\rho_1;\cdots;\cap\rho_n}$, so $\Theta \cup \{(\sigma :_L \neg\varphi)\}$ is faithful to $\mathcal{M}$. Otherwise, $f(\sigma)$ is not in the domain of $\mathcal{M}^{\cap\rho_1;\cdots;\cap\rho_n}$, so $\Theta \cup \{(\sigma :_L \times)\}$ is faithful to $\mathcal{M}$.

(*Back*): Since $(\sigma :_{L;\varphi} \psi) \in \Theta$, we obtain $\mathcal{M}^{\cap\rho_1;\cdots;\cap\rho_n;\cap\varphi}, f(\sigma) \models \psi$, which implies $\mathcal{M}^{\cap\rho_1;\cdots;\cap\rho_n}, f(\sigma) \models \varphi$. Hence, $\Theta \cup \{(\sigma :_L \varphi)\}$ is faithful to $\mathcal{M}$.

A.4 Remaining Proof of Lemma 5

Here we show case (iii) fully and the cases for φ of the form p, $[\psi]\gamma$ of Lemma 5.

First consider the case (iii). If L is empty, the statement of (iii) becomes vacuously true since Θ is open. Otherwise, $L \equiv \rho_1; \cdots; \rho_n$, and the saturatedness of Θ and the rule $(\times Back)$ imply either $(\sigma :_{\rho_1;\cdots;\rho_{n-1}} \neg\rho_n) \in \Theta$ or else $(\sigma :_{\rho_1;\cdots;\rho_{n-1}} \times) \in \Theta$. By induction hypothesis, either $(\mathcal{M}^{\Theta})^{\cap\rho_1;\cdots\cap\rho_{n-1}}, \sigma \not\models \rho_n$ or else σ is not in $(\mathcal{M}^{\Theta})^{\cap\rho_1;\cdots\cap\rho_{n-1}}$. In both cases, σ is not in $(\mathcal{M}^{\Theta})^{\cap\rho_1;\cdots\cap\rho_n}$.

Second, let $\varphi \equiv p$. For (i), suppose $(\sigma :_L p) \in \Theta$; since Θ is saturated, rule $(\downarrow \epsilon)$ implies $(\sigma :_\epsilon p) \in \Theta$ so, by definition, $\sigma \in V^{\Theta}(p)$. Moreover, rule $(Back)$ and induction hypothesis imply that σ is in $(\mathcal{M}^{\Theta})^{\cap\rho_1;\cdots\cap\rho_n}$; hence, $(\mathcal{M}^{\Theta})^{\cap\rho_1;\cdots\cap\rho_n}, \sigma \models p$. For (ii), use a similar argument now with $(\downarrow \epsilon\neg)$ and $(Back)$.

Third, let $\varphi \equiv [\psi]\gamma$. For (i), suppose $(\sigma :_{\rho_1;\cdots;\rho_n} [\psi]\gamma) \in \Theta$ and, further, that $(\mathcal{M}^{\Theta})^{\cap\rho_1;\cdots\cap\rho_n}, \sigma \models \psi$; we show $(\mathcal{M}^{\Theta})^{\cap\rho_1;\cdots\cap\rho_n;\cap\psi}, \sigma \models \gamma$. Since Θ is saturated, rules $([\cdot])$ and (Back) imply either $(\sigma :_L \neg\psi) \in \Theta$ or else both $(\sigma :_L \psi) \in \Theta$ and $(\sigma :_{L;\psi} \gamma) \in \Theta$. But from assumption and induction hypothesis, $(\sigma :_L \neg\psi) \notin \Theta$ and thus $(\sigma :_L \psi) \in \Theta$ and $(\sigma :_{L;\psi} \gamma) \in \Theta$. Then, again by induction hypothesis, $(\mathcal{M}^{\Theta})^{\cap\rho_1;\cdots\cap\rho_n;\cap\psi}, \sigma \models \gamma$. For (ii), suppose $(\sigma :_{\rho_1;\cdots;\rho_n} \neg[\psi]\gamma) \in \Theta$. Since Θ is saturated, rule $(\neg[\cdot])$ implies both $(\sigma :_L \psi) \in \Theta$ and $(\sigma :_{L;\psi} \neg\gamma) \in \Theta$. By induction hypothesis, both $(\mathcal{M}^{\Theta})^{\cap\rho_1;\cdots\cap\rho_n}, \sigma \models \psi$ and $(\mathcal{M}^{\Theta})^{\cap\rho_1;\cdots\cap\rho_n;\cap\psi}, \sigma \not\models \gamma$ so $(\mathcal{M}^{\Theta})^{\cap\rho_1;\cdots\cap\rho_n}, \sigma \not\models [\psi]\gamma$.

A.5 Execution of the Tableau Method

When we run Lotrecscheme with the tableau method for intersection semantics for the formula $(p \rightarrow \Box[p]q) \wedge \neg[p]\Box q$ we obtain the following closed branch at some point:

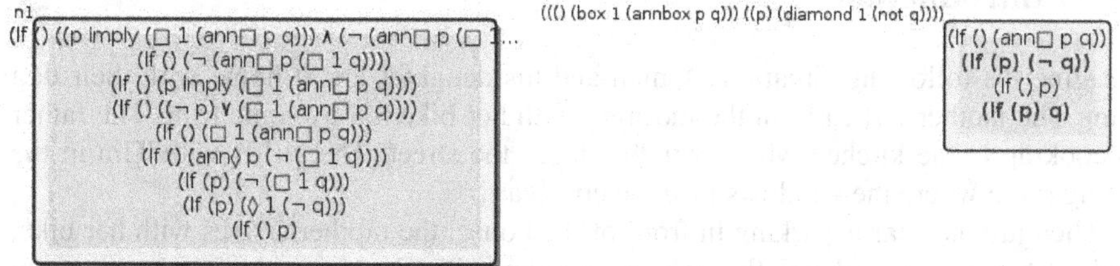

The branch contains two world symbols (that are the two nodes above). As the node `n1` contains `If () p` means that the term $(n1 :_\epsilon p)$ is in the current branch.

A Pragmatistic Approach to Propositional Knowledge Based on the Successful Behavior of Belief

Aránzazu San Ginés[1] and Rohit Parikh[2]

[1] IFS-CSIC and Granada University, Spain
[2] City University of New York, New York, USA
aransangines@gmail.com

Abstract. Every belief has a life that goes from the agent having the belief now, the transmission of the belief to other agents, and the persistence of the belief through time. In this article we propose the idea that the belief can be said to be successful in relation to any of these respects. We will call them, respectively, the first, second, and third person perspective on knowledge and investigate the requisite properties of these three perspectives.

We do not base our approach on the notion of truth as is common, or on the notion of justification, which is another basis. Our concern is not with knowledge as corresponding to truth but knowledge as corresponding to stable belief.

Keywords: Propositional Knowledge, Successful belief, Pragmatism.

1 Introduction

Imagine the following situation. A man and his daughter are at home with their dog Tim. The mother left early in the morning with her bike to buy some fruit. The father is cooking in the kitchen whose window faces the street. The girl is with Tim in the living room where the windows face the courtyard.

Then just as a car is parking in front of the house, the mother arrives with her bike, and waves to her husband through the window. She heads to the garage (a room separated from the house) to leave the bike there. From the car alights Mike, a friend of the family who is going to have lunch with them.

The daughter comes to the kitchen. - *Look dad*, she says, *Tim is excited! He doesn't stop wagging his tail! - Ha ha ha*, the dad answers, *he knows that your mother has arrived.- Oh, is that right? Where is she?- In the garage, setting the bike.* Then, someone knocks. - *Open the door*, says the father, *it's Mike*. The girl opens the door with Tim at her side, and Mike walks in. The dog at once stops wagging his tail, and goes to lie down in the living room apparently disappointed. - *What is it dad? What is it with Tim now? - Mmmm, it seems that he actually didn't know before, that your mother had arrived.*

In the previous example we can appreciate two interesting and revealing uses of the verb 'to know'. In the first one, the father says that the dog "knows ..." because he observes that Tim *acts* in the way he would expect the dog to do if Tim were

M. Banerjee and S.N. Krishna (eds.): ICLA 2015, LNCS 8923, pp. 146–157, 2014.

genuinely convinced of the (true) belief (ascribed to the dog) that the mother has arrived.[1] Tim's belief is successful for the father in this sense, so he says that the dog knows. The second time, however, he says that Tim "actually didn't know ...". The father is now looking at the situation in a diachronic way. If the dog had actually known, Mike's entry would not have disappointed him as it apparently did. The father sees in the reaction of the dog proof that the new datum, namely Mike entering the house, undermines Tim's belief that the mother has arrived. From the father's perspective the belief fails to remain strong through time, thus being the belief unsuccessful for the father in this new sense.

(Parikh, 1994, p.530) proposes a criterion of knowledge which emphasizes the idea of successful behavior. He gives the example of a mouse in a maze which is given a choice between two boxes, one of which contains cheese, but not the other. He says:

> Suppose now that the mouse invariably goes to that box which contains the cheese. We would then say, "he somehow knows where the cheese is", and would look for some evidence that might have given him a clue. Perhaps we would find such a clue, but our judgment that he knew where the cheese was would not depend on such a clue, but rather on our perception of successful behaviour on the part of the mouse which we were unable to otherwise understand.

The idea exemplified above contrasts, as Parikh pointed out in his paper (pp. 529-530), with the usual way to think of knowledge in terms of Justified True Belief + Gettier Blocker (Meeker, 2004, p.157), where the Gettier Blocker is a clause devised to prevent a justified true belief to be considered knowledge when the belief is only accidentally true.

In our present paper, we modify Parikh's idea of successful behavior, and try to develop it to provide a more general understanding of what is usually called 'to know that ...'. In order to do this we will distinguish among three possible respects in which a belief can be said to be successful at a time t, that depends upon the relative position of the agent who evaluates the belief. We will distinguish among

1) an agent who evaluates someone's belief according to internal conditions at a particular time (*first person perspective*),
2) an agent who evaluates someone's belief according to external conditions at a particular time (*second person perspective*),
3) and finally an agent who evaluates someone's belief according to external conditions over an interval of time (*third person perspective*).

Our criterion to determine whether someone knows something or not at t will depend on the perspective we take, and thus on the particular context. If the answer is positive, if we claim at a time t that someone knows that φ (for the belief has proved at t to be successful from the perspective we are taking), this claim may still be

[1] Observe the difference between 'being genuinely convinced of φ' and 'fully believe that φ', where φ is a belief. A person who fully believes that φ at a time t is allowed a change of mental state at a time $t'>t$. On the other hand, we are using 'being genuinely convinced of φ at t' to express that a person fully believes that φ at t, and nothing can change this mental state.

corrected if the belief proves at some point to be unsuccessful at t from any of the other perspectives (e.g. the dog's example: Tim didn't *actually* know). In the same line, the corpus of knowledge of an individual B at a time t consists of all the beliefs that having been seen to be successful from the relative perspective taken by B at different times $t' \leq t$, have not revealed themselves yet for B, up to time t, as unsuccessful from any other perspective.[2]

Our view is a pragmatistic[3] one. Observe that, as with Levi (1980, p.1), for us knowledge is not a question of pedigree. When A says that B knows that φ, A is not making a reference to the quality of those premises and derivations that are supposed to be the basis of the acceptance of φ as knowledge. In fact, most of the time A will even admit that he has not the faintest idea of the story of φ, that is, of the reasons why φ seems to count now as a belief for B.

We will introduce in the following pages, the first, second, and third person perspectives to propositional knowledge, and by doing that we will refer to some of the examples that have flooded, and continue to flood, the literature on propositional knowledge of the last decades. In particular, our approach will fit in nicely with difficult cases such as Harman's or Stanley's bank examples.

2 Three Points of View to Propositional Knowledge

2.1 The First Person Perspective

Let us begin with a look at one particular situation. In a playground two teachers observe a group of ten boys and one girl running towards the back of the school main building. After the children have disappeared behind the building, the teachers hear someone falling down and crying. They have then the following conversation:

(A) Oh, I believe that one of the boys has fallen down.
(B) How could you possibly know that?
(A) Well, I don't *know* it but it seems pretty likely, don't you think?

In this example (A) believes at a time t_0 that one of the boys has fallen down. However he also admits to not knowing it. The belief *fails* to be identified as knowledge at t_0 by the person having it although, as (A) points out to (B), from (A)'s perspective (which he assumes to be a common perspective) the belief seems also justified at t_0. Observe as well that even if one of the boys had actually fallen down, and (A) discovered it afterwards at another time t_1, he still would not say at t_1 that he knew at t_0 that a boy had fallen down. That is, (A) may have had a justified true belief that φ at t_0 that he confirms at t_1, and still not say at t_1 that he knew at t_0 that φ. The

2 It could be the case that a belief that was unsuccessful for an individual from one perspective at a time t, was successful for the same individual and from the same perspective at $t'>t$, and vice versa. When that happens we are only considering the most recent evaluation of the belief.

3 Observe that we are following Peirce's own term (pragmaticism) which he introduced to distinguish himself from that of William James.

questions we want to answer along this section are the following: When would a person having a belief say that the belief is justified? When would this same person say that he knows it?

In order to answer these questions, we will introduce four definitions, and before that some nomenclature.

Nomenclature. Let A be an individual, L the smallest language (containing a complete set of Boolean connectives) in which all the beliefs that A can have are expressible, and $\bar{L}$ the set of formulae (formulated over L) of some logic. We suppose (with Grove, 1988, p.157) that $\bar{L}$ contains the propositional tautologies and that it is closed under modus ponens. $\bar{L}$ must also be compact. Γ is the belief-system of A at a particular time t. We do not assume that Γ is logically closed, i.e., a theory. Rather Γ is a consistent belief base. $\emptyset$ is the empty set.

Now imagine that we have a belief that is not tautological, φ, and we want to evaluate its naturalness at a time t against those others of our beliefs, γ, that are related to φ, and are salient at the moment of the evaluation, salient in the sense that we are particularly conscious of them at t. For instance, imagine that φ is about trains, then intuitively γ would contain all of my relevant beliefs about trains, but not my belief about the beautiful hair of my neighbor. Definition 1. assumes that we already have some such notion of salience.

Definition 1. (Acceptability) Let $\gamma \subseteq \Gamma$ be the salient set of beliefs related to φ that does not include φ. The belief φ is *acceptable* for A at t if Con(γ, φ), that is γ and φ do not imply a contradiction, and one of the following cases holds:

1.1. $\gamma = \emptyset$;

1.2. $\gamma \neq \emptyset$ and $\gamma \vDash \varphi$. We say here that $\emptyset$ is a *witness* of the acceptability of φ relative to γ;

1.3. $\gamma \neq \emptyset, \gamma \nvDash \varphi$, and $\exists\{p_1, ..., p_n\}$ non-empty, finite set of potential beliefs (not necessarily in Γ) expressed as literals in L, such that $\gamma \cup \{p_1, ..., p_n\} \nvDash \bot$, $\{p_1, ..., p_n\} \nvDash \varphi$, and $\gamma \cup \{p_1, ..., p_n\} \vDash \varphi$. As before, we call $\{p_1, ..., p_n\}$ a *witness* of the acceptability of φ relative to γ.

Observe that if φ is acceptable then $\neg\varphi$ could also be acceptable, as acceptability is a rather weak notion. We will clarify this notion in Definition 4. when we would introduce the notion of justification which, as we will see, is stronger than the notion of acceptability.

In the case thus of Con(γ, φ) and $\gamma \neq \emptyset$, we say that we have an *acceptable* belief, φ, if we are able to semantically infer φ either from γ alone, or from γ together with a set of *simple assumptions* satisfying certain properties (specified above).[4] But, how much effort, or otherwise said, how much *credulity* do we need in order to accept, along with γ, these *simple assumptions*? Definition 2. answers that question in terms of a comparison of *credulities*. The higher the value of the utility $\nu(\cdot,\cdot)$, the more credulity we need. (Note that we are not defining the credulities or the Grove spheres we make use of below, but assuming that these are provided in some plausible manner and are making use of them.)

Before introducing Definition 2. we need some preliminaries. It is well known in the field of belief revision that some beliefs are more plausible than others. The belief, for example, that 'John is at the beach' is more plausible than the belief that 'John is in Antarctica' given my current belief-system. We need then to take this fact into account. In this sense, we will use in the definition a system of spheres, *S*, of the kind proposed by Grove (1988), that is, brought to the context of the present article, a collection of sets of possible worlds (maximal consistent subsets of *L*) such that (see Grove, 1988, pp. 158-159):

1. *S* is totally ordered by inclusion. The elements of *S* are called spheres.
2. The set of possible worlds that contain Γ as a subset (models of Γ), $|\Gamma|$, is the minimum of *S*. The set of all possible worlds is the largest sphere, *M*.
3. If X is a finite, consistent set of literals in *L*, there is a smallest sphere in *S* containing a model of X.

Observation. Let *S* be our assumed system of Grove spheres. X a finite, consistent set of literals in *L*. Then there is a smallest sphere in *S* containing $|X|$. We will call this minimum e(X).

Definition 2. Let be $\gamma, \gamma', \delta \subseteq \Gamma$ not empty. Let also be X and Y any finite set of potential beliefs expressed as literals in *L*, and *S* a system of Grove spheres. We assume an ordinal utility $\nu(\cdot,\cdot)$ with the following properties:

a) $\nu(\emptyset, \delta) \leq \nu(X, \gamma)$ (thus $\nu(\emptyset, \delta)$ is a lowest possible value of ν - adding nothing, $\emptyset$, to a set of beliefs does not require any credulity)

b) If $\gamma \models p\ \forall p \in X$, then $\nu(X, \gamma) = \nu(\emptyset, \delta)$

c) If $\neg Con(\gamma, X)$, then $\nu(Y, \gamma') \leq \nu(X, \gamma)$ (this yields a highest possible value of ν - adding new beliefs inconsistent with our previous beliefs does require a great credulity)

d) If $X, Y \neq \emptyset$, $Con(\gamma, X)$, $Con(\gamma', Y)$ and $e(|\gamma| \cap |X|) = e(|\gamma'| \cap |Y|)$, then

$$\nu(X, \gamma) \leq \nu(Y, \gamma') \text{ iff } \#\overline{X} \leq \#\overline{Y},$$

where # refers to the cardinal of the set next to it, and $\overline{X}$ and $\overline{Y}$ are respectively the subset of elements of X and Y such that $\gamma \not\models p\ \forall p \in \overline{X}$, and $\gamma' \not\models q\ \forall q \in \overline{Y}$.

e) If $X, Y \neq \emptyset$, $Con(\gamma, X)$, $Con(\gamma', Y)$ and $e(|\gamma'| \cap |Y|) \subset e(|\gamma| \cap |X|)$, then

$$\nu(X, \gamma) < \nu(Y, \gamma')$$

[4] Definition 1. owes inspiration to Aliseda-Llera (1997). Rohit Parikh has a slightly different version of Definition 2. The two versions will be reconciled in the journal version of the paper.

We use in e) the fact that the more the conditions we add, the smaller the set of possible worlds satisfying the conditions is. Let us explain how we use it. Imagine that I add to my set of beliefs 'John is in Antarctica', then together with it other items will be also added, items that somehow were implied by the previous one (this is the case, for example, of 'John must have access to this kind of expeditions', 'John is brave', and so on). On the other hand, if I rather decide to add to my set of beliefs 'John is at the beach' the additional set of items that I will also need to include will be smaller (observe for instance, that I already believed before that 'John has access to this kind of expeditions to the beach', and that John does not really need to be brave to go to the beach). It seems then that the less plausible a belief is, the more conditions it adds, and thus the smaller the set of possible worlds will be, and this is what we are using here. So if X = {'John is at the beach'} and Y = {'John is in Antarctica'}, what we should expect to get is that $e(|\gamma| \cap |Y|) \subset e(|\gamma| \cap |X|)$, where γ is my salient set of beliefs at the moment.

$\nu(\cdot,\cdot)$ is of special importance in order to introduce, in Definition 3., our so called *deficit of a belief* φ *given the salient set of beliefs* γ *related to it*, which in its turn, Definition 4., will play an important role to give an answer to the first of the questions to which we devoted this section: When would a person having a belief say that the belief is justified?

Definition 3. Let be $\varphi, \psi \in \Gamma, \gamma, \gamma' \subseteq \Gamma$, not-empty salient sets of beliefs related to but not containing, respectively, φ and ψ, and S a system of Grove spheres. We call $d(\varphi, \gamma)$ the *deficit of* φ *given* γ. We establish the following relation between deficits:

a) If φ and ψ are *acceptable* beliefs, then:

$$d(\varphi, \gamma) \leq d(\psi, \gamma') \text{ iff } \exists \mathrm{X}\ \forall \mathrm{Y} \text{ s.t. } \nu(\mathrm{X}, \gamma) \leq \nu(\mathrm{Y}, \gamma'), \text{ where}$$

X is a *witness* of the acceptability of φ relative to γ, and Y is a witness of the acceptability of ψ relative to γ'.

b) If φ is a not an *acceptable* belief, then:

$$d(\psi, \gamma') \leq d(\varphi, \gamma)$$

Definition 4. Let be $\varphi \in \Gamma, \gamma \subseteq \Gamma$ the salient set of beliefs related to φ which does not include φ, and S a system of Grove spheres. $\gamma \neq \emptyset$. φ is *justified* for A at t if φ is *acceptable* for A at t, and $d(\varphi, \gamma) < d(\neg\varphi, \gamma)$. We will say at t that 'φ is justified', if φ is justified for us at t.

Intuitively, the deficit of φ is the amount of credulity one needs to infer φ from one's salient set of beliefs related to φ, γ. We will come back again to this idea below when we discuss the third person perspective on knowledge. For now, let us just see how this definition works given the example we introduced at the beginning of the section. In it, φ is the belief that 'one of the boys has fallen down'. Let γ be the set {'ten boys (b_1, ..., b_{10}) and one girl were running', 'now someone is crying', 'one of the children fell down'}, and let S be a system of Grove spheres. Observe that given this γ, φ and $\neg\varphi$ are acceptable, since $\gamma \cup X \models \varphi$ and $\gamma \cup Y \models \neg\varphi$, where X = {'the girl did not fall down'} and Y = {'b_1 did not fall down', ..., 'b_{10} did not fall down'}. Observe also that given the similarity of the items in X and Y, either $e(|\gamma| \cap |Y|) \subset e(|\gamma| \cap |X|)$ or $e(|\gamma| \cap |Y|) = e(|\gamma| \cap |X|)$, so we have that in any case $d(\varphi, \gamma) < d(\neg\varphi, \gamma)$. φ is hence justified for (A) at the time it is uttered.

Finally, the belief that φ of an agent *A* at a time *t* is knowledge at *t* for *A* from the first person perspective if, and only if, φ is *acceptable* for *A* at *t*, and $\neg\varphi$ is not acceptable for *A* at *t*, where the salient set of beliefs related to $\neg\varphi$ is assumed to be that of φ. Observe that when the evaluator, *B*, and the person having the belief, *A*, are different, the mechanism is very similar. In this case, *A* ascribes to *B* a belief-system, a salient set of beliefs and so on, and carries on the calculations there.

The idea is that in order for the belief that φ to be knowledge for *A* from the point of view of the first person perspective, φ (but not $\neg\varphi$) should *naturally* follow from those salient elements of the current belief-system that are connected to φ. We have opted for an abductive approach (cf. Aliseda-Llera, 1997) because of the importance that we claim for the creative component. When I have a belief, in order for it to be acceptable for me, I need to be able to naturally imagine a context out of the constraints given by my current salient beliefs, from which φ can be semantically inferred. The acceptable belief will be knowledge for me from the first person perspective if I cannot naturally imagine, out of the same constraints, a context in which its negation ($\neg\varphi$) can also be semantically inferred. In the previous example, for instance, (A) believes that a boy has fallen down. It is perfectly possible for (A) though, to imagine also a context in which the girl would have been the one to fall down. The two scenarios are naturally possible, this being the reason why (A) finds it hard to say that he knows φ. He is considering the belief from the first person perspective.

2.2 The Second Person Perspective

Suppose that *A* has the belief that φ, and that *A* shares it with *B* at a time *t*. *A*'s belief will be successful for *B* from the second person perspective if *B* is convinced at *t* that *A* will act in agreement with φ. Imagine, for instance, the following situation. We are in Spain. Today is the morning of December 22nd, the day of the Christmas lottery. Most people have the radio on, hoping for their numbers to be winners. (A) and (B) are respectively a woman and her boyfriend who are having a coffee in a cafeteria. (A) receives a call. After she hangs up the following dialog takes place:

(A) We have won the lottery!!
(B) What? How do you know that?
(A) It was my father. He seemed quite excited. He said he has good news. He wants us to be at home in half an hour to tell us the news and celebrate. You see? I know it! We have won the lottery!

(B) Really? The lottery? Would you call your boss right now and tell her that you resign?
(A) Mmmmm. OK, OK, you're right. I don't know for certain ...

(B) challenges the belief. He needs to check that (A) has no doubts about it. In this line, if (A) had answered the challenge taking it, (B) would have accepted and even said that '(A) really thinks that she knows that her father has won the lottery'. (B) however could still remain skeptical about the belief, not seeing it as certain yet. In fact it is quite possible that (B) would have prevented (A) from calling her boss. While he would reckon that she *must* know it, he would not endorse the belief. Observe that in order for (B) to say that '(A) knows that we have won the lottery', (A)'s belief not only should be successful for (B) from the second person perspective, but it also should be part of the corpus of knowledge of (B).

In the example, however, what actually happens is that (A) refuses to resign, which makes (A)'s belief *unsuccessful* for (B) from the second person perspective. One important thing to observe here is the magnitude of the challenge. We claim that this magnitude will depend on the specific situation in which the belief is uttered, and on how coherent the belief is with (B)'s salient set of beliefs. The less the coherence, the higher the challenge. In our example, winning the lottery seems highly implausible to (B), hence the difficult challenge he throws to his girlfriend.

Another issue should be discussed in relation with the lottery example. Observe that when the evaluator of the belief coincides with the person having the belief, the first and the second person perspectives get highly connected. In our example, when (A) says that she knows that "we have won the lottery", her salient set of beliefs, γ_1, related to the belief under evaluation would be something similar to the following {'today's the day of the lottery', 'Dad has good news', 'Dad plays lottery every year', ('good news' and 'day of the lottery') → 'we won the lottery'}. Seconds later, however, when the challenge is thrown and the reconsideration of the belief begins, the salient set of belief changes from the original γ_1 to γ_2, where γ_2 is closer to the following set of beliefs {'today's the day of the lottery', 'today's a normal day', 'Dad has good news', 'Dad plays lottery every year', ('good news' and 'day of the lottery' and 'normal day') → ('we won the lottery' or 'Dad got a new job' or ...)}. That is, the challenge leads (A) to reconsider the items in γ_1 in order to increase her degree of reliance on the considered belief. What we observe, nevertheless, is that she fails, she cannot increase her degree of reliance on the belief since given the new set of salient beliefs γ_2, the belief that "we have won the lottery" is no longer successful from the first person perspective. She realizes then that she cannot rely on her belief enough to act in agreement with it. Her belief is neither successful from the second person perspective.

An interesting question then is: What does the level of reconsideration of the belief, and thus of the salient set of beliefs, depends on? The answer is clear: it depends on the needed degree of reliance on the belief, and so on the magnitude of the challenge. Let us see that through the example, proposed by Stanley (2005, pp. 3-4) of the 'bank cases'. This time the challenge is to wait until Saturday to deposit a certain amount of money into the account, instead of doing it today. In the *low stakes* case, the couple has no pending payments to make. In the *high stakes* case, an important payment is due and there will be a heavy penalty if the payment is not made. If we express the situation in a game-theoretical way, what we have is something similar to this:

Low stakes case

	Deposit today	Try to deposit on Saturday
Bank open on Saturday	5	10
Bank not open on Saturday	5	5

High stakes case

	Deposit today	Try to deposit on Saturday
Bank open on Saturday	5	10
Bank not open on Saturday	5	- 90

Let α be the action 'Try to deposit on Saturday', and $0 \leq p \leq 1$ the probability that the bank were open on Saturday. Consider also the expected value E[α] of α.

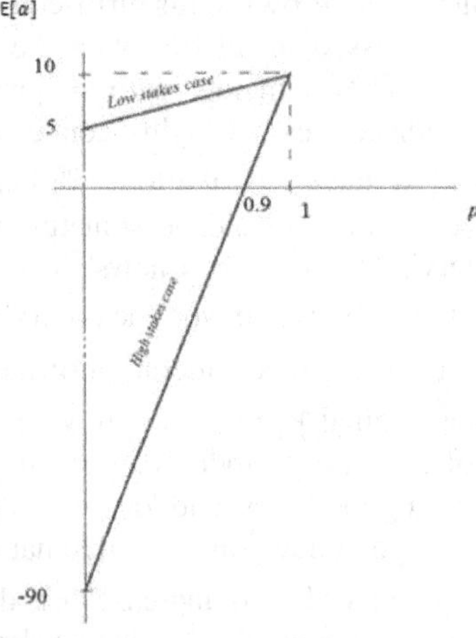

Fig. 1.

In the *low stakes* case E[α] (the red line) is always positive (Fig. 1.). The challenge is not hard to take. As a consequence, no revision needs to be made and γ does not change. Hannah's current degree of reliance on the belief is already enough and Hannah says that she knows that the bank is open the following day. In the *high stakes* case, however, in order for E[α] to be positive the probability, *p*, of being open on Saturday must be higher than 0.9 (see the blue line). But is Hannah sure enough to consider *p* to be higher than 0.9? The challenge is harder to take. The revision needs to take place and with it the original γ will have to be reconsidered before Hannah may say again that she knows that 'the bank is open the following day *for sure*'.

To sum up, the belief that φ of an agent A at a time t is knowledge at t for some agent B from the second person perspective if, and only if, B is convinced at t that A will behave according to φ.

2.3 The Third Person Perspective

The third person perspective is that of an observer who draws attention to the dynamic of an individual A's belief-system (i.e. what the salient set of beliefs related to a belief are at any time) given the evolution of the situation in which the particular belief is elicited. In particular, we have in mind two sort of contexts: one in which the observer has actually witnessed the evolution of the situation and the reactions of A (witness-case), and another in which the observer predicts what is going to happen (prediction-case). We say that A's belief that φ shows a successful behavior for B from the third person perspective at a time t if A's belief that φ is successful for B at t from the first person perspective (given the idea held by B of A's belief-system), and the evolution (witness-case) or *natural* evolution (prediction-case) of the situation does not present a challenge to it, that is, the situation evolves in a way in which none of the new items added to the set of salient beliefs undermines A's belief that φ.

Let us introduce an example in order to clarify our proposal:

A is a teacher in a school. He witnesses the following situation in the playground: At the corner of the main building is student *S1*. Running towards him is *S2*. After a couple of minutes *S1* and *S2* have moved out from the sight of the teacher who then hears someone crying. The teacher believes that '*S1* is crying' (φ).

Suppose that the salient set of beliefs of A related to φ at the time the teacher sees the scene t_0 is: γ= {'*S2* was running towards *S1*', 'Someone is crying', '*S2* hates *S1*', '*S2* is a violent boy', '*S1* has reported *S2*'s bad behavior', 'if *S2* heard about the report, he would hit *S1*', 'if *S1* were hit, he would cry'}. In this case, on the one hand $\gamma \cup${'*S2* has heard about the report'} $\models \varphi$, and hence φ is *acceptable* (see section 2.1), and on the other hand, $\neg\varphi$ is not *acceptable* for A at the same time t_0 given γ. Thus the belief φ is knowledge for A at t_0 from the first person perspective. But is A's belief also knowledge for A at t_0 from the third person perspective?

The situation evolves in such a way that, when A hears the cry, he runs towards it. Once there, at t_1, A observes that *S1* lies on the ground crying, while *S2* is calmly talking to someone else. A is puzzled. The fact that *S2* is calmly talking to another student causes the introduction into γ at t_1 of at least two new beliefs: $\gamma' = \gamma \cup$ {'*S2* seems to be calmly talking to another student', 'after hitting a person one is excited'}. The teacher would have needed more *credulity* at t_0 in order to get φ out of the new salient set of beliefs: $\gamma' \cup$ {'S2 has heard about the report', 'S2 is pretending to be calm'} $\models \varphi$. That is, $d(\varphi, \gamma) < d(\varphi, \gamma')$. In this case (witness-case) the belief that φ, was not knowledge at t_0 for the teacher from the third person perspective. The teacher would not say at t_1 that he knew at t_0 that *S1* was crying.

The previous example is, as we have just pointed out, a witness-case example. Let us talk now for a while about the prediction-cases. In these cases the observer, the evaluator of the belief, predicts the evolution of the situation. So the question to raise is: What is the evolution which the observer will predict? When is a situation said to evolve *naturally*? The answer seems to have clearly a social nature, which brings us to Harman's well known examples (1968), and Meeker's idea that "there seems to be an inescapably *normative* dimension to social defeaters" (Meeker, 2004, p.162). Imagine, for instance that in our previous example the observer is a friend of *S2* who has witnessed how after hitting *S1* and warning him against denouncing him again, *S2* goes to have a chat with another of his friends. Suppose also that while going away from there, the observer hears how the teacher says to a colleague 'Oh no, *S1* is crying'. Observe that also in this case, for the observer the teacher does not know φ at t_0 from the third person perspective, and the reason is that the situation *naturally* evolves in a way that will eventually undermine his belief that φ. The observer sees *A* as belonging to a particular community. This circumstance determines the *natural* evolution of the situation. The teacher *ought* to run towards the cry, but once he is there no one will tell him what has actually happened. The new information he will get there, according to the prediction the observer makes, will only have the effect of undermining his previous belief. Observe that from the observer's perspective it is the socially expected way in which the situation evolves, and not the actual way in which it does which is relevant.

To sum up: The belief that φ of an agent *A* is knowledge for *B* at a time *t* from the third person perspective if, and only if, φ is knowledge for *B* at *t* from the first person perspective, and the situation in which φ was elicited does not evolve (witness-case) or *naturally* evolve (prediction-case) in a way in which the new items added to the set of salient beliefs undermine, from *B*'s perspective, *A*'s belief that φ.

3 Conclusion

We propose an approach to propositional knowledge based on the idea of successful behavior. Our pragmatistic approach finds truth, seen as transcendental, as too heavy a burden to knowledge. Because, how are we suppose to prove that we have attained the truth? We have no idea. Our intention, following Peirce, has been to show that we do not need to care.

Acknowledgements. We would like to thank especially Adriana Renero for the time and interest she devoted to discuss 'knowledge' with us. We are also thankful to Juliet Floyd for her comments on a previous version of this article, and to the three reviewers. This work has been partly supported by the projects FFI2013-44836-P of the Spanish Ministry of Economy and Competitiveness, and P12-HUM-1216 of the Junta de Andalucía, each of which is gratefully acknowledged. This paper would not have been possible either without the Fulbright/Ministerio de Educación postdoctoral grant. The research of Rohit Parikh was supported in part by a grant from CUNY.

References

1. Aliseda-Llera, A.: Seeking Explanations: Abduction in Logic, Philosophy of Science and Artificial Intelligence. ILLC Dissertation Series 1997-4 (1997)

2. Gettier, E.: Is justified true belief knowledge? Analysis 23, 121–123 (1963)
3. Grove, A.: Two Modellings for Theory Change. Journal of Philosophical Logic 17, 157–170 (1988)
4. Harman, G.: Knowledge, Inference, and Explanation. American Philosophical Quarterly 5(3), 164–173 (1968)
5. Levi, I.: The Enterprise of Knowledge. An Essay on Knowledge. Credal Probability, and Chance. The MIT Press, Cambridge (1980)
6. Meeker, K.: Justification and the Social Nature of Knowledge. Philosophy and Phenomenological Research LXIX 1, 156–172 (2004)
7. Parikh, R.: Vagueness and Utility: The Semantics of Common Nouns. Linguistics and Philosophy 17, 521–535 (1994)
8. Parikh, R., Renero, A., San Ginés, A.: Justified True Belief, Plato, Gettier and Turing, slides of a talk in Montreal (2013), `https://cuny.academia.edu/RohitParikh`
9. Peirce, C.S.: The Fixation of Belief. In: Hartshorne, C., Weiss, P. (eds.) Collected Papers of Charles Sanders Peirce, 6 vols., pp. 1–15. Harvard University Press, Cambridge (1931-5). Also in: Popular Science Monthly 12, 1-15 (1877)
10. Stanley, J.: Knowledge and Practical Interests. Oxford University Press, New York (2005)

Büchi Automata Optimisations Formalised in Isabelle/HOL

Alexander Schimpf[1,*] and Jan-Georg Smaus[2]

[1] Institut für Informatik, Universität Freiburg, Germany
[2] IRIT, Université de Toulouse, France
smaus@irit.fr

Abstract. In applications of automata theory, one is interested in reductions in the size of automata that preserve the recognised language. For Büchi automata, two optimisations have been proposed: bisimulation reduction, which computes equivalence classes of states and collapses them, and α-balls reduction, which collapses strongly connected components (SCCs) of an automaton that only contain one single letter as edge label. In this paper, we present a formalisation of these algorithms in Isabelle/HOL, providing a formally verified implementation.

1 Introduction

Model-checking is an important method for proving systems correct, and is applied in industrial practice [1]. In previous work [2], we present a reference implementation for an LTL (linear temporal logic) model checker for finite-state systems à la SPIN [5]. The model checker follows the well-known automata-theoretic approach. Given a finite-state program P and an LTL formula ϕ, two Büchi automata are constructed: the *system automaton* that recognises the executions of P, and the *property* or *formula automaton* expressing all potential executions that violate ϕ, respectively. Then the product of the two automata is computed and tested on-the-fly for emptiness. This implementation is realised and verified using Isabelle/HOL [7].

One important part of automata-based model checking is the translation of an LTL formula into a Büchi automaton. The standard algorithm for this problem has been proposed by Gerth *et al.* [4]. Previously to [2], we have implemented and verified this algorithm in Isabelle/HOL [8]. In this paper, we consider two of the optimisations proposed by Etessami and Holzmann [3] to reduce the size of the formula automaton.

In model checking, the system automaton is usually much larger than the property automaton, but since the size of the property automaton is a multiplicative factor of the overall complexity, it is worthwhile to put substantial effort into its optimisation [3].

The first optimisation is *bisimulation* reduction, which computes equivalence classes of states and collapses them. The algorithm of [3] uses a so-called colouring.

* Supported by DFG grant CAVA, Computer Aided Verification of Automata.

M. Banerjee and S.N. Krishna (eds.): ICLA 2015, LNCS 8923, pp. 158–169, 2015.

proc *BasicBisimReduction*$(A) \equiv$
  /* Init: $\forall q \in Q.\ C^{-1}(q) := 1$, and $\forall q \in F.\ C^0(q) := 1$, $\forall q \in Q \setminus F.\ C^0(q) := 2$.*/
  $i := 0$;
  while $|C^i(Q)| \neq |C^{i-1}(Q)|$ **do**
    $i := i + 1$
    foreach $q \in Q$ **do**
      $C^i(q) := \langle C^{i-1}(q), \cup_{(q,a,q') \in \delta} \{(C^{i-1}(q'), a)\} \rangle$
    od
    Rename colour set $C^n(Q)$, with $\{1, \ldots, |C^i(Q)|\}$, using lexicogr. ordering.
  od
  $C := C^i$; **return** $A' := \langle Q' := C(Q), \delta', q'_{\mathcal{I}} := C(q_{\mathcal{I}}), F' := C(F) \rangle$;
  /* δ' defined so that $(C(q_1), a, C(q_2)) \in \delta'$ iff $(q_1, a, q_2) \in \delta$ for $q_1, q_2 \in Q$*/

Fig. 1. Basic Bisimulation Reduction Algorithm [3]

Our formalisation has revealed that there is a mistake in the initialisation of the algorithm, which we have corrected in our implementation.

The second optimisation is *α-balls reduction*, which collapses strongly connected components (SCCs) that only contain one single letter as edge label.

The rest of the paper is organised as follows: Section 2 gives some preliminaries. Section 3 recalls the two optimisations of [3] in turn. Section 4 presents our Isabelle formalisations of those algorithms, and Sec. 5 concludes.

2 Preliminaries

We recall the basic notions of automata as used by [3]; for more details see [10].

Usually, Büchi (or finite) automata have transitions labelled with characters from an alphabet Σ. In [3], a generalisation of such labellings is considered, but for our purposes, this is not necessary and so we assume simple characters. We assume that a Büchi automaton A is given by $\langle Q, \delta, q_{\mathcal{I}}, F \rangle$. Here Q is a set of states, $\delta \subseteq (Q \times \Sigma \times Q)$ is the transition relation, $q_{\mathcal{I}} \in Q$ is the initial state, and $F \subseteq Q$ is the set of final states. The language $L(A)$ is defined as the set of those ω-words which have an accepting run in A, where a *run* on word $w = a_1 a_2 \ldots$ is a sequence $q_0 q_1 q_2 \ldots$ such that $q_0 = q_{\mathcal{I}}$ and $(q_i, a_{i+1}, q_{i+1}) \in \delta$ for all $i \geq 0$, and it is *accepting* if $q_i \in F$ for infinitely many i.

Isabelle/HOL [7] is an interactive theorem prover based on Higher-Order Logic (HOL). You can think of HOL as a combination of a functional programming language with logic. Isabelle/HOL aims at being readable by humans and thus follows the usual mathematical notation, but still the syntax needs some explanations which we provide when we come to the examples. In our presentation of Isabelle code we have stayed faithful to the sources.

3 The Original Algorithms

3.1 The Bisimulation Reduction Algorithm

Fig. 1 shows the basic bisimulation reduction algorithm in pseudo-code. The letter C with a superscript refers to the iterations of the computation of a colouring.

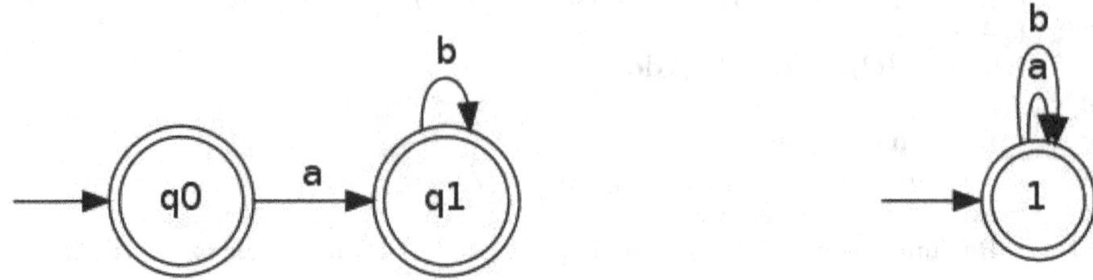

Fig. 2. An automaton and its incorrect reduction according to [3]

The idea is that in the beginning ($i = 0$) accepting states have colour 1 and non-accepting states have colour 2, and in each step, the colour of a state is obtained by its old colour and a combination of the successor state colours and the corresponding edge labels. This means that if two states have the same colour but they differ in the colours of their successors (taking into account the edge labels), then those two states must be distinguished; we say that the colouring is *refined.* In the end, states with the same colour can be joined.

The algorithm initialises not only C^0 but also C^{-1} (we might call this "pre-initialisation") which is a trick making the formulation of the algorithm more concise, by allowing for a loop condition that makes a comparison between the current and the previous colouring, even for $i = 0$.

However, our formalisation of the reduction algorithm, to be shown later, has revealed that there is a mistake in this pre-initialisation. This is illustrated in Figure 2. Here we have $Q = \{q0, q1\}, F = Q, C^0(q0) = 1, C^0(q1) = 1, C^{-1}(q0) = 1, C^{-1}(q1) = 1$. Just before the **while** we have $|C^0(Q)| = |C^{-1}(Q)| = 1$; even stronger, we have $C^0(Q) = C^{-1}(Q) = \{1\}$. What matters is that the loop condition is false and hence the loop is not entered at all. Therefore the result $C = C^0$ is computed, yielding the automaton shown on the right in Figure 2.

Our explanation is as follows: The pre-initialisation $\forall q \in Q.\ C^{-1}(q) := 1$ is conceptually wrong. It expresses that "pre-initially" ($i = -1$), there is only one colour. If by coincidence the input automaton has only accepting or only non-accepting states, then "initially" (index $i = 0$), there is also just one colour. The loop condition will then wrongly say "we have done enough refinement steps".

The problem really manifests itself for the case that $F = Q$, i.e., all states are accepting: each refinement step takes into account the edge labels and not just whether a state is accepting or not. The initialisation however only considers whether a state is accepting or not, and so not doing any refinement wrongly results in identifying all states ($q0$ and $q1$ in the example).

The conceptual mistake happens to cause no harm in the case $F = \emptyset$, since the initialisation of the algorithm establishes the property $F = \emptyset$ and trivially maintains it since no refinement is done. The accepted language is then empty.

Our solution is to replace the condition $|C^i(Q)| \neq |C^{i-1}(Q)|$ with $i \leq 0 \lor |C^i(Q)| \neq |C^{i-1}(Q)|$ (in the actual formalisation: $i > 0 \longrightarrow |C^i(Q)| \neq |C^{i-1}(Q)|$)

so that the loop body will definitely be entered for $i = 0$ at least once. This is shown in Fig. 3 and will be discussed in Sec. 4.1.

3.2 α-Balls Reduction

This optimisation may appear simple and rather specialised, but in fact, it is quite effective in our context of model checking, more precisely, on Büchi automata that are the result of a translation from *generalised* Büchi automata which in turn are the output of the formula translation [4]. Note also that the reduction does not work for finite automata; it only works for Büchi automata.

The idea of the reduction is that, if in a Büchi automaton we are ever stuck in a component, and the only transition labels in this component are α, and there is some accepting state in the component, then we can treat the entire component as a single accepting state with a self-transition labelled by α.

Definition 1. *For $\alpha \in \Sigma$, a* fixed-letter α-ball[1] *inside a Büchi automaton A is a set $Q' \subseteq Q$ of states such that:*

1. *$\alpha \in \Sigma$ is the unique letter which labels the transitions inside Q';*
2. *the nodes of Q' form an SCC of A;*
3. *there is no transition leaving Q' , i.e., no $(q', b, q) \in \delta$ where $q' \in Q'$ and $q \notin Q'$.*
4. *$Q' \cap F \neq \emptyset$.*

Proposition 1. *Given a Büchi automaton $A = \langle Q, \delta, q_\mathcal{I}, f \rangle$, suppose $Q' \subseteq Q$ is a fixed-letter α-ball of A. Let $A' = \langle (Q \setminus Q') \cup \{q_{new}\}, \delta', q'_\mathcal{I}, (F \setminus Q') \cup \{q_{new}\} \rangle$ where*

$$\delta = \{(q_1, b, q_2) \mid q_1, q_2 \in Q \setminus Q'\} \cup \{(q_1, b, q_{new}) \mid (q_1, b, q_2) \in \delta, q_1 \in Q, q_2 \in Q'\} \cup \{(q_{new}, \alpha, q_{new})\},$$

and $q'_\mathcal{I} = q_{new}$ if $q_\mathcal{I} \in Q'$, else $q'_\mathcal{I} = q_\mathcal{I}$. Then $L(A) = L(A')$.

4 Isabelle Formalisation

The work presented in this paper is a fragment of a bigger library being developed on automata in the context of model checking, in particular the construction of the *property automaton* (see Sec. 1). Modularity, generality and reuse are important concerns in this project, which is why the Isabelle code chunks presented here exhibit some aspects that we do not discuss in all detail.

Generally, automata are represented as *record types* that are parametrised by the type of the states and the type of the alphabet, among others. E.g., in line 2 in Fig. 3, $'q$ is the type of the states. The fields of these records, mostly denoted by calligraphic letters, refer to the states, the final states, etc. E.g., in line 5 in Fig. 3, $\mathcal{Q}$ gives the state set, and in line 7, $\mathcal{F}$ refers to being accepting.

[1] [3] defines more generally: a *fixed-formula α-ball*, i.e., α is a formula.

```
1: definition LBA_bpr_C ::
2:       "('q, 'l, 'more) LBA_scheme ⇒ ('q ⇒ nat) nres"
3:  where
4:    "LBA_bpr_C 𝒜 ≡ do {
5:       let Q = 𝒬 𝒜;
6:       let C = (λq. 1);
7:       let C' = (λq. if ℱ 𝒜 q then 1 else 2);
8:       let i = 0;
9:       (_, C', _) ←
10:      WHILEIT
11:        (LBA.LBA_bpr__whilei 𝒜 Q)
12:        (λ(C, C', i). i>0 ⟶ card (C' ‘ Q) ≠ card (C ‘ Q))
13:        (λ(C, C', i). do {
14:          let i = Suc i;
15:          let C = C';
16:          let f = (λq. (C q, ℒ 𝒜 q,  C ‘ successors𝔄 𝒜 q));
17:          let R = f ‘ Q;
18:          fi ← set_enum R;
19:          let C' = fi o f;
20:          RETURN (C, C', i)
21:        }) (C, C', i);
22:      RETURN C'
23:    }"
```

Fig. 3. BPR colouring

In the formalisation we present here, we use automata that follow [4] but differ from the standard definition in one important aspect: we assume that not the *edges*, but rather the *states* of automata are labelled. We call those automata *labelled Büchi automata* (*LBA*), as opposed to *BA*. Moreover, in our representation we have a *set* of initial states rather than a unique initial state. Instead of a set of accepting states we use a predicate to express whether a state is an accepting state or not. This kind of automata representation is suitable in our context, since we formalise the algorithm in the context of the Büchi automaton construction from an LTL formula according to [4], where the output automaton corresponds to a state labelled rather than a transition labelled automaton.

Big parts of our library concern BAs, however, and technically, LBAs are defined as an extension of the BA record type where labels for the states are added and the labels for the edges are "disabled".

4.1 The Bisimulation Reduction Algorithm

The Isabelle formalisation of the algorithm from Fig. 1 is shown in Figure 3. The formalisation uses the Isabelle refinement framework [6] for writing what resembles imperative code. We explain some of the syntactic constructs.

Lines 5 to 8 accomplish the initialisation. The *let* should be understood as imperative assignment. We assign: Q is the state set of the input automaton $\mathcal{A}$; C

```
1: definition
2:   "set_enum S ≡ do {
3:     (_, m) ← FOREACHi
4:                 (set_enum__foreachi S)
5:                 S
6:                 (λx (k, m). RETURN (Suc k, m(x↦k)))
7:                 (1, empty);
8:     RETURN (λx. case m x of None ⇒ 0 | Some k ⇒ k) }"
```

Fig. 4. Numbering of sets

is the $(i-1)$th colouring and is initialised to the function that colours each state as 1; C' is the ith colouring and is initialised to the function that colours each accepting state as 1 and all others as 2. We need C and C' in order to determine whether the current iteration has actually refined the current colouring.

Line 9 performs a kind of nondeterministic assignment: the term following it is essentially a set of values, and an element of this set is assigned to $(\_, C', \_)$ nondeterministically. The refinement framework allows us to specify algorithms with such nondeterminism, prove theorems about them, and replace the nondeterminism by a deterministic implementation later and independently.

Back to the code: line 10 contains the **while**-construct in this language. Its first argument (line 11) consists of a loop invariant that must be provided by the programmer and that is used in correctness proofs. The second argument (line 12) is the loop condition corrected as explained in Sec. 3.1. The third argument (lines 13 to 21) is the loop body which takes the form of a λ-term with an abstraction over (C, C', i), applied to the argument (C, C', i) (line 21) which corresponds to the initial values explained two paragraphs above.

During each iteration, a new colouring is computed in the form of a function f that assigns a certain triple to each state; R is then defined as the image of f on Q, i.e., R is the set of all the colours of the new colouring. In order to convert those complicated triples into simple numbers, an auxiliary function shown in Fig. 4 is used; *fi*, obtained by a nondeterministic assignment as in line 9, is then the numbering of R. The new colouring is then the composition of *fi* and f. It assigns a number to each state.

The function *set_enum* computes the set of all unique numberings for a set S, so that *set_enum* S is the set of bijections between S and $\{1, \ldots, \#S\}$. The definition of the function starts with a **foreach**-construct (lines 3-7). In line 3 one possible result of the loop is chosen non-deterministically and is assigned to $(\_, m)$. In the following line a loop-invariant is provided in order to be able to prove correctness properties of the definition. The second parameter in line 5 represents the iterated set. For each element of S the body of the loop (line 6) is applied sequentially in an arbitrary order, where x is an element of S and (k, m) an intermediate result of the loop, that is propagated through the iterations starting with $(1, \mathit{empty})$. Such results correspond to a pair consisting of the next number to assign and an already constructed mapping of numbers to elements of S. An empty mapping is denoted by *empty* and for an existing mapping m a new mapping is constructed with $m(x \mapsto k)$, that behaves like m with the exception that it maps x to k. In the last line the above constructed mapping is turned into a function: elements not in S are mapped to 0. The following lemma states the correctness of the construction:

```
lemma set_enum_correct:
  assumes "finite S"
  shows "set_enum S ≤ SPEC (λf. bij_betw f S {1..card S})"
```

The term *bij_betw* f S $\{1..card\ S\}$ says "f is a bijection between S and $\{1, \ldots, \#S\}$", and the entire lemma says that the results of *set_enum* S in terms of functions f fulfil the specification "f is a bijection between S and $\{1, \ldots, \#S\}$".

Thus we have a bijection between $f(Q)$ and $C'(Q)$ in the actual procedure, i.e., line 9 in the pseudo-code in Fig. 1 is correctly implemented.

Termination After each execution of the body of the loop, the number of colours according to C' increases compared to C, or the iteration stops after that execution of that loop body. At the same time, the number of possible colours in C' is bounded by the number of states in the input automaton. Hence the number of colours cannot increase indefinitely and therefore the iteration stops eventually.

Correctness The proof of correctness of the procedure is based on the characterisation of the ith colouring. For $i > 0$, we have the following loop invariant:

```
definition (in LBA)
  "LBA_bpr_C_inv Q ≡ λ(C, C', i).
     ∀q∈Q. ∀q'∈Q. C' q = C' q'
                     ⟷ (C q = C q'
                          ∧ ℒ 𝒜 q = ℒ 𝒜 q'
                          ∧ C ‘ successors q = C ‘ successors q')"
```

This invariant corresponds exactly to the modification of C' after each iteration and is thus simple to prove based on the bijectivity of f and C'. For the general case, i.e., including $i = 0$, the following holds: whenever $C'(q) = C'(q')$ for two states $q, q' \in Q$, then either both q, q' are in F or they are both not in F.

When the iteration of the loop stops, i.e., the number of colours in C' does not change anymore compared to C, we obtain a bijection between C and C'. Considering that the invariant relates C to C', we obtain that C and C' yield the same equivalence classes, i.e., for any states $q, q' \in Q$, we have $C'(q) = C'(q')$ if and only if $C(q) = C(q')$. In Fig. 5, we have defined the characteristics that are needed in the subsequent proofs and that *LBA_bpr_C* indeed fulfils. We have chosen to introduce a definition for this characterisation because it occurs in several places in our development.

All in all, we obtain the characterisation of the resulting colouring shown by the lemma in Figure 6. Indeed, we need to assume that the set of states is

```
definition (in LBA)
  "LBA_bpr_C_char
    ≡ λC. ∀q∈𝒬 𝒜. ∀q'∈𝒬 𝒜.
             C q = C q' ⟶ (ℱ 𝒜 q ⟷ ℱ 𝒜 q')
                          ∧ ℒ 𝒜 q = ℒ 𝒜 q'
                          ∧ C ‘ successors q = C ‘ successors q'"
```

Fig. 5. Characterisation of the colouring

```
lemma (in LBA) LBA_bpr_C_correct:
  "LBA_bpr_C 𝒜 ≤ SPEC LBA_bpr_C_char"
```

Fig. 6. Correctness property of the colouring

finite. Otherwise we could not apply the above shown properties about *set_enum*. Finiteness is given here by an implicit assumption denoted by "(**in** *LBA*)".

This characterisation turns out to be a sufficient condition for proving the correctness of the resulting coloured automaton. To obtain the automaton, we apply a renaming function *LBA_rename* which takes an LBA and a colouring function C and returns the LBA where each state has been renamed using the colouring function. The renaming function has properties shown in Figure 7.

For example the second line of the lemma states that the initial states of the renamed automaton are exactly the renamings of the initial states of the original automaton. The other lines state similarly that the transition function, the final states, and the labels are preserved by the renaming.

The term *inv_into* $Q\ f\ q$ gives "the"[2] inverse element of q under f in Q. In the case that this inverse is unique, i.e., f is injective on Q, it is straightforward to show that the renaming preserves language equivalence. However, the very purpose of the colouring is to *reduce* the number of states, hence not to be injective! But even in this general case, language equivalence holds, as is expressed by the following lemma:

```
lemma (in LBA) LBA_bpr_C_rename_accept_iff:
  assumes "LBA_bpr_C_char C"
  shows "∀w. LBA_accept (LBA_rename 𝒜 C) w ⟷ LBA_accept 𝒜 w"
```

The proof of the lemma works constructively. The "⟵" direction consists of taking a run (sequence of states) r for word w of the input automaton and colouring r componentwise using C, i.e., r is mapped to a run $C \circ r$ in the coloured automaton.

The "⟶" direction requires an auxiliary function:

[2] This is the famous ϵ operator of HOL: It represents an arbitrary but fixed term fulfilling a given property.

```
lemma LBA_rename_simps:
  "δ (LBA_rename 𝒜 C) q a = C ‘ δ 𝒜 (inv_into (𝒬 𝒜) C q) a"
  "ℐ (LBA_rename 𝒜 C) = C ‘ ℐ 𝒜"
  "ℱ (LBA_rename 𝒜 C) q ⟷ ℱ 𝒜 (inv_into (𝒬 𝒜) C q)"
  "ℒ (LBA_rename 𝒜 C) q = ℒ 𝒜 (inv_into (𝒬 𝒜) C q)"
```

Fig. 7. Renaming function for LBAs

Fig. 8. An LBA and its colouring

```
fun bpr_run
where
  "bpr_run 𝒜 C r q0 0 = q0"
| "bpr_run 𝒜 C r q0 (Suc k)
    = (SOME q'. C q' = C (inv_into (𝒬 𝒜) C (r (Suc k)))
                ∧ q'∈δ_L 𝒜 (bpr_run 𝒜 C r q0 k))"
```

This function is needed for the following reason: if we start from a run r for w in the coloured automaton and use *inv_into* Q f q to compute a corresponding state in the input automaton for each state q in r, then these states do not necessarily “fit together”, i.e., they may not form a run in the input automaton.

Example 1. Figure 8 shows an LBA accepting the word $aaa\ldots$ on the left, and the simplified automaton obtained by colouring on the right. Obviously, the state sequence in the coloured automaton is always simply a sequence of 1’s. The inverse of “1” is either q_1 or q_2, so simply translating $11\ldots$ back into the original LBA would give either $q_1q_1\ldots$ or $q_2q_2\ldots$. In neither case this would correspond to a run of the original LBA.

As the example shows, constructing a run in the original LBA, given a run in the coloured LBA, is not so simple and has to be done step by step starting from the initial state. This is what the function *bpr_run* is good for. It starts by taking an inverse of the initial state of the run r of the coloured automaton, and then always picks an appropriate inverse for each next state in r. The existence of such an inverse is guaranteed by the assumption that the colouring C fulfils the characterisation according to Figure 5.

The entire procedure for constructing the coloured automaton is then given in Figure 9. By the considerations given above, this construction is correct:

```
lemma (in LBA) LBA_bpr_correct:
  "LBA_bpr 𝒜
    ≤ SPEC (λ𝒜_C. LBA 𝒜_C ∧ (∀w. LBA_accept 𝒜_C w ⟷ LBA_accept 𝒜 w))"
```

```
definition LBA_bpr :: "('q, 'l, _) LBA_scheme ⇒ (nat, 'l) LBA nres"
where
  "LBA_bpr 𝒜 ≡ do {
     C ← LBA_bpr_C 𝒜;
     RETURN (LBA_rename 𝒜 C)
   }"
```

Fig. 9. Procedure for constructing the coloured automaton

In addition to language equivalence we also show that the coloured automaton is well-formed, i.e. *LBA* $\mathcal{A}_C$ holds.

4.2 α-Balls Reduction

Most of the Isabelle development on this topic uses in fact BA, not LBA, and the formalisation of α-balls reduction for LBA is based on the one for BA. However, we focus here on LBA because the use of LBA required certain adaptations that partly make the contribution of our work.

An α-ball for LBA has, of course, an Isabelle definition. However, this is not very readable and so we prefer to present the following characterisation:

```
lemma αball_L_full_def:
  "αball_L 𝒜 α Q ≡
     𝒢 𝒜 ↾ Q ∈ sccs 𝒜 ∧ Q ≠ {}
     ∧ (∀q∈Q. successors 𝒜 q ⊆ Q)
     ∧ (∀q∈Q. ℒ 𝒜 q = α)"
```

The lemma says that Q is an α-ball iff the following four conditions hold: (1) the graph of $\mathcal{A}$, restricted to Q, is an SCC of $\mathcal{A}$; (2) $Q \neq \emptyset$; (3) there is no edge leading out of Q; (4) all states in Q are labelled with α.

Fig. 10 gives an auxiliary definition for balls reduction. There is a loop for working through the SCCs. Trivial balls or balls without accepting states are removed. One-element balls are left untouched. Finally, non-trivial balls are collapsed to a single state. Based on the computation of SCCs according to Tarjan's algorithm [9] we obtain a function that computes the SCCs and then reduces, as shown in Figure 11. The ball reduction on LBAs thus returns a well-formed equivalent automaton and a set $SCCs_R$ representing its SCCs.

The correctness follows from the correctness of *LBA_reduce_ball_aux* and *tarjan*:

```
lemma (in LBA) LBA_reduce_balls_correct:
  "LBA_reduce_balls 𝒜 ≤ SPEC(λ(𝒜_R, SCCs_R).
     LBA 𝒜_R ∧ (∀w. LBA_accept 𝒜_R w ⟷ LBA_accept 𝒜 w)
     ∧ 𝒬 𝒜_R = ⋃SCCs_R
     ∧ pairwise_disjoint SCCs_R
     ∧ (∀Q∈SCCs_R. 𝒢 𝒜_R ↾ Q∈sccs 𝒜_R ∧ Q ≠ {})
   )"
```

```
definition LBA_reduce_balls_aux ::
  "[('q, 'l, 'more) LBA_scheme, 'q set set]
     ⇒ (('q, 'l, 'more) LBA_scheme × 'q set set) nres"
where
  "LBA_reduce_balls_aux 𝒜 SCCs
   ≡ FOREACHi
         (LBA_reduce_balls_aux_foreach_inv 𝒜 SCCs)
         SCCs
         (λQ (𝒜_R, SCCs_R).
           if ¬ ball_L 𝒜_R Q then RETURN (𝒜_R, {Q} ∪ SCCs_R)
           else if scc_trivial 𝒜_R (𝒢 𝒜_R ↾ Q)
                then RETURN (remove_states 𝒜_R Q, SCCs_R)
           else if ∀q∈Q. ¬ ℱ 𝒜_R q
                then RETURN (remove_states 𝒜_R Q, SCCs_R)
           else if (∃q. Q = {q}) then RETURN (𝒜_R, {Q} ∪ SCCs_R)
           else
             case ball_L_get_α 𝒜_R Q of
               None ⇒ RETURN (𝒜_R, {Q} ∪ SCCs_R)
             | Some α ⇒
                 do { (𝒜_R, Q_R) ← LBA_remove_αball 𝒜_R Q;
                      ASSERT (Q_R⊆Q ∧ Q_R ≠ {});
                      RETURN (𝒜_R, {Q_R} ∪ SCCs_R) }
           )
         (𝒜, {})"
```

Fig. 10. Auxiliary definition for α-balls reduction

```
definition LBA_reduce_balls ::
  "('q, 'l, 'more) LBA_scheme
    ⇒ (('q, 'l, 'more) LBA_scheme × 'q set set) nres"
where
  "LBA_reduce_balls 𝒜 ≡ do {
     SCCs ← tarjan 𝒜 (ℐ 𝒜);
     (𝒜_R, SCCs_R) ← LBA_reduce_balls_aux 𝒜 SCCs;
     RETURN (𝒜_R, SCCs_R)
   }"
```

Fig. 11. Ball reduction on LBAs

5 Conclusion

We have presented an Isabelle/HOL formalisation of two Büchi automata optimisations proposed by [3]. The context of this work is explained in detail in [2]: implementing full-fledged model checkers verified in Isabelle/HOL. Within this endeavour, it is worthwhile to invest effort in the optimisation of the property automaton. While the optimisations are particularly relevant for model checking, they are abstract enough to be applicable to Büchi automata in general.

The difficulty lay in making the right design decisions for the formalisation. We generally tried to model the Isabelle proofs on the paper-and-pencil proofs of the literature, but especially for the colouring, it turned out to be better to develop a new constructive proof from scratch. In that particular case, as mentioned above, the original code contained a mistake, which we discovered during our vain efforts to prove the code correct.

Concerning the first optimisation, there is also a more sophisticated algorithm [3]. However this algorithm relies on the *edge* labels in an automaton and thus it is not immediately possible to implement this algorithm in our scenario.

Our formalisation consists of around 1500 lines of code (approximately 3000 lines of code if the formalisation of Tarjan's SCC algorithm is counted).

References

1. Clarke, E.M., Grumberg, O., Peled, D.A.: Model Checking, 5th print. MIT Press (2002)
2. Esparza, J., Lammich, P., Neumann, R., Nipkow, T., Schimpf, A., Smaus, J.-G.: A fully verified executable LTL model checker. In: Sharygina, N., Veith, H. (eds.) CAV 2013. LNCS, vol. 8044, pp. 463–478. Springer, Heidelberg (2013)
3. Etessami, K., Holzmann, G.J.: Optimizing Büchi automata. In: Palamidessi, C. (ed.) CONCUR 2000. LNCS, vol. 1877, pp. 153–167. Springer, Heidelberg (2000)
4. Gerth, R., Peled, D., Vardi, M.Y., Wolper, P.: Simple on-the-fly automatic verification of linear temporal logic. In: Dembinski, P., Sredniawa, M. (eds.) Proceedings of the 15th International Symposium on Protocol Specification, Testing, and Verification. IFIP Conference Proceedings, vol. 38, pp. 3–18. Chapman and Hall (1996)
5. Holzmann, G.J.: The SPIN Model Checker - primer and reference manual. Addison-Wesley (2004)
6. Lammich, P., Tuerk, T.: Applying data refinement for monadic programs to Hopcroft's algorithm. In: Beringer, L., Felty, A. (eds.) ITP 2012. LNCS, vol. 7406, pp. 166–182. Springer, Heidelberg (2012)
7. Nipkow, T., Paulson, L.C., Wenzel, M.: Isabelle/HOL. LNCS, vol. 2283. Springer, Heidelberg (2002)
8. Schimpf, A., Merz, S., Smaus, J.-G.: Construction of Büchi automata for LTL model checking verified in Isabelle/HOL. In: Berghofer, S., Nipkow, T., Urban, C., Wenzel, M. (eds.) TPHOLs 2009. LNCS, vol. 5674, pp. 424–439. Springer, Heidelberg (2009)
9. Tarjan, R.E.: Depth-first search and linear graph algorithms. SIAM J. Comput. 1(2), 146–160 (1972)
10. Thomas, W.: Automata on infinite objects. In: Handbook of Theoretical Computer Science, Volume B: Formal Models and Semantics (B), pp. 133–192. Elsevier and MIT Press (1990)

Nyāya's Logical Model for Ascertaining Sound Arguments

Jaron Schorr

Tel Aviv University, Ramat-Aviv, Israel
jaron.schorr@gmail.com

Abstract. The logical debate in India of the first millennia AD revolved around the concept of *pramāna*. The term *pramāna* was taken to mean 'the criterion of knowledge'. Current researchers of Indian philosophy are certain that Indian logicians all agreed that *pramāna* is the sound operation of the mental processes which produce mental knowledge episodes. Conversely, according to my research, Nyāya thinkers believed the criteria of knowledge are the rules of the use of things in everyday habitual behaviors. The issue which stood at the center of the Indian logical debate, I wish to suggest, was the following: On the one hand, there were thinkers who believed the rules of logic were prior to and independent of habitual everyday human behaviors. On the other hand, there were the Naiyāyikas who believed the rules of logic were derived from habitual everyday human behaviors and the rules of usage they provided.

Keywords: Syllogism, inference, criteria, *pramāṇa*, *anumāṇa*, *upamāṇa*, *pratyakṣa*.

1 Introduction

My purpose in what follows is to present several points that came to my attention during the research I have conducted concerning one of Indian philosophy most important schools of thought – the Nyāya. Indian philosophical debate concerning logic revolved around the concept of *pramāṇa*. In colloquial usage, *pramāṇa* means a measuring device, a criterion or a standard. The meaning of *pramāṇa* in the context of the Indian philosophical debate resembled greatly to its everyday usage. It was taken to mean 'criteria of knowledge'.

It follows from the above that the philosophical debate concerning logic in India revolved around the question 'what the criteria of knowledge are?' Up until now, researchers of Indian philosophy agreed that all Indian thinkers of the first millennia AD presupposed knowledge to be mental episodes which mental processes produce and accordingly, that the criteria of different types of knowledge are the sound operation of the

M. Banerjee and S.N. Krishna (eds.): ICLA 2015, LNCS 8923, pp. 170–182, 2014.

mental processes which produce them.[1] These researchers were certain that the Indian philosophical debate concerning logic was conducted within the confines of the above conceptual framework.

Conversely, my research shows that at least one Indian school of thought, the Nyāya, did not operate within the confines of the said framework. The criteria of knowledge Nyāya thinkers proposed were entirely different. Their proposed criteria were based on the following principle: The rules of grammar allow us to speak as objects of knowledge only of things we know how to use, things which play a role and fulfill a function in human life. The meaning of a tree is given to it by the various ways in which it is used. Tree is that thing we built stuff out of, that thing we use to warm our houses with, that thing the fruits of which we eat and under the shade of which we take refuge. The meaning of things is given to them by the rules of their use or to state the same thing differently, the rules of the use of things which we learn when we learn human behavior determine their meanings. The criteria of knowledge are then, according to Nyāya, the rules of the use of things in habitual behaviors. This means that according to Nyāya, it makes no sense to speak as objects of knowledge of things we don't know how to use, things there are no habitual behaviors in which they participate and accordingly, no rules for their use.

It is important to note that Indian logic was not formal.[2] Its subject matters were not abstract variables, logical symbols and propositions whose forms have been separated from their contents. The subject matters of Indian logic were everyday entities and concepts. The issue which stood at the center of the Indian philosophical debate regarding the *pramāṇa*, I suggest, was the following: On the one hand, there were thinkers who believed the rules of logic were prior to and independent of habitual everyday human behaviors. On the other hand, there were the Naiyāyikas who believed the rules of logic were derived from habitual everyday human behaviors and the rules of usage they provided. Nyāya was and still is a school of thought that claims that arguments which are not grounded in well known, well familiar courses of action, are not logical, not rational. Since Nyāya logic was not formal, since its subject matters were everyday entities and concepts, it considered only arguments which were sound and meaningful, arguments which conformed to rules of usage in everyday life, to be strictly speaking, logical and rational.

The main purpose of Nyāya philosophical project was to develop a logical model for ascertaining whether arguments were sound. This logical model was designed to determine whether arguments conformed to the rules of the use of things in everyday habitual behaviors. The model operated as follows: It examined whether arguments satisfied the

---

[1] MATILAL 2002, pp. 368 – Matilal states that: "*Prāmaṇa* is what 'makes' knowledge. Since knowledge is always an episode (an inner event in Indian philosophy, in fact, a sub-category of mental occurrence), *prāmaṇa* has also a causal role to play. It is the "most efficient" cause of the knowledge episode. Knowledge yields determination of an object *x* or a fact that *p* (*artha-parichchedda*) as the result, and *prāmaṇa* is "instrumental" in bringing about that result. This is the causal role of a *prāmaṇa*. In MOHANTY 1992, pp. 229, a similar position is presented by Mohanty. He states that " A sort of causal theory of knowledge is built into the *prāmaṇa* theory: a true cognition must not only be true to its object, but must also be generated in the right manner, i.e. by the appropriate causes.

[2] For further reading on the matter see MOHANTY, pp. 100-127, MATILAL 1986, pp. 118-127 and MATILAL 1990, pp. 23 -28, MATILAL 1998, pp. 1-22

criteria of soundness. Arguments which satisfied all the required criteria were determined to be sound. Arguments which failed to satisfy the criteria were rejected, however convincing they may have seemed.

2 The Four Criterions of Knowledge, the Four *Pramāṇas*

2.1 The *Pratyakṣa-pramāṇa,* the Criterion of Perceptual Knowledge

Uddyotakara, the author of the Nyāyavārtika, the second commentary of the Nyāyasūtra, which was composed probably during the 6th century AD, presents the outlines of his theory of *pramāṇa* already in the introduction of his work. The discussion in the introduction is divided to three parts, each one of which is dedicated to the clarification of a key concept Vātsyāyana uses in the opening sentence of the Nyāya-bhāṣya. The opening sentence of the Bhāṣya states that "*pramāṇa* is useful and it is fully compatible with its objects *(arthavat)* because the capability to act *(pravṛṛti-sāmarthya)* reveals that the object of perception was obtained by *pramāṇa*". Uddyotakara first turns to examine the relation between *pramāṇa* and *samartha-pravṛṛti* – meaningful, compatible and competent activity. It seems Uddyotakara considers the clarification of the relation between *pramāṇa* and *samartha-pravi,* which he takes to be a relation of mutual-interdependence, to be crucial to the understanding of Vātsyāyana's statement '*pramāṇa* is *arthavat*'.

Uddyotakara does not consider the relation between *pramāṇa* and *samartha-pravṛṛti* to be a causal relation. He makes this point perfectly clear in his response to the following objection: The *pūrvapakṣin* notes that whereas it is perfectly reasonable to maintain that *pramāṇa* conditions *samartha-pravṛṛti*, it is senseless to maintain that *samartha-pravṛṛti* conditions *pramāṇa*, since the latter temporally precedes the former.[3] In causal relations, time plays a crucial role. If relata *x* conditions relata *y* it follow that relata *x* temporally precedes relata *y*; and if relata *y* is temporally posterior to relata *x*, it follows that it is impossible for relata y to condition relata *x*. Under such circumstances, it is indeed impossible for *samartha-pravṛṛti* to condition *pramāṇa*. Uddyotakara responds by saying that time plays no role in the analysis he presented of the relation between *pramāṇa* and *samartha-pravṛṛti*.[4] If the relation between *pramāṇa* and *samartha-pravṛṛti* is a kind of relation in which time plays no role, it cannot be a causal relation, and if it is not a causal relation, there is no reason to withdraw the claim that *pramāṇa* and *samartha-pravṛṛti* are mutually interdependent. If Uddyotakara holds time to have no bearing on the relation between *pramāṇa* and *samartha-pravṛṛti*, he must be holding *pramāṇa*

---

[3] NV 1.1.1, pp. 3 line 4 - परस्परापेक्षित्वादुभयासिद्धिरिति चेत् – न, अनादित्वात्. यदि प्रमाणतोऽर्थप्रतिपत्तौ प्रवृत्तिसामर्थ्यम्, यदि वा प्रवृत्तिसामर्थ्यात् प्रमाणतोऽर्थप्रतिपत्तिः, किं पूर्वं किं वा पश्चादिति वाच्यम्. यदि तावत् प्रमाणतः पूर्वमर्थप्रतिपत्तिः, प्रवृत्तिसामर्थ्यमन्तरेण किमिति प्रतिपद्यते? अथ पूर्वं प्रवृत्तिसामर्थ्यमनवधार्यार्थं किमिति प्रवर्तते? तस्मात् प्रवृत्तेः प्रमाणतोऽर्थप्रतिपत्तेरवा पूर्वापरभावो न कल्प्यते इति.

[4] NV 1.1.1, pp. 3 line 9 -तच्च नैवम्. कस्मात्? अनादित्वात्. अनादिरयं संसार इति पूर्वाभ्यस्तसूत्रे प्रतिपादयिष्यामः. आदिमति च संसारे एष दोषः, किं पूर्वं प्रमाणतोऽर्थप्रतिपत्तिः, आहोस्वित् पूर्वं प्रवृत्तिसामर्थ्यमिति.

and *samartha-pravṛṛti* to be logically, grammatically related. The purpose of his analysis must have been to provide an account of the grammatical relations between the concepts of *pramāṇa* and *samartha-pravṛṛti* and not an account of the causal relation between *pramāṇa* and *samartha-pravṛṛti*.

Uddyotakara's analysis of the grammatical relations between *pramāṇa* and *samartha-pravṛṛti* consists of three short passages. First, he points out that there is a grammatical relation between what we understand to be an 'object of interest' and the activity which accomplishes it - *samartha-pravṛṛti*. He states:

> When the thought I want to 'obtain that thing' propels a man to act, and he indeed 'obtains that thing', the activity (*pravṛṛti*) he performs is said to be fruitful (*samartha*). Similarly, when the thought I want to 'avoid that thing' propels a man to act, and he indeed 'avoids that thing', the activity he performs is said to be fruitful. Otherwise, his activity is said to be non-fruitful.[5]

The grammatical point Uddyotakara tries to turn our attention to in the above passage is as follows: The terms we use to express the things we are interested in pursuing are the same terms we use to express the activities which accomplish these things. One's desire to "drink coffee", to "climb a tree" or to "wear a coat", could be fulfilled only by him "drinking coffee", "climbing a tree" or "wearing a coat". This grammatical point can be formulated in the following manner: The thing p one is interest in pursuing can be accomplished only by the activity we describe as p and the activity we describe as p is the activity which is capable of accomplishing only the thing p one is interest in pursuing. It follows from the fact that the things we are interested in pursuing and the activities which accomplish them are expressed by the same terms, that for each activity p corresponds one and only one object it accomplishes, i.e., the object p we are interested in pursuing; and that for each object p we are interested in pursuing corresponds one and only one activity, i.e., the activity p which accomplishes it.

In the subsequent passage, Uddyotakara states that the classification of activities to *samartha* and *asamartha* corresponds to the classification of *pramāṇa* to real and unreal.[6] The real *pramāṇa*, Uddyotakara explains, is that which correctly endows things with their meanings.[7] A meaningful thing is by definition a thing which serves a purpose, a thing which plays a role and which people find interest in. Hence, saying that the real *pramāṇa* correctly determines the meaning of something to be p is tantamount to saying that the real *pramāṇa* correctly identifies p as something people find interest in. Since the group of objects the real *pramāṇa* provides is the group of things people find interest in, and since for each such thing p corresponds one and only one

---

[5] NV 1.1.1, pp. 2 line 17 - प्रवृत्तेरपि द्वैविध्यं भवति समर्थासमर्थभेदात्. या खलु रागादिमत्प्रवृत्तिः सा समर्था चासमर्थ च भवति. इष्टमाप्स्यामीति प्रवर्तमानो यदा प्राप्नोति तदा समर्था, अनिष्टं हास्यामीति प्रवर्तमानो यदा जहति तदापि समर्था. यदा विपर्ययस्तदासमर्थेति.

[6] NV 1.1.1, pp. 2 line 21 - तत् पुनः प्रवृत्तेर्द्वैविध्यं प्रमाणस्यार्थवदनर्थकत्वात्.

[7] NV, pp. 2 line 21 - प्रमाणं तावदर्थपरिच्छेदकम्. प्रमाणसामान्यात् प्रमाणप्रतिरूपकमपि प्रमाणमित्युपचर्यते.

activity, i.e., the activity *p* which accomplishes it, the group of objects the real *pramāṇa* provides is logically equivalent to the group of fruitful activities.

Uddyotakara concludes his analysis of the relation between *pramāṇa* and *samartha-pravṛṛti* by noting that:

> When a man acts according to an object the real *pramāṇa* provides, the activity he performs is fruitful and the fact that it is fruitful reveals him to be a knower. On the other hand, when a man acts according to an object the unreal *pramāṇa* provides the activity he performs is not fruitful (and the fact that it is not fruitful reveals him not to be a knower).[8]

It follows from the fact that for every fruitful activity *p* corresponds one and only one meaningful thing *p*, that if there is a fruitful activity we express as *p* we can unequivocally determine that *p* is meaningful. It further follows from the fact that the group of all possible fruitful activities is logically equivalent to the group of all possible meaningful things, that we can determine a certain thing *p* to be meaningful only if there is a fruitful activity we describe as *p*. Fruitful activities are, then, that by which we can determine whether a certain thing *p* is meaningful and without which determining whether things are meaningful would have been impossible. In other words, fruitful activities determine the bounds of sense, they are the criteria of meaning; and since knowing is nothing but grasping the correct meaning of a thing, fruitful activities are the criteria of knowledge, they are the *pramāṇa*.

Further on in the introduction, Uddyotakara examines the issue of *kāraka-śabdas*. The term *kāraka-śabda* means 'word which expresses activity'. All the words which denote the agents, the instruments or the objects of activities fall under the category of *kāraka-śabda*. An account of the way *kāraka-śabdas* operate is, of course, of pivotal importance to a position which regards the linguistic expression of activities to be the criterion of meaningfulness and knowledge. Uddyotakara explains how he takes *kāraka-śabdas* to operate when he responds to an objection raised by the *pūrvapakṣin*. The *pūrvapakṣin* claims it to be inconceivable for *kāraka-śabdas* to derive their meaning from something other than the on-going activity to which they refer and that *kāraka-śabdas* can obtain their meaning and become operative only after the on-going activity to which they refer is completed.[9] To this Uddyotakara replies that if a 'cook' was a cook only while he was cooking, this would mean that he became a cook only once he began cooking and that he would cease to be a cook as soon as he completed cooking, which is preposterous.[10] Similarly, if a 'stove' and a 'dish' were a stove and a dish only during a cooking session, it would mean that they became a stove and a dish

---

[8] NV 1.1.1, pp. 2 line 2 - सोSयं प्रमाता यदा प्रमाणेनार्थमवधार्य प्रवर्तते तदास्य प्रवृत्तिः समर्था भवति. यदा पुनः प्रमाणाभासेनावधार्य प्रवर्तते तदासमर्था. तस्याः पुनरर्थवत्त्वम्

[9] NV 1.1.1, p. 5 line 21 - न च कारकशब्दाः क्रियासंबन्धमन्तरेणात्मानं लभन्त इति. न च द्रव्यमात्रं कारकम्, न च क्रियामात्रमिति. कारकशब्दो हि प्रवर्तमानः क्रियासाधने क्रियाविषेशयुक्ते प्रवर्तते. प्रमातृप्रमेयशब्दौ च कारकशब्दौ तावन्तरेण क्रियां न प्रवर्तेयातामिति.

[10] NV 1.1.1, pp. 6, line 3 - न, पाचकादिशब्दवत् त्रिकालविषयत्वात्. न ब्रूमः क्रियासंबन्धेनैव कारकशब्दाः प्रवर्तन्त इति. अपि तु त्रिकालविषया एते. यदि क्रियासंबन्धनिमित्ता स्युः, न क्रियामन्तरेण प्रवर्तेरन्. क्रियासंबन्धमन्तरेण तु प्रवर्तन्ते.

only once the cooking session began and that they would cease to be a stove and a dish as soon as the cooking session completed, which is equally preposterous.

Kāraka-śabdas, Uddyotakara concludes, do not refer to the constituents of an ongoing activity but rather, they express the role a thing plays, the purpose it fulfills or the use we make of it in a habitual behavior. The word 'cook' expresses the role that the agent of cooking plays in the habitual behavior of 'cooking', the word 'stove' expresses the function the instrument of cooking fulfills in 'cooking' and the word 'dish' expresses the thing which the activity of cooking produces. *Kāraka-śabdas*, the terms we use to describe the activity which accomplishes a thing *p*, express as a matter of fact the use we make of *p* in a certain habitual behavior. So, instead of saying that by pointing out an activity we describe as *p* and only by pointing out such activity, we can unequivocally determine that *p* is meaningful, we can now say that by pointing out a use we can make of *p* in a habitual behavior and only by pointing out such usage, we can unequivocally determine that *p* is meaningful. In other words, usage determines the meaning of things or, the meaning of things is their usage. The meaning of a tree is that which we can build staff out of, cut down and climb on. Knowing what a 'tree' is, grasping the true meaning of 'tree', depends on our acquaintance with the habitual behaviors in which trees are used. We become acquainted with the usage of things in habitual behaviors by learning these habitual behaviors - a process which requires much exercise, train and drill, the development of skills and techniques. The true meaning of things is given by their usage in habitual behavior and being acquainted with the usage of things in habitual behavior requires learning these behaviors.

Uddyotakara reveals another important aspect of his theory of *pramāṇa* during a discussion in which he explains Vātsyāyana use of the term *'artha-pratipatti'*, 'perception of an object'. The *pūrvakṣin* asks Uddyotakara to explain why Vātsyāyana uses the word *artha*, 'object'. Perception is always the perception of an object, so the explicit mention of the word 'object' seems to be redundant. The word *artha* is included in the statement, Uddyotakara replies, for the purpose of precluding the perception which has the *pramāṇa* as its object. The objects which *pramāṇa* provides, Uddyotakara explains, reveal purposes that human beings are interested in pursuing. Contrarily, there can be no purpose in pursuing *pramāṇa*.[11] *Pramāṇa* pertains to habitual ways in which things are used. These habitual ways are courses of action people fixed because they have repeatedly proven their capacity to accomplish the purposes they were supposed to accomplish. Doubting that things can be used in the ways they were over and over successfully used before is senseless. It is senseless for people who are capable of climbing trees or cutting trees down to question whether a tree can be climbed on or cut down. Someone who disagrees that a tree can be climbed on or cut down does not fully understand what a tree is. People do not doubt the habitual ways in which they use things because doubting the successful ways in which they use things is senseless. Since the habitual ways in which things are used are not doubted, no purpose can be achieved by pursuing the *pramāṇa*. The *pramāṇa* is, therefore, not fit to serve as an object of perception. More importantly, the fact that the habitual ways in which we use things are beyond doubt establishes them as *pramāṇa*.

---

[11] NV 1.1.1, p. 7, line 1 - अर्थग्रहणं प्रमाणविषयप्रतिपत्तिप्रतिषेधार्थम्. यतो न प्रमाणविषयाप्रतिपत्तिः पुरुषमितिकर्तव्यतायां प्रयोजयति, किं त्वर्थस्य तथाभावविषया. यदायमर्थस्य तथाभावं प्रतिपद्यते, अथ प्रवर्तत इति.

2.2 The *Anumāna-Pramāṇa,* the Criterion of Inferential Knowledge

The guiding principles which Uddyotakara presents in the introduction of the vārtika pertain to all forms of *pramāṇa*. The *pratyakṣa pramāṇa* pertains to ways of usage which are immediately available to us as a result of our capability to follow habitual behaviors. This interpretation of the *pratyakṣa* is in accord with its definition in *sūtra* 1.1.4: A habitual way of using a thing is directly grasped, it can be grasped without the assistance of words and it is unerring and unequivocal. The *anumāna-pramāṇa,* is said to provide knowledge which is over and above the knowledge which *pratyakṣa* provides. The obtainment of inferential knowledge is said be dependent on two conditions: Firstly, it depends on the perception of the *pakṣa,* the object of the inference, as qualified by the inferential sign, the *liṅga*. The question of whether 'the *pakṣa* as qualified by the *liṅga*' is meaningful, i.e., a case of knowledge, can be resolved by the *pratyakṣa-pramāṣa*. If there is a use we can make of 'the *pakṣa* as qualified by the *liṅga*' in a certain habitual behavior, then 'the *pakṣa* as qualified by the *liṅga*' is a case of knowledge. Secondly, inferential knowledge depends on the perception of *avinabhāva*, a relation of invariable concomitance between the *liṅga* and the *sādhya.*[12] In order to figure out what Uddyotakara takes the *avinabhāva* relation to be, we must turn to an especially illuminating discussion he conducts with the *pūrvapakṣin* in vārtika 1.1.5

Uddyotakara and the *pūrvapakṣin* attempt in this discussion to get to the bottom of the relation of invariable concomitance between fire and smoke which renders the inference of the former from the latter possible. Uddyotakara argues that fire and smoke are not causally related, inherently related or even generally related.[13] Smoke, Uddyotakara points out, is sometimes perceived apart from fire - as when smoke rises from ember - and fire is sometimes perceived apart from smoke - as when no smoke is seen to arise from a modern gas stove (that is of course not Uddyotakara's example but mine). The unescapable conclusion, Uddyotakara proceeds to argue, is that there is no constant companionship between smoke and fire.[14] Since there is no constant companionship between smoke and fire, it is wrong to argue that 'wherever there is smoke there is fire' and since maintaining that 'wherever there is smoke there is fire' is impossible, it must be admitted that the presence of fire is not what is inferred from the perception of smoke! The *pūrvapakṣin* responds by saying that this argument of Uddyotakara goes directly against a fact accepted by all men.

---

[12] NV 1.1.5, pp. 41, line 10 - बुभुत्सावतो द्वितीयात् लिङ्गदर्शनात् संस्काराभिव्यक्त्युत्तरकालं स्मृतिः, स्मृत्यनन्तरं च पुनर्लिङ्गदर्शनमयं धूम इति. तदिदमन्तिमं प्रत्यक्षं पूर्वाभ्यां प्रत्यक्षाभ्यां स्मृत्या चानुगृह्यमाणो लिङ्गपरामर्शोऽनुमानं भवति.

[13] NV 1.1.5, pp. 47, line 9 – अस्तु तावत् पूर्वः कार्यकारणभाव इति. तन्न, अतद्वृत्तित्वात्. न हि धूमोऽग्नौ वर्तते नाप्यग्निधूमे, स्वकारणवृत्तित्वात्. अतो न कार्यकारणभावः. नैकार्थसमवायोऽपि, ताभ्यामन्यस्यानारम्भात्. न हि भिन्नजातीयाभ्यां द्रव्यमारभ्यत इति. न च तावप्यन्यत्र वर्तेते, स्वकारणवृत्तित्वादित्युक्तम्. संबन्धमात्रं तत्र वर्तत इति चेत्? तदप्यनुमातुं न शक्यते. कथम्? यदि तावदेवं कुरुते अस्ति संबन्धोऽग्निधूमयोरिति – तन्न, अप्रतीतत्वात्. अनग्निकस्यापि धूमस्य दर्शनात् न संबन्धानुमानम्.

[14] NV 1.1.5, p. 47, line 16 - न, उभयोर्व्यभिचारो दृष्टः. अनग्निधूमो दृष्टः, अधूमश्चाग्निरिति उभयं व्यभिचारि. तस्मात् न साहचर्यमपि.

The following examples may help us understand what Uddyotakara is driving at. Assume one sees smoke rising from her house. Assume, further, that in previous cases she rushed over to her house upon seeing smoke, she found out that no fire was burning there. Would this prevent her from rushing over to her house this time? The answer is absolutely not. She would again rush over, most probably with the same sense of urgency. Now, assume one sees dark clouds approaching. Assume, further, that in some previous cases she saw dark clouds approaching, she rushed outside to take the laundry in, only to find out later that the clouds produced no rain. Would this prevent her from rushing to take the laundry in this time? The answer is probably not. She would again rush outside fearing that the coming rain would wet her clothes.

What the above examples show is that we don't assume the presence of fire upon perceiving smoke because in previous cases we always perceived smoke alongside fire. Rather, we react to smoke as if we see fire because we were taught to act this way.[15] Reacting to fire upon seeing only smoke has repeatedly proven itself useful in saving lives, property, time and effort. The inference of fire from smoke is a pattern of behavior we were taught to follow. It is this pattern of behavior from which the relation of *avinabhāva* between fire and smoke is derived from. A relation of *avinabhāva* is therefore, a relation of meaning. We have learned to associate the meaning of 'smoke' with the meaning of 'fire', we have learned that part of what 'smoke' means is that it indicates 'fire', when we have learned habitual behaviors which involve 'smoke' and 'fire'. The invariable concomitance of fire and smoke is based on the fact that there is a use we can make of 'qualifying smoke by fire' and not on a relation between smoke and fire which is prior to and independent of habitual human behaviors. Generally speaking, the question whether maintaining that 'the *liṅga*' is invariably concomitant with the *sādhya*' is meaningful, whether it is a case of knowledge, can be resolved by the *pratyakṣa-pramāṇa*. If there is a use we make of 'the *liṅga* as invariably qualified by the *sādhya*' in habitual behaviors, then maintaining that 'the *liṅga* is invariably concomitant with the *sādhya*' is meaningful, it is a case of knowledge.

So, provided one is acquainted with the habitual behaviors in which the *pakṣa*, the *liṅga* and the *sādhya* are used, the fact that 'the *pakṣa* is qualified by the *liṅga*' and the fact that 'the *liṅga* is invariably qualified by the *sādhya*' are immediately given to him by the *pratyakṣa-pramāṇa*. However, the inference 'the *pakṣa* is qualified by the *sādhya*' does not 'naturally' follow from the above two perceptual facts. For inference is also a form of habitual behavior that one needs to learn, by exercise, train and drill and by developing skills and techniques. Only one who knows how to infer would accept that 'the *pakṣa* is qualified by the *sādhya*' after he is presented with the facts that 'the *pakṣa* is qualified by the *liṅga*' and that 'the *liṅga* is invariably qualified by the *sādhya*'. It is possible that one would be presented with the facts that 'the *pakṣa* is qualified by the *liṅga*' and that 'the *liṅga* is invariably qualified by the *sādhya*' and yet, refuse to admit that 'the *pakṣa* is qualified by the *sādhya*'. What this refusal demonstrates, however, is this person's incapacity to infer.

---

[15] My above interpretation of inference is inspired by Baker and Hacker commentary on Wittgenstein's Philosophical investigations. See Baker, G. P. & Hacker P.M. S. (2009), p. 88-90.

2.3 The *Upamāṇa-pramāṇa,* the Criterion of Analogical Knowledge

Uddyotakara clarifies best his position regarding the *upamāna-pramāṇa* in a discussion he conducts with the *pūravpakśin* in Vārtika 1.1.39. The question Uddyotakara and the *pūravpakśin* are in disagreement over is how the similarity of things is determined. The two things the similarity of which is examined in this particular discussion are 'the production of dishes', on the one hand, and 'the production of sounds' on the other.

The *pūrvapakṣain* first rejects the possibility that 'the production of dishes' is similar to 'the production of sounds' because dishes and sounds are identical, for the obvious reason that dishes and sounds are very different things. Next, he rejects the possibility that 'the production of dishes' is similar to 'the production of sounds' because dishes and sounds are produced in the same way. Dishes are produced, for example, by putting vegetables, meat and spices in a casserole and by putting the casserole in the oven. Sounds, on the other hand, are produced, for example, by playing a music instrument or by using the vocal chords. Since the above two methods for determining similarity must be rejected, the *pūrvapakṣain* concludes that the similarity between 'the production of dishes' and 'the production of sounds' must be the result of the following procedure: We abstract the property 'being a product' from actual cases in which dishes and sounds are produced, and conclude from the fact that this property qualifies both that 'the production of dishes' and 'the production of sounds' are similar.[16]

Uddyotakara replies that there is no criterion for determining similarity irrespective of the habitual behaviors in which the two things whose similarity is examined are involved.[17] Irrespective of the activity of playing chess, the actions of moving wooden pieces and tapping the computer's keyboard would seem different; in the context of playing chess, however, they could be regarded as similar – making a move in a game of chess. Contrarily, two identical actions of putting a piece of paper in an envelope may seem similar irrespective of the context in which they are performed. In the proper context, however, one action could be a vote placed during an election and the other, a love letter sent from a husband to his wife. There are no general rules prior to and independent of habitual behaviors for determining whether two things *p* and *q* are similar. Determining whether *p* and *q* are similar is possible only with respect to the rules of a certain behavior.[18] The *upamāna-pramāṇa*, the criterion of analogical knowledge, is that *p* and *q* are similar if the rules of the habitual behavior in which they participate determine them to be similar.

[16] NV 1.1.39, pp. 130, line 7 – यदपि, यथा तथेत्युपसंहारे कृते त्थेत्यनेन शब्देन सर्वसामान्यं वा, कृतकत्वसामान्यं वाभिधीयते? कृतकत्व विशेषो वा? सर्वसामान्यं तावत् न युक्तम्, तथेति व्यपदेशाशक्यत्वादिति. कृतकत्वविशेषोऽपि न युक्तः, अन्यथा शब्दस्य कृतकत्वात्. परिशेषात् कृतकत्वसामान्यम्. तच्च हेतुनैवोक्तमिति – तदप्ययुक्तम्, उपनयस्योपमानार्थत्वात्.

[17] NV 1.1.39, pp. 130, line 12 – कृतकत्वसामान्यं तु शब्दसन्निधावभिधीयत इति चित्रमिदम्. कृतकत्वसन्निधावभिधीयत इति. शब्देन च विशिष्यमाणं कथं सामान्यं भविष्यति?

[18] The idea to interpret Uddyotakara's point thus I have taken from Baker's and Hacker's essay 'Doing the right thing and doing the same thing' in BAKER-HACKER, pp.145 - 149

2.4 The *Śabda-pramāṇa,* the Criterion of Verbal Knowledge

The *śabda-pramāṇa,* the criterion of verbal knowledge, would not concern us in the present paper. It is sufficed to say that an assertion *p* is a piece of verbal knowledge if it expresses the perceptual, inferential or analogical knowledge it is supposed to express in accord with the rules of everyday language and the rules of grammar. Similarly to the other *pramāṇas,* there are no general rules prior to and independent of habitual behaviors for determining whether an assertion is a piece of knowledge. An assertion is a piece of knowledge if what it expresses was provided by the *pratyakṣa,* the *anumāna* or the *upamāna-pramāṇas* and if it expresses it in accord the rules of everyday language and the rules of grammar.

3 Nyāya's Five-Limbs Logical Model for Ascertaining Sound Arguments

The *pramāṇas,* the criteria of knowledge, are peculiar and specific rules of usage derived from habitual behaviors. There are no general rules prior to and independent of habitual behaviors for determining whether a piece of perceptual, inferential, analogical or verbal information is meaningful and can qualify as a piece of knowledge. However, the fact that the criteria of knowledge are particular and behavior-dependent, by no means renders them impractical. For, albeit being particular and behavior-dependent, they are well familiar and accepted by all.

Let us try to illustrate the above point. Disputants may disagree whether the argument 'ants are intelligent because they are capable of making decisions' is sound. But they must agree that 'humans are intelligent because they are capable of making decisions' and that there is 'nothing which is, one the hand, capable of making decisions and, on the other, not intelligent'. Disagreement by one of the disputants over one of the propositions presented above would entail that he does not fully understand what 'intelligence' and 'making decisions' are; and it is impossible to conduct a fruitful discussion with someone who does not fully understand the meaning of the concepts he wish to discuss. On the other hand, an all-embracing agreement by all the disputants over the above propositions establishes the *avinabhāva* relation between 'the capability of making decisions' and 'intelligence'.

Similarly, disputants may disagree whether the argument 'penguins are birds because they have feathers' is sound. But they must agree that 'parrots are birds because they have feathers' and that 'nothing is known to have feathers and yet, not be a bird'. Disagreement by one of the disputants over the above propositions necessarily entails that he does not fully understand what 'birds' and 'feathers' are and it is impossible to conduct a fruitful discussion with someone who does not fully understand the meaning of the concepts he wishes to discuss. On the other hand, an all-embracing agreement by all the disputants over the above propositions establishes the *avinabhāva* relation, the relation of meaning between 'feathers' and 'birds'.

The principle which underlies Nyāya's five-limbs logical model is, then, that arguments are sound only if they accord with the criteria set by the *pramāṇas*. That is, arguments are sound only if the concepts they employ accord with the rules of the use of the things they denote in everyday life. If an argument fails to satisfy the said criteria, it follows that it employs concepts contrary to the rules of their use, that is, it follows that the concepts it employs assign to the things they denote, meaning which is in discord with their meaning as everyday life determines it to be. Such arguments which fails to satisfy the criteria of soundness must be dismissed as incoherent and nonsensical, however convincing they may seem.[19]

To illustrate how the model he proposes operates, Uddyotakara chooses the philosophically-loaded argument 'sound is not eternal'. Uddyotakara's choice of the argument 'sound is not eternal' is not incidental. Demonstrating beyond doubt that the above argument is sound, demonstrates *ipso facto* that the opposite argument 'sound is eternal' is nonsensical. The argument that 'sound is eternal' is an exceptionally important and an entrenched position in Indian culture. Showing it to be nonsensical is a volatile and subversive move on behalf of Uddyotakara.

Let us now observe how the five-limbs logical model operates:

3.1 The Statement of the Proposition – 'Sound is Not Eternal'.

This piece of verbal knowledge must satisfy the conditions set by the *śabda-pramāṇa*. These conditions are that the perceptual fact which the proposition expresses is sound and that is expresses it in accord with the rules of grammar. At this stage, the truthfulness of the proposition is known only to its proponent. The fact that its truthfulness is unknown at this stage to the other disputants does not mean that it does not satisfy the conditions set by the *śabda-pramāṇa*. For whether or not it satisfies these conditions depends solely on the truthfulness of the perceptual fact the propositions expresses and on the fact that it expresses it in accord with the rules of grammar.

3.2 The Statement of the Reason – 'Sound is Not Eternal because it is Produced'

The statement of the reason must satisfy the conditions set by the *anumāna pramāṇa*, the criterion of inferential knowledge. According to Uddyotakara's account of *anumāna* we have presented above, if the rules of habitual behaviors determine that sounds are produced and that whatever is produced is not eternal, the statement 'sound is not eternal' must be a piece of inferential knowledge. The purpose of the subsequent two stages of the logical model would be to provide conclusive evidence that sounds are produced and that whatever is produced is not eternal. If such conclusive evidence would be provided, inferring that 'sound is not-eternal' would be irresistible, for the following simple reason: if sound is shown beyond doubt to be the kind of thing which, if it is produced, it is non-eternal, it makes no sense not to infer that sound is non-eternal.

3.3 The Statement of a Familiar Instance – 'Dishes are Produced and are Not Eternal'

The familiar instance depends on the *pratyakṣa-pramāṇa*, the criterion of perceptual knowledge. The role of the familiar instance is first, to establish the existence of a group

[19]The idea that assertions must be dismissed if they are incompatible with everyday known facts I have borrowed from Descombes. In DESCOMBES, pp.1 he states: "Phenomenon" here means whatever may contradict our speculations and lead us to correct out initial descriptions. There are phenomena if there are facts that could result in the overthrow of even the most entrenched dogmas or in the rejection of the conclusions of even the soundest lines of reasoning.

of objects about which it makes sense to say that they are produced and non-eternal and second, to make sure that there is not even a single known object about which it makes sense to say that it is produced and yet, eternal. The familiar instance Uddyotakara suggests is 'a dish is produced and not eternal'. The acceptance of the said familiar instance by all the disputants is a precondition for conducting the discussion. If a disputant disagrees that dishes are understood in habitual behaviors to be a kind of thing which is produced and non-eternal, this merely shows that he does not fully understand the meaning of the concepts 'produced' and 'non-eternal', the very concepts the discussion is about. On the other hand, if all the disputants agree that dishes are produced and non-eternal, the existence of a group of objects which are produced and non-eternal is established. If the disputants moreover fail to come up with an example of an object which is produced and yet, is eternal, the relation of *avinabhāva* between production and non-eternality is established. The only question left now is whether the *avinabhāva* relation between production and non-eternality which pertains to dishes pertains also to sounds.

3.4 The Statement of the Application – 'Like Dishes, Sounds are also Produced and are not Eternal'.

The question whether the *avinabhāva* relation between production and non-eternality pertains to sounds can be answered by an appeal to the *upamāna-pramāṇa*, the criterion of analogical knowledge. The *upamāna* set one important condition: Two things are similar if they perform the same role in the context of a certain habitual behavior. In our particular case, dishes and sounds are similar if they are both produced, that is, if they did not exist prior to a certain point of time and began to exist since that point of time. Since sound is a kind of thing which does not exist prior to a certain point of time and begins to exist since that point of time, sounds are similar to dishes in this respect. Like dishes which are produced and not eternal, sounds are also produced and not eternal.

At this stage, anyone who is capable of inferring cannot resist accepting the conclusion that 'sound is not eternal', for the conditions necessary for inference are fully met. It was demonstrated that the rules of the use of 'sound' in habitual behaviors determine its meaning to be a kind of thing which is produced and non- eternal. The deeply entrenched position that 'the sounds of the vedas are eternal' is shown to be nonsensical when it is confronted with the meaning everyday habitual behaviors assign to 'sound'. The said traditional and deeply entrenched position is shown to employ the words 'sound' and 'eternality' contrary to the meaning determined to them by everyday parlance.

3.5 The Statement of the Conclusion – 'Sound is not Eternal'

The same statement that was presented in stage one is presented again now as a conclusion. The truthfulness of the proposition is now known not only to its proponent but also to the other disputants.

To conclude, the purpose of the logical model described above is to confront philosophical arguments with everyday facts, i.e., with the meaning which everyday habitual behaviors determine to the things which constitute human experience. The logical model of Nyāya determines that philosophical arguments are sound and meaningful, if the concepts they employ conform with the meaning that everyday usage determine to the things they denote. This is in accord with Nyāya position that only arguments which conform to the rules of everyday life are strictly speaking, logical and rational.

References

BAKER-HACKER = Baker, G.P., Hacker, P.M.S.: Wittgenstein: Rules, Grammar and Necessity. Analytical Commentary on the Philosophical Investigations, vol. 2. Wiley-Blackwell, Chichester (2009)

DESCOMBES = Descombes, V.: The Mind's Provisions. A Critique of Cognitivism. Princeton University Press, Princeton and Oxford (2001)

JHA = Jhā, G. (trans.): The Nyāya-Sūtras of Gautama with the Bhāṣya of Vātṣyāyana and the Vārtika of Uddyotakara Translated into English (4 Vols). Motilal Banarsidass, Delhi (reprint 1984)

MATILAL 1986 = Matilal, B.K.: Perception: an Essay on Classical Indian Theories of Knowledge. Clarendon Press, Oxford (1986)

MATILAL 1990 = Matilal, B.K.: Logic, Language, and Reality: an Introduction to Indian Philosophical Studies. Motilal Banarsidass, Delhi (1990)

MATILAL 1998 = Matilal, B.K.: The Character of Logic in India. In: J. Ganeri & H. Tiwari (eds.). State University of New York Press, Albany (1998)

MATILAL 2002 = Matilal, B.K.: On the Concept of Philosophy in India. In: Ganeri, J. (ed.) The Collected Essays of Bimal Krishna Matilal. Mind, Language and World, vol. 1. Oxford University Press, New Delhi (2002)

MOHANTY = Mohanty, J.: Reason and Tradition in Indian thought: an Essay on the Nature of Indian Philosophical Thinking. Clarendon Press, Oxford (1992)

NV = Thakur, A. (ed.): Nyāyabhāṣyavārtika of Bhāradvāja Uddyotakara. Indian Council of Philosophical Research, New Delhi (1997)

Negative Existentials and Non-denoting Terms

Paul Schweizer

Institute for Language, Cognition and Computation
School of Informatics, University of Edinburgh
paul@inf.ed.ac.uk

Abstract. Logical and semantical issues surrounding non-denoting terms have been investigated since ancient times, in both the Western and Indian philosophical traditions. And in a more applied formal setting, such issues have also gained importance in constructive mathematics, as well as computer science and software engineering. The paper first presents a strategic exploration of logical treatments of reference failure in Western thought, and then goes on to provide a comparative examination of the issue in the Indian tradition, particularly with respect to the dispute between the Yogācāra-Sautrāntika school of Buddhism and the Nyāya school of Hinduism. The paper concludes by advancing a formalization of the Buddhist *apoha* semantical theory in terms of a dual-domain Free logic.

1 The Analysis of Non-existence in Western Logic

It is a distinctive feature of human language and thought that we can introduce terms purporting to designate some object or entity in the world, but where no such object or entity exists. And we can then go on to use such terms to make grammatically well formed assertions which appear to be *meaningful*, and indeed many of these meaningful assertions about non-existent objects seem to be either *true* or *false*. This phenomenon poses some deep challenges for philosophy and logical theory which have been recognized and investigated since ancient times, in both the Western and Indian traditions. And in a more applied formal setting, such issues have also gained importance in constructive mathematics, as well as computer science and software engineering. In the context of ancient Greek philosophy, a well known version of the problem is articulated in Plato's riddle of non-being, often referred to as the predicament of 'Plato's beard'. Let us suppose that Plato was always a clean shaven individual and never sported facial hair. In such a case, we would seem to be asserting a true proposition with the negative existential statement 'Plato's beard did not exist'. But if Plato's beard did not exist, then exactly what are we talking about when we say that he didn't have one? And how can we make any coherent assertion involving the term 'Plato's beard' when, by hypothesis, it fails to denote? Even the cogency of the seemingly innocent 'Plato did not have a beard' seems threatened.

M. Banerjee and S.N. Krishna (eds.): ICLA 2015, LNCS 8923, pp. 183–194, 2015.

1.1 Sense, Reference and Definite Descriptions

Frege's [1] distinction between sense and reference provides a powerful and far reaching response to scenarios such as Plato's beard. In accord with Kant's maxim that existence is not a predicate applying to individual objects, Frege analyzed assertions of existence in terms of the extensions of concepts. Hence to say that aardvarks exist is, in effect, to say that the 'cognitive content' or sense (*Sinn* in Frege's terminology and intension in Carnap's) expressed by the term 'aardvark' is true of at least one object in the universe of discourse. So there are individuals in the range of the existential quantifier that satisfy or 'fall under' the concept 'aardvark'. And conversely, to say that unicorns do not exist is to say that no individuals in the range of the existential quantifier fall under this concept, and hence that its extension is empty. This makes existence a second order claim about concepts rather than objects. 'Aardvark' and 'unicorn' are general terms to which singular terms can attach to form atomic statements. Frege applies the distinction between sense and reference to singular terms as well, such as 'Pegasus' or 'Sherlock Holmes'. To say that Pegasus does not exist is again to say that no individual in the domain of discourse falls under the 'Pegasus' concept. In other words, 'Pegasus' has a sense but no reference.

This dual level analysis provides an elegant explanation of why terms with empty extensions can still contribute to *meaningful* discourse. At the level of intension or sense, there is still semantic content associated with terms such as 'Pegasus' and 'unicorn'. According to Frege's principle of compositionality, the semantic value of a complex whole is a function of the semantic values of its respective parts and their mode of combination. Propositions (or 'complete thoughts') are the intensions of declarative sentences, and the sense of a non-denoting term such as Pegasus can still contribute to the intensional level of sentences in which it occurs, to yield a meaningful proposition. And indeed, this supplies a very elegant explanation of the semantic content conveyed by literature and other forms of fictional discourse.

A proposition is the intension of a declarative sentence, while for Frege its extension or reference is a truth value. In accord with the above principle of compositionality, failure of reference for singular terms must turn the method of designating the reference of a sentence involving such terms into a partial function on the range {True, False}. Since Pegasus has a sense but no reference, the sense can contribute to a proposition, while the lack of reference entails that functional combination at this level fails, and sentences involving Pegasus will lack a truth value. If 'Pegasus' has no referent then neither does the sentence 'Pegasus is winged', so that its truth value is undefined or '*u*'. Bivalence must be sacrificed if genuinely non-denoting terms are allowed, and the principle of strict compositionality requires lack of reference to recursively propagate in the manner of an infectious disease. If the referent of 'Pegasus is winged' is *u*, then the value of, e.g. 'Pegasus is winged or snow is white' must also be *u*, because if there is an input missing to the disjunctive truth function then there can be no output. This yields a version of Kleene's system of weak 3 valued logic.

In Russell's [2] response to Frege the level of intension is not invoked, and instead Russell focuses purely on referential considerations. His 'logically proper name' is a pure indexical referring to immediate aspects of raw sensation, while standard and logically improper names are analyzed along the lines of definite descriptions. On Russell's account, expressions involving the definite article, such as 'the ϕ', are treated according to the standard existence and uniqueness constraints, $\exists x(\phi x \wedge \forall y(\phi y \leftrightarrow y = x))$. This analysis yields a formula rather than a singular term, and to make a further assertion *about* 'the ϕ', requires an appropriate augmentation of the base formula. Hence 'The ϕ is Ψ' is formalized as $\exists x(\phi x \wedge \forall y(\phi y \leftrightarrow y = x) \wedge \Psi x)$. If there is no object in the domain of discourse satisfying both the existence and uniqueness constraints, then 'the ϕ' is a vacuous description and the corresponding formula above will be false, as will any further formula attempting to assert something about 'the ϕ'. There is no present King of France, and if we let Kx symbolize the property in question, then $\exists x(Kx \wedge \forall y(Ky \leftrightarrow y = x))$ is rendered false by the falsity of the first conjunct. Consequently 'The present King of France is just' and 'The present King of France is not just' both turn out false (on both narrow and wide readings of negation), and now uniform falsity, rather than lack of truth value, propagates through the account.

But, contra both Frege and Russell, there is an intuitive sense in which we might want to make *true* assertions using non-denoting terms, such as those involving basic logical properties like self identity: 'The present King of France is identical to the present King of France', or statements using fictional names that affirm details of the literary context, like 'Sherlock Holmes was a brilliant detective'. It is also convenient to retain the logical form of a genuine *singular term* for both proper names and definite descriptions. But this won't work in classical logic for expressions that don't refer. If **t** is a singular term standing, say, for 'Plato's beard', then the negative existential mentioned above, *viz.*, $\neg\exists x(x = \mathbf{t})$ is a *contradiction* in classical first-order logic with identity, since it's a basic requirement of the model theory that **t** be assigned some object in the domain. This highlights a crucial asymmetry in the classical approach, where general terms are allowed to have empty extensions while singular terms are not.

1.2 Free Logic

As Lambert [3] perspicuously observes, the branch of non-classical logic known as Free logic is largely motivated in response to this asymmetry. The traditional logic of general terms supposed that the inference from $\forall y(\phi y \rightarrow \Psi y)$ to $\exists y(\phi y \wedge \Psi y)$ was valid, because the terms ϕ and Ψ were thought to have *existential import*. But this imposes an unwanted restriction on the range of applicability of formal reasoning, and on the modern and broader approach no such import is presupposed. The general terms ϕ and Ψ are allowed to be true of nothing, and hence the inference is invalidated. For example, since there are no unicorns, the actual world is a model of the sentence 'Every unicorn is an aardvark', formalized as $\forall y(Uy \rightarrow Ay)$, while it is false that $\exists y(Uy \wedge Ay)$, so the actual world serves as a counterexample to the inference. On the modern approach, an *additional premise* of the form $\exists y(Uy)$ is required to restore existential import and yield the valid (but unsound) piece of reasoning: $\forall y(Uy \rightarrow Ay)$, $\exists y(Uy)$ $\therefore$ $\exists y(Uy \wedge Ay)$.

However, classical first-order logic with identity retains a somewhat curious exception to the need for an additional premise. If the (potentially complex) 1-place predicate expression ϕy is replaced with the complex 1-place predicate $y = \mathbf{t}$, then the original inference pattern $\forall y(y = \mathbf{t} \rightarrow \Psi y)$ $\therefore$ $\exists y(y = \mathbf{t} \wedge \Psi y)$ goes through on its own.

The expression ' $= \mathbf{t}$' has existential import in the traditional sense, while in general the expressions ϕy and Ψy do not. This traditional residue derives from the asymmetrical fact that singular terms are required to denote while general terms can be empty. $\exists y(y = \mathbf{t})$ is a truth of classical logic for every singular term $\mathbf{t}$ in the language, and hence does not need to be introduced as an extra premise. This can itself be viewed as an undue restriction on the range of applicability of formal reasoning, since it is not possible to carry out intuitively plausible inferences concerning objects that do not or might not exist in the actual world. And in the same manner as above, the natural strategy is to devise a logic free of existence assumptions with respect to its terms, both *singular* and general (Lambert [4]).

In Free logic, the quantifiers are interpreted in the normal way, as ranging over some domain of discourse $\boldsymbol{D}$, normally construed as the set of 'existent objects'. But the singular terms may denote objects outside of $\boldsymbol{D}$, or fail to denote altogether. This de-coupling of singular reference from the range of the quantifiers undermines two fundamental inference patterns of classical logic, namely Universal Instantiation (UI) and Existential Generalization (EG). According to UI, $\forall y\phi y \;\therefore\; \phi\mathbf{t}$ is a valid inference. But it fails in Free logic because the quantifier $\forall y$ only ranges over objects $e \in \boldsymbol{D}$, whereas '$\mathbf{t}$' may not refer to any such e. So from the fact that every $e \in \boldsymbol{D}$ has property ϕ, it does not follow that $\mathbf{t}$ does. And according to EG, $\phi\mathbf{t} \therefore \exists y\phi y$ is a valid inference. But similarly this fails in Free logic because, e.g., $\mathbf{t}$ may denote a nonexistent object not in the range of $\exists y$, thus allowing for the possibility of true premise and false conclusion.

Analogous to the foregoing transition from traditional to modern logic in the case of general terms, now that singular terms are also free of existence presuppositions, an *additional premise* is required to restore validity. Existential import with respect to singular terms is expressed via an existence predicate for individuals (in violation of Kantian notions), normally using Russell's '*E!*' notation. With the use of identity, the existence predicate can be defined as $E!(\mathbf{t}) \;:=_{\mathrm{def}}\; \exists y(y = \mathbf{t})$. In the case of both UI and EG, $E!(\mathbf{t})$ is the suppressed premise required to yield an inference pattern valid in the context of Free logic. Hence $\mathrm{UI}_{\mathrm{Free}}$ has the form $\forall y\phi y,\, E!(\mathbf{t}) \;\therefore\; \phi\mathbf{t}$, and $\mathrm{EG}_{\mathrm{Free}}$ has the form $\phi\mathbf{t},\, E!(\mathbf{t}) \;\therefore\; \exists y\phi y$. It is now possible to directly articulate the fact that Pegasus does not exist with the formula $\neg E!(\mathbf{t})$, letting $\mathbf{t}$ denote the mythical flying horse. And while it's true that neither Plato's beard nor Pegasus exist, it's nonetheless *false* that $\exists x\neg E!(x)$.

1.3 Definite Descriptions Revisited

As noted earlier, Russell's 1905 theory of definite descriptions analysed expression such as 'the ϕ' in terms of a formula rather than a singular term. However, it is often convenient to be able to render such expressions as genuine terms, and have a uniform treatment of simple terms such as individual constants or proper names, along with complex singular terms such as definite descriptions and function terms. In *Principia Mathematica*, Russell [5] introduced his variable-binding, term-forming 'iota' operator to do just that. If it's provable that the existence and uniqueness conditions are satisfied, then a Russellian iota operator '$\imath$' yields a complex singular term as follows: if $\vdash \exists x(\phi x \wedge \forall y(\phi y \leftrightarrow y = x))$ then $\imath x\phi x$, read as 'the x such that ϕx', or simply 'the ϕ' is defined (contextually) as *that* unique x. The definite description '$ix\phi x$' can then be used as a legitimate complex singular term for making assertions such as $\exists y(y = ix\phi x)$, $\Psi(ix\phi x)$, and the seemingly innocuous $\phi(ix\phi x)$. In the special case of definite descriptions, a 1-place predicate ϕx is used to define a 0-place function, i.e. a singular term. In the general case, an n-ary relation $R^n(x_1,..., x_{n-1},y)$ can be used to define an $(n - 1)$-ary *total* function f^{n-1}, if R^n satisfies the corresponding existence and uniqueness constraints $\forall x_1 ... \forall x_{n-1}\exists y\ \forall z[R^n(x_1,...x_{n-1},y) \wedge (R^n(x_1,...x_{n-1},z) \rightarrow z = y)]$ in which case $f^{n-1}(x_1,..., x_{n-1}) = y$ and the set of $(n - 1)$-ary total functions can be viewed as a proper subset of the set of n-place relations.

However, not all functions that we might wish to consider are total, and this can be due to a failure of either constraint. Furthermore, such failures might not be known to us at the time the function term is introduced. For example, 0-place definite descriptions are often vacuous, as in 'the greatest prime number', although prior to Euclid's proof the semantic status of this description was not definitively known. The function $f(x) = x^{-1}$ on the reals is partial, since it is not defined in the case of $x = 0$, and the description 'the x such that $x^2 = 2$' fails the uniqueness constraint. Nonetheless it is often expedient to perform logical and mathematical manipulations involving partial functions, and thus in the general case Russell's constraints seem unduly restrictive. For example, on Russell's account, it is a logical truth that $\exists y(y = ix\phi x)$. However, it might be useful to be able to introduce the term $ix\phi x$ *without* first proving that the existence condition is satisfied, *a la* Free logic, and then employ the term to articulate the discovery that $\neg E!(ix\phi x)$, if it's later found that no such object exists.

In the context of providing a foundation for mathematics, Frege sought to avoid the truth value gaps mentioned above that result from descriptions that fail to denote, and his solution was to assign a 'dummy value' from the realm of existents. This is akin to the current strategy in computer science of assigning an 'error object' in such cases (see Gumb [6]). The (generic) Free logic approach is to dispense with existence assumptions for such terms and use the existence predicate to preserve valid patterns of inference. This is also the intuitive strategy adopted by Troelstra and van Dalen [7] with their E-logics in the context of constructive mathematics. Within Free logic there are various choices regarding descriptions that fail to denote. Making all atomic formulas containing empty descriptions *false* yields a 'negative' free description theory equivalent to Russell. In contrast, making all identities between empty descriptions *true* yields a 'positive' description theory analogous to Frege's solution above, although instead of taking the 'dummy value' from the realm of existents, it is now more natural to use a nonextistent object, *as per* the semantics outlined below. So called 'neutral' Free description theories constitute yet a third option, where bivalence is sacrificed and statements involving empty terms lack a truth value, as on Frege's strictly compositional approach.

1.4 Inner and Outer Domains

From the point of view of classical semantics, there are two distinct ways in which singular terms can fail to denote. First, a term can be genuinely empty, in the sense that it maps to nothing at all, in which case the semantical interpretation function on the set of terms is itself partial. Second, it can map to something, but this 'something' is not in the realm of actual or proper existents, and hence is outside the range of the (classical) quantifiers. In this case the semantical interpretation function can be total, but with a range that exceeds the scope of the quantifiers. This is in broad accord with Meinong's [8] famous and influential distinction between existentent and subsistent objects. Subsistence is a wide ontological category that includes both concrete and abstract objects, where concrete objects both exist and subsist, while abstract entities merely subsist. Meinong's idea serves as an inspiration behind a standard version of Free logic in which the semantic structures have both an 'inner' and 'outer' domain, and where the inner domain $\boldsymbol{D}_i$ specifies the universe of existent objects over which the quantifiers range. There are technical choices to be made concerning the relation between $\boldsymbol{D}_i$ and the outer domain $\boldsymbol{D}_o$, and it's possible to make them disjoint, or to adopt the Meinongian picture and let $\boldsymbol{D}_i \subseteq \boldsymbol{D}_o$. In the current exposition the latter option will be selected, and we will allow $\boldsymbol{D}_i$ (although not $\boldsymbol{D}_o$) to be empty, thereby evading yet another philosophically dubious presupposition of classical logic, namely that at least one object must exist, which presupposition is embodied in the logical truth $\exists y(y = y)$. A straightforward semantics for this type of dual-domain Free logic can be specified as a direct extension of the classical approach, where the objects not belonging to the inner domain cannot be accessed by the quantifiers, but where such objects *can* be accessed by the interpretation function, both to serve as the referents of singular terms, and to appear in the extensions of predicate expressions.

A Free logic interpretation for the respective first-order language with identity $\mathcal{L}$, is a triple $< \boldsymbol{D}_i, \boldsymbol{D}_o, \boldsymbol{f} >$, where $\boldsymbol{D}_i$ is a (possibly empty) set of existent objects, $\boldsymbol{D}_o$ is a (non-empty) set of subsistent objects, and $\boldsymbol{D}_i \subseteq \boldsymbol{D}_o$. $\boldsymbol{f}$ is an interpretation function such that for every individual constant $\boldsymbol{c}$ of $\mathcal{L}$, $\boldsymbol{f}(\boldsymbol{c}) \in \boldsymbol{D}_o$, and for every n-place predicate $\boldsymbol{P}^n$ of $\mathcal{L}$, $\boldsymbol{f}(\boldsymbol{P}^n) \subseteq \boldsymbol{D}_o{}^n$. Given an interpretation $< \boldsymbol{D}_i, \boldsymbol{D}_o, \boldsymbol{f} >$, the valuation function V assigns truth values to formulas Θ of $\mathcal{L}$ in the following manner (truth functional combinations are evaluated as normal):

(i) if Θ is of the form $\boldsymbol{P}^n\, c_1,\ldots, c_n$, then $V(\Theta)$ = True iff $< \boldsymbol{f}(c_1),\ldots,\boldsymbol{f}(c_n) > \in \boldsymbol{f}(\boldsymbol{P}^n)$. $V(\Theta)$ = False otherwise;

(ii) if Θ is of the form $\boldsymbol{c}_1 = \boldsymbol{c}_2$, then $V(\Theta)$ = True iff $\boldsymbol{f}(\boldsymbol{c}_1)=\boldsymbol{f}(\boldsymbol{c}_2)$. $V(\Theta)$ = False otherwise;

(iii) if Θ is of the form $E!(\boldsymbol{c})$, then $V(\Theta)$ = True iff $\boldsymbol{f}(\boldsymbol{c}_1) \in \boldsymbol{D}_i$. $V(\Theta)$ = False otherwise;

(iv) if Θ is of the form $\forall \boldsymbol{v} \phi \boldsymbol{v}$, then $V(\Theta)$ = True iff for every $e \in \boldsymbol{D}_i$, $V_e^{\boldsymbol{a}}(\Phi \mathbf{v}/\boldsymbol{a}) = 1$, where $\boldsymbol{a}$ is a *new* individual constant, $\Phi\mathbf{v}/\boldsymbol{a}$ is the result of substituting $\boldsymbol{a}$ for every free occurrence of $\mathbf{v}$ in Φ, and $V_e^{\boldsymbol{a}}$ is the valuation function on the interpretation $< \boldsymbol{D}_i, \boldsymbol{D}_o, \boldsymbol{f}^* >$ which is exactly like $< \boldsymbol{D}_i, \boldsymbol{D}_o, \boldsymbol{f} >$ except that $\boldsymbol{f}^*(\boldsymbol{a}) = e$. $V(\Theta)$ = False otherwise.

In this 'positive' Free logic, predications involving nonexistent objects can be evaluated as *true* on the basis of set membership, in the typical Tarskian fashion. For example, suppose the merely subsistent Pegasus is an element of $\boldsymbol{D}_o$ but not $\boldsymbol{D}_i$, the 1-place predicate Wx stands for the property of 'being winged', $\boldsymbol{f}(\boldsymbol{c}_1)$ = Pegasus, and $\boldsymbol{f}(W)$ = {Pegasus, ... }. Then 'Pegasus is winged' is formalized as Wc_1 and is evaluated as True, while $E!(c_1)$ comes out False. Leblanc and Thomason [9] provide a

7-schema axiomatization of Free logic with identity which is sound and complete with respect to this semantics. It incorporates the UI_{Free} rule previously discussed, as well as the axiom $\forall xE!x$.

1.5 Actual versus Possible

Another well developed framework for dealing with objects that can be referred to but do not actually exist is supplied by modal logic, and the discussion of Western logic will finish with a brief examination of possible world semantics. Although in the actual world Plato did not possess a beard, it's nonetheless *possible* that he could have grown one, say like Aristotle's, and so there's a plausible sense in which Plato's beard 'exists' in alternative possible worlds. Similarly, there have never been flying horses in this world, but if biological evolution had taken a somewhat different course then there *might* have been. Indeed, the possibility of a winged horse seems no more outlandish than the palaeontological fact of flying dinosaurs, and thus Pegasus is a possible though non-actual creature.

There are a number of options and technical choices that must be made when providing a semantics for quantified modal logic, and Kripke's [10] groundbreaking work adopts some key choices that embody principles of Free rather than classical logic. The most distinctive of these concerns the extensions of predicates. Each world w in a modal structure has a domain $\boldsymbol{D}_w$ of objects that exist at that index. Let $\boldsymbol{UD}$ be the union of all domains $\boldsymbol{D}_w$ for worlds in the structure. Then the binary interpretation function $\mathcal{I}(w, \boldsymbol{P}^n)$ can assign an object e to the extension of the predicate $\boldsymbol{P}^n$ at some world w, even though $e \notin \boldsymbol{D}_w$ and hence e does not exist at that world. The only restriction is that $\mathcal{I}(w, \boldsymbol{P}^n) \subseteq \boldsymbol{UD}^n$. Conversely, a predication can turn out to be *false* in a world w, when evaluated with respect to an object $e \notin \boldsymbol{D}_w$, but where *e does exist* at another world w' which has access to w. In addition, Kripke upholds the principle that the quantifiers have existential import and are thereby restricted at each world to the set $\boldsymbol{D}_w$. This combination of features is in harmony with the positive dual-domain semantics for Free logic described above, where $\boldsymbol{UD}$ corresponds to the outer domain D_o, while $\boldsymbol{D}_w$ constitutes the inner domain $\boldsymbol{D}_i$ of locally existent objects over which the quantifiers range. One of the prime advantages of this combination of choices is that it allows both the Barcan formula, $\forall x\Box\Psi x \rightarrow \Box\forall x\Psi x$, and its equally implausible converse to be refuted, thereby yielding the maximum degree of articulation with respect to scope interactions between the quantifiers and the modal operators. Neither the Barcan formula nor its converse are derivable in Free logic, whereas both are valid in straightforward modal extensions of classical logic (see Schweizer [11] for further discussion).

2 The Analysis of Non-existence in Classical Indian Philosophy

In classical Indian philosophy, the riddle of non-being was a historical focal point of controversy, particularly between rival Buddhist and Hindu schools. The remainder of the paper will explore the polemical exchange between the Yogācāra-Sautrāntika school of Buddhism and the orthodox Nyāya *darśana* of Hinduism. The exposition relies primarily on Matilal [12,13], Siderits [14,15] and Tillemans [16] as sources.

2.1 The *Apoha* Semantics of Dharmakīrti

In the 7th century A.D. the Yogācāra-Sautrāntika philosopher Dharmakīrti provided an extended development of the nominalistic theory of his predecessor Dinnāga. *Apoha* nominalism emerged within an ontological framework of radical particularism, in which each existent is held to be absolutely unique and distinct from every other, and thus it is not strictly true to say that two objects have a property in common. Not only are there no universals or abstract entities crowding the metaphysical heavens, but there are not even genuine similarities or resemblances between distinct objects to underwrite our everyday use of property terms. On this beautifully self-consistent analysis, the conventional use of property terms is explained in purely negative fashion. Every object differs absolutely from every other, but objects differ from each other in different ways, and these assorted modes of differing sustain, through two applications of negation, our use of ordinary language predicates.

The *apoha* analysis is based on the idea that the conventionally correct use of a term is acquired through various learning episodes, where encounters with particular objects give rise to a mental paradigm which guides the language user's verbal behavior. This paradigm serves as an internal representation or conceptual 'image', whose primary function is to exclude incompatible representations and thereby specify some portion of the term's *anti-extension*. For example, use of the term 'cow' is based on a particular mental paradigm, which does not encode the abstract features which all cows (are mistakenly supposed to) share, but rather which guides our ability to exclude other objects, and hence judge that a given table or chair is a non-cow. So the particular paradigm or conceptual construct is first used to exclude non-cows, and the extension of the general term 'cow' is obtained through a second application of negation, as the set of all things which are not non-cows. In this manner, the extension of the general term is obtained without commitment to any genuine properties or positive similarities shared by members of the set of cows.

Of course, this immediately leads to the question, 'On what basis does the paradigm exclude some objects and not others, if not by tacit appeal to relevant similarities?' Dharmakīrti's consistent, though semantically somewhat unsatisfying answer, is that exclusion is a *causal* property of the representation as an actual cognitive structure, so that incompatibility between representations is not a logical or semantical trait, but rather is more akin to a repulsive mechanical force. Dharmakīrti's view is not an 'idea' theory of meaning, and the mental paradigm is not a type of pictorial image accessible to consciousness. According to Siderits [14], 'for the Buddhists the psychological machinery that explains our use of words to refer... is purely causal in nature and semantically invisible' (p. 99). Thus unlike the Fregean model, *apoha* semantics is predominantly concerned with reference, and we discover nothing about sense or conceptual *content* when we discover that, say, two terms 'A' and 'B' are co-extensive. Instead this is always a quasi-empirical finding.

As one of the basic metaphysical tenants of Buddhism, the Yogācāra-Sautrāntika school embraced the principle that existence is purely momentary. On this view, the world is not materially preserved from one moment to the next, but rather consists of a series of discrete 'instants' (*kṣaṇas*) of existence, followed by complete annihilation before the next instant occurs. In combination with this view of reality as a kind of Heraclitean flux, the Yogācāra-Sautrāntikas also supported the widespread Indian distinction between brute sensation (*nirvikalpika*) and determinate perception (*savikalpika*). According to this distinction, the raw data supplied by sensory contact with the world must be ordered with respect to a verbal/conceptual scheme, before various objects can be perceived *as* members of their respective categories. This imposition of a conceptual framework on the chaotic field of raw sensation is required to provide the propositional content of ordinary perceptual experience, while the basal level of indeterminate sensation is strictly ineffable. Thus the ordinary objects which we experience in propositionally structured perception do not exist independently of our conceptual activities. Only the instantaneous and ultimately unique particulars are real, and, for reasons quite analogous to Russell's arguments concerning logically proper names, are not referred to with ordinary singular terms, while the enduring and composite objects which we perceive and talk about in everyday speech are diagnosed as conceptual constructs.

2.2 Negative Existentials and Non-denoting Terms

When the foregoing analysis of the objects of perception and reference is combined with *apoha* nominalism, the result is an elegant treatment of negative existentials, which the Yogācāra-Sautrāntikas defended against rivals, especially those of the Nyāya school. The Naiyāyikas held that some absences, *viz.* those which can be associated with existing counter-positive instances, are real and can be directly perceived. Thus when I say I can see that, for example, there is no gorilla in the doorway, this absence itself is said to be directly perceived, because there is a clearly defined counter-positive phenomenon, namely, the way the doorway would look if there were any particular gorilla standing in it. In contrast, the Yogācāra-Sautrāntikas maintain that absences are never perceived but only inferred. And the inferential mechanisms involved stem directly from the two-step negation of *apoha* semantics.

On the Yogācāra-Sautrāntika view, my non-perception of the gorilla is nothing other than my perception of the actual doorway in question. When I judge that I see a doorway, this is an instance of determinate perception, and as such it necessarily involves the mental paradigm governing my use of the term 'doorway'. This paradigm is a conceptual construction which enables me to apply the term under the correct assertability conditions. Thus the perceptual data with which I am now presented must be such that it is not excluded by the doorway paradigm, *i.e.* it must be such that it is not a non-doorway. But since a gorilla is included in the non-doorway class, I can rightly judge that there is no gorilla present in my immediate visual field, simply on the basis of my determinate perception of this doorway. Because of the exclusionary machinery of *apoha* semantics, the perception of the doorway *simplicitor* is a sufficient condition for inferring the non-presence of a gorilla (or any other non-doorway construction).

The *apoha* semantic analysis applies to singular as well as general terms (see, e.g., Tillemans [16]), and can be uniformly extended to cases where the subject of the assertion has no counter-positive instance, either because the subject is purely fictional (but possible), or because it is impossible. The feature which distinguishes conceptual constructions which are 'actual' is that the assertability conditions for terms denoting actual objects are constrained by direct causal interactions with ultimately existing particulars, while in the case of fictional objects the assertability conditions are governed purely by linguistic conventions. Thus to make the statement 'Kripke exists' is to hold that the conceptual construction designated by the term 'Kripke' (which we would normally take to be an actual individual in our naive, pre-theoretic belief that enduring and composite entities such as human individuals are real) is causally tied to the non-linguistic world of ultimate particulars, in such a way that stimuli from this world, combined with salient linguistic conventions, yield the result that the statement is at present correctly assertable, while at some undetermined future point it will not be. In a related vein, to assert that 'Pegasus does not exist' involves holding that the 'Pegasus construction' is not directly tied to the world of ultimate particulars, and the rules governing its use are constrained purely by discourse conventions. In this case, even though Pegasus does not exist, the discourse conventions warrant the assertion that 'Pegasus is winged', since the Pegasus concept is psychologically generated in response to the story in which Pegasus is presented as a flying horse. This allows statements such as 'Pegasus is winged' to have the same type of subject as factually grounded assertions, since in both cases the subject is a conceptual construction. Thus, in a manner analogous with Meinong's view that 'being so' is independent of 'being', the statement 'Pegasus is winged' is construed as both 'true' and *about* a genuine 'object'.

In the vernacular of the dispute between the Buddhist and Nyāya schools, a 'horned hare' is a stock example of a fictitious object, and according to the Buddhists the predication 'The hare's horn is sharp' is a normal sentence that we may employ in our discourse for various purposes. In contrast, advocates of the realist Nyāya school such as Vācaspati argue that the subject term of a sentence must refer to something actual, and if not, then the sentence is in need of philosophical paraphrasing in a manner strikingly akin to Russell's 1905 view. In order to cogently assert that 'the hare's horn does not exist', this must be analysed as the claim that 'each thing that is a horn does not belong to a hare' (Matilal [12], p. 81). As in Russell, this analysis (implicitly) relies on quantified variables rather than singular terms to express the lack of reference. So as in Russell, predications involving fictitious objects turn out uniformly false: both 'The hare's horn is sharp' and 'The hare's horn is not sharp' are evaluated as falsehoods. At this level there is nothing paradoxical about the analysis, and Russell's theory provides an explicit formalization of the basic idea. However, the Naiyāyikas did acknowledge a subtle but 'superficial' self-contradiction when expressed as the general principle that 'nothing can be truly affirmed or denied of a fictitious entity', since this is itself presumably intended as a true statement about fictitious entities.

The Yogācāra-Sautrāntika analysis of statements involving possible but non-actual objects is then carried over to statements about impossible objects, where the stock example is 'Devadatta, the son of a barren woman'. The fact that the specification of such an object is not self-consistent does not prevent the formation of an attendant conceptual construct (since the construction itself does not possess the incompatible traits), and thus we can make comprehensible assertions about Devadatta, wherein these assertions will have a constructed subject in many ways comparable to Kripke or Pegasus. Attempted *application* of the 'Devadatta' concept will result in the discovery of a null extension, since everything must be excluded. However, I would argue that the Yogācāra-Sautrāntika semantical theory does not seem to possess the resources needed to distinguish merely possible but non-actual entities like a hare's horn or Meinong's Golden Mountain, from impossible objects such as the son of a barren woman. The impossibility and hence non-existence of the latter is due to the mutual incompatibility of the *meanings* involved. Appeal to the level of sense or intension reveals that the description can perforce be satisfied by no object, and hence the purported individual is impossible. But the purely exclusionary mechanisms of the *apoha* account are not sufficient to distinguish cases of contingent non-existence from the analytically unsatisfiable, since the two are extensionally identical. To capture the definitional impossibility of 'the son of a barren woman' would require the introduction of something like Carnap's 'meaning postulates' to specify the salient natural language *content* carried by these terms.

2.3 *Apoha* Semantics and Free Logic

Siderits contrasts the Yogācāra-Sautrāntika view with Meinong's account, and argues that the Buddhist view has all of the virtues of Meinongianism with none of its vices. Assertions about nonexistent objects are given subjects and truth conditions in accord with common sense (as in Meinong), but not at the price of an 'ontological slum', bloated with subsistent but nonexistent objects. This latter claim is far from clear however, since the objects of predication do exist *qua* 'conceptual constructions'. Thus according to Dharmakīrti, 'Pegasus' does not refer to some attenuated individual residing in the nether world of abstract entities, but rather designates a private *mental* object of some kind. Furthermore, there is now not just *one* salient object of reference for the entire linguistic community, but instead there is one for every linguistic agent, just as there is an idiosyncratic Kripke concept, Everest concept, Zeus concept, etc. Thus the Buddhist view seems to constitute a type of psychologically instantiated Meinongianism, where the objects of reference are multiplied rather than decreased. This approach is perhaps more realistic than a logically idealized account with a single semantical structure for an entire linguistic community, although the sense in which it is genuinely 'nominalist' is in need of clarification. From an externalist point of view it is nominalist, since terms do not refer to external, mind-independent entities. But from an internalist perspective, linguistics expressions are interpreted as referring to conceptual constructs, which are the psychological analogues of ordinary, everyday objects.

Hence, I would propose a dual-domain Free logic as an appropriate way to *formally model* the individual nominalistic ideolects. In contrast to the division between existent and non-existent oblects underlying the Free logic domains, on the *apoha* view all cognitive representations exist as mental structures and hence are ontologically commensurate as such. So the demarcation between actual and 'non-actual' must be delineated as above – the inner domain $\boldsymbol{D}_i$ is comprised of the conceptual constructions such as Kripke and Everest which are causally tied to the non-linguistic world of ultimate particulars, while the outer domain $\boldsymbol{D}_o$ is comprised of conceptual constructs such as Pegasus and Zeus which are not directly tied to the world of particulars, and where the rules governing their use are constrained purely by linguistic conventions. On the *apoha* account, 'impossible objects' such as 'Devadatta' and Meinong's 'the round square' are countenanced as well, and should be mapped to constructs inhabiting the outer domain. As noted above, the constructs themselves do not possess the incompatible attributes in question and hence are not themselves impossible. So it is both fitting and necessary to block the deducibility of the aforementioned and seemingly innocuous notion that $\phi(\imath x\phi x)$, which in fact is not benign and will lead to paradox in the case of inconsistent descriptions. This is achieved by adopting Lambert's Law $\forall y(y = ix\phi x \leftrightarrow \forall x(\phi x \leftrightarrow x = y))$ as a basic principle of Free description theory. In terms of the semantics for definite descriptions, the following clause is added to the foregoing specification of a Free logic interpretation:

(v) if $\forall x(\phi x \leftrightarrow x = \boldsymbol{c})$ is true, where $\boldsymbol{c}$ is an individual constant, then $\boldsymbol{f}(ix\phi x) = \boldsymbol{f}(\boldsymbol{c})$, and if $\neg\exists x(\forall y \phi y \leftrightarrow y = x)$, then $\boldsymbol{f}(ix\phi x) = e$, where $e \in \boldsymbol{D}_o \setminus \boldsymbol{D}_i$, and where the interpretation function $\boldsymbol{f}$ on the set of individual constants must be surjective with respect to $\boldsymbol{D}_o$.

The nominalism of the Buddhist view indicates that the predicate extensions given by the interpretation function $\boldsymbol{f}$ and the attendant set membership conditions used in the formal definition of truth for atomic formulas should *not* be seen as reflecting some literal correspondence theory of truth. Instead, they simply encode the internalized discourse conventions of the speaker's linguistic community. In this manner it is possible to provide the basics of a formal semantics for natural language that models the *apoha* view, and hence provides a structure that can reflect the conventional truth conditions for sentences involving non-denoting terms.

References

1. Frege, G.: On Sense and Meaning. In: Geach, P., Black, M. (eds.) Translations from the Philosophical Writings of Gottlob Frege. Basil Blackwell, Oxford (1960)
2. Russell, B.: On Denoting. Mind 14, 479–493 (1905)
3. Lambert, K.: Free Logic: Selected Essays. CUP, Cambridge (2003)
4. Lambert, K.: Existential Import Revisited. Notre Dame Journal of Formal Logic 4, 288–292 (1963)
5. Russell, B., Whitehead, A.: Principia Mathematica. CUP, Cambridge (1910)
6. Gumb, R.: Free Logic in Program Specification and Verification. In: Morscher, E., Hieke, A. (eds.) New Essays in Free Logic, pp. 157–189. Kluwer, London (2001)
7. Troelstra, A.S., van Dalen, D.: Constructivism in Mathematics. Elsevier, Dordrecht (1999)
8. Meinong, A.: On the Theory of Objects. In: Chisholm, R. (ed.) Realism and the Background of Phenomenology, pp. 76–117. Free Press, Glencoe (1960)
9. Leblanc, H., Thomason, R.: Completeness Theorems for Some Presupposition-free Logics. Fundamenta Mathematicae 62, 125–164 (1968)
10. Kripke, S.: Semantic Considerations on Modal Logic. Acta Philosophica Fennica 16, 83–94 (1963)
11. Schweizer, P.: Free Logic and Quantification in Syntactic Modal Contexts. In: Morscher, E., Hieke, A. (eds.) New Essays in Free Logic, pp. 69–83. Kluwer, London (2001)
12. Matilal, B.K.: Logic, Language and Reality. Motilal Banarsidass, Delhi (1985)
13. Matilal, B.K.: The Character of Logic in India. SUNY, Albany (1998)
14. Siderits, M.: Indian Philosophy of Language. Kluwer, Dordrecht (1991)
15. Siderits, M.: Śrughna by Dusk. In: Siderits, M., Tillemans, T., Chakrabarti, A. (eds.) Aphoha: Buddhist Nominalism and Human Cognition, pp. 283–304. Columbia University Press, New York (2011)
16. Tillemans, T.: Dharmakīrti. In: E. Zalta (ed.) The Stanford Encyclopedia of Philosophy (2014), `http://plato.stanford.edu/archives/spr2014/entries/dharmakiirti/`

Ordinals in an Algebra-Valued Model of a Paraconsistent Set Theory

Sourav Tarafder[1,2]

[1] Department of Commerce,
St. Xavier's College,
30, Mother Teresa Sarani, Kolkata, 700016, India
[2] Department of Pure Mathematics,
University of Calcutta,
35, Ballygunge Circular Road, Kolkata, 700019, India
souravt09@gmail.com

Abstract. This paper deals with ordinal numbers in an algebra-valued model of a paraconsistent set theory. It is proved that the collection of all ordinals is not a set in this model which is dissimilar to the other existing paraconsistent set theories. For each ordinal α of classical set theory α-like elements are defined in the mentioned algebra-valued model whose collection is not singleton. It is shown that two α-like elements (for same α) may perform conversely to validate a given formula of the corresponding paraconsistent set theory.

Keywords: non-classical set theory, ordinal numbers, paraconsistent logic.

1 Introduction

Boolean-valued models of classical set theory were introduced by Dana Scott, Robert M. Solovay and Petr Vopěnka in the 1960s. If $\mathbb{B}$ is a *complete Boolean algebra* then $\mathbf{V}^{(\mathbb{B})}$ is a model of Zermelo Fraenkel set theory with the Axiom of Choice, ZFC (cf. [1]). If the complete Boolean algebra is replaced by a *complete Heyting algebra*, $\mathbb{H}$ then essentially the same proof shows that $\mathbf{V}^{(\mathbb{H})}$ becomes a model of Intuitionistic Zermelo Fraenkel set theory, IZF [4]. Later Takeuti, Titani, Kozawa and Ozawa generalised the development to some appropriate *lattice-valued model* of *quantum set theory* or *fuzzy set theory* [6–8] [10, 11]. We describe the construction of these algebra-valued models and the notion of validity of a formula in these models in §2.

It can be proved that for Heyting-valued models $\mathbf{V}^{(\mathbb{H})}$, the validity of the Axiom of Choice AC in $\mathbf{V}^{(\mathbb{H})}$ is equivalent to the *law of excluded middle* $a \vee a^*$ in $\mathbb{H}$. This is a remarkable fact linking set theoretic properties in algebra-valued models of set theories to algebraic properties of the corresponding algebra. This observation was used in [5] to define a class of algebras $\langle \mathbf{A}, \wedge, \vee, \Rightarrow, \mathbf{1}, \mathbf{0} \rangle$ called reasonable implication algebras. An algebra $\langle \mathbf{A}, \wedge, \vee, \Rightarrow, \mathbf{1}, \mathbf{0} \rangle$ is called a *reasonable implication algebra* if the following hold:

M. Banerjee and S.N. Krishna (eds.): ICLA 2015, LNCS 8923, pp. 195–206, 2015.

1. $\langle \mathbf{A}, \wedge, \vee, \mathbf{1}, \mathbf{0} \rangle$ is a complete distributive lattice,
2. $x \wedge y \leq z$ implies $x \leq y \Rightarrow z$,
3. $x \leq y$ implies $z \Rightarrow x \leq z \Rightarrow y$, and
4. $x \leq y$ implies $y \Rightarrow z \leq x \Rightarrow z$.

A reasonable implication algebra is called *deductive* if

$$(x \wedge y) \Rightarrow z = x \Rightarrow (y \Rightarrow z).$$

If $\mathcal{L}$ is any first-order language then by NFF (the *negation-free fragment*) we mean the closure of the atomic formulas of $\mathcal{L}$ under $\wedge$, $\vee$, $\rightarrow$, $\exists$, and $\forall$. The elements of NFF will be called *negation-free formulas.*

For the axiom schemes in the axiom system for ZF (i.e., Separation and Replacement), we write NFF-Separation and NFF-Replacement for the subscheme where we only allow instances of negation-free formulas in the scheme.

One of the main results in [5] is the following.

Theorem 1. *Let* $\mathbb{A} = \langle \mathbf{A}, \wedge, \vee, \Rightarrow, \mathbf{1}, \mathbf{0} \rangle$ *be a deductive reasonable implication algebra. Then* Extensionality, Pairing, Infinity, Union *and* Powerset *and the schemes* NFF-Separation *and* NFF-Replacement *are valid in* $\mathbf{V}^{(\mathbb{A})}$.

If $\mathbb{A}$ is a deductive reasonable implication algebra and $\mathcal{L}_{\mathbb{A}}$ is a logic that is sound and complete with respect to $\mathbb{A}$, then $\mathcal{L}_{\mathbb{A}}$ plays the role of the propositional fragment of the logic of the set theoretic model $\mathbf{V}^{(\mathbb{A})}$.

Below, we shall give an example PS3 of a deductive reasonable implication algebra which is neither a Heyting nor a Boolean algebra. Its logic is a *paraconsistent logic*[1] which then gives rise to a model of *paraconsistent set theory.*

The three-valued matrix $\mathbb{PS}_3 = \langle \{1, 1/2, 0\}, \wedge, \vee, \Rightarrow \rangle$ is a deductive reasonable implication algebra, where the truth tables for the operators are given below:

$\wedge$	1	1/2	0
1	1	1/2	0
1/2	1/2	1/2	0
0	0	0	0

$\vee$	1	1/2	0
1	1	1	1
1/2	1	1/2	1/2
0	1	1/2	0

$\Rightarrow$	1	1/2	0
1	1	1	0
1/2	1	1	0
0	1	1	1

Let us now introduce a unary operator * in $\mathbb{PS}_3$ having the following truth table:

*	
1	0
1/2	1/2
0	1

We use the symbol PS3 to refer to the augmented structure $\langle \mathbb{PS}_3, ^* \rangle$. The designated set is $\{1, 1/2\}$. The logic which is sound and complete with respect to PS3 is a paraconsistent logic [9] and we observed in [5] that $\mathbf{V}^{(\mathrm{PS}_3)}$ is a model of paraconsistent set theory. In [2] and [3] the same truth tables have appeared

[1] A logic is called paraconsistent if there exist formulas φ and ψ such that $\{\varphi, \neg\varphi\} \nvdash \psi$. Semantically, the premises get the designated values but the conclusion does not.

but along with these tables there is one more table for an unary operator called the *inconsistency operator*. Also the approaches there are totally different from [9] which was developed independently.

In this paper, we study the set theory of the model $\mathbf{V}^{(\mathrm{PS}_3)}$, in particular the notion of ordinals in this particular paraconsistent set theory. Ordinals have been studied in non-classical set theories; e.g., Titani used a notion of ordinal number in *lattice-valued set theory* [10] where the definition of ordinal is not exactly the one used in classical set theory. On the other hand let us consider the paraconsistenet set theory described by [12] where the classical definition of ordinal is used.

In §4, we stick to the classical definition of ordinals. However, our paraconsistent set theory differs considerably from Weber's (as noted in [5]) since the axiom scheme of full comprehension is valid in Weber's set theory and invalid in $\mathbf{V}^{(\mathrm{PS}_3)}$.[2]

Usually in the literature of paraconsistent set theory general comprehension is taken as valid. In [5] it is proved that $\mathbf{V}^{(\mathrm{PS}_3)}$ does not validate the general comprehension axiom scheme. Also theorem 14 of this paper shows that there does not exist a set of all ordinal numbers which disproves the general comprehension. One of the motivations of this paper in connection with the work done in [5] is that without the comprehension axiom scheme a set theory may behave well enough as a paraconsistent set theory. There is no need to think that paraconsistent set theories are only built to deal with the set theoretic paradoxes e.g., *Russell's paradox.* Rather this paper will show a set theory being paraconsistent may agree with important classical set theoretic results. This fact is corroborated in Intuitionistic set theory as viewed in Heyting algebra-valued models [4]. It may also be mentioned that the identity as used here is not exactly the classical one. In future we shall deal with these issues in more detail.

2 Definitions and Preliminaries

2.1 Some Classical Definitions

We develop the classical theory of *transitive sets*, *well-ordered sets*, and ordinal numbers in the setting of $\mathbf{V}^{(\mathbb{A})}$. The following definitions 2–4 are stated in classical metalanguage whereas the formalizations will take place in §4.

Definition 2. *A set x is said to be transitive if every element of x is a subset of x, or equivalently, if $y \in z$ and $z \in x$ implies $y \in x$.*

Definition 3. *A set* A *is said to be well-ordered by a relation* R *if* R *is a linear order on* A *and any non-empty subset of* A *has a least element with respect to* R.

Definition 4. *An ordinal number is a transitive set well-ordered by $\in$.*

[2] Comprehension axiom scheme: if $\mathrm{P}(x)$ is a property then $\{x : \mathrm{P}(x) \text{ holds}\}$ is a set.

2.2 The Algebra-Valued Model Construction and $\mathbf{BQ}_\varphi$

The following account is taken from [5] with the permission of the first author of that paper.

We start with a model of classical set theory $\mathbf{V}$ and an algebra $\langle \mathbf{A}, \wedge, \vee, \Rightarrow, {}^*, \mathbf{1}, \mathbf{0}\rangle$. A universe of *names* is constructed by transfinite recursion:

$$\mathbf{V}_\alpha^{(\mathbb{A})} = \{x;\ x \text{ is a function and } \mathrm{ran}(x) \subseteq \mathbf{A} \text{ and there is } \xi < \alpha \text{ with } \mathrm{dom}(x) \subseteq \mathbf{V}_\xi^{(\mathbb{A})}\} \text{ and}$$

$$\mathbf{V}^{(\mathbb{A})} = \{x;\ \exists\alpha(x \in \mathbf{V}_\alpha^{(\mathbb{A})})\}.$$

Let $\mathcal{L}_\in$ be the language of set theory having the propositional connectives $\wedge$, $\vee$, $\rightarrow$, and $\neg$. $\mathcal{L}_\mathbb{A}$ stands for the extended language of $\mathcal{L}_\in$ extended by adding constants corresponding to each element in $\mathbf{V}^{(\mathbb{A})}$. Similar to the case in Boolean-valued model (cf. [1]) the following *(meta-)induction principle* for $\mathbf{V}^{(\mathbb{A})}$ will be used in this paper whenever it is needed: for every property Φ of names, if for all $x \in \mathbf{V}^{(\mathbb{A})}$, we have

$$\forall y \in \mathrm{dom}(x)(\Phi(y)) \text{ implies } \Phi(x),$$

then all names $x \in \mathbf{V}^{(\mathbb{A})}$ have the property Φ.

Following the Boolean-valued model construction a map $[\![\cdot]\!]$ is defined from the class of all formulas in $\mathcal{L}_\mathbb{A}$ to the set $\mathbf{A}$ of truth values as follows. If $u, v \in \mathbf{V}^{(\mathbb{A})}$ and φ, ψ are any two formulas in $\mathcal{L}_\mathbb{A}$ then

$$\begin{aligned}
[\![u \in v]\!] &= \bigvee_{x\in\mathrm{dom}(v)} (v(x) \wedge [\![x = u]\!]),\\
[\![u = v]\!] &= \bigwedge_{x\in\mathrm{dom}(u)} (u(x) \Rightarrow [\![x \in v]\!]) \wedge \bigwedge_{y\in\mathrm{dom}(v)} (v(y) \Rightarrow [\![y \in u]\!]),\\
[\![\varphi \wedge \psi]\!] &= [\![\varphi]\!] \wedge [\![\psi]\!],\\
[\![\varphi \vee \psi]\!] &= [\![\varphi]\!] \vee [\![\psi]\!],\\
[\![\varphi \rightarrow \psi]\!] &= [\![\varphi]\!] \Rightarrow [\![\psi]\!],\\
[\![\neg\varphi]\!] &= [\![\varphi]\!]^*,\\
[\![\forall x\varphi(x)]\!] &= \bigwedge_{u\in\mathbf{V}^{(\mathbb{A})}} [\![\varphi(u)]\!], \text{ and}\\
[\![\exists x\varphi(x)]\!] &= \bigvee_{u\in\mathbf{V}^{(\mathbb{A})}} [\![\varphi(u)]\!].
\end{aligned}$$

Let $\mathrm{D} \subseteq \mathbb{A}$ be a set of *designated truth values* (in short: *designated set*). A formula φ of $\mathcal{L}_\mathbb{A}$ is said to be D-valid in $\mathbf{V}^{(\mathbb{A})}$ if $[\![\varphi]\!] \in \mathrm{D}$ and is denoted by $\mathbf{V}^{(\mathbb{A})} \models_\mathrm{D} \varphi$. We omit D in the notation whenever it is clear from the context.

The following equations (for a formula φ of the language of set theory) played a crucial role in the development of the theory in [5]:

$$[\![\forall x(x \in u \rightarrow \varphi(x))]\!] = \bigwedge_{x\in\mathrm{dom}(u)} (u(x) \Rightarrow [\![\varphi(x)]\!]). \qquad (\mathrm{BQ}_\varphi)$$

In order to show that an instance of bounded quantification over the formula φ behaves properly in $\mathbf{V}^{(\mathbb{A})}$, one needs the validity of the formula BQ_φ in $\mathbf{V}^{(\mathbb{A})}$. Unfortunately, the fact that $\mathbb{A}$ is a deductive reasonable implication algebra is not sufficient to prove this for all formulas φ; but it is sufficient to prove it for all negation-free formulas φ. Details can be found in [5, § 3.1].[3]

2.3 Definition of α-like Elements

As in the theory of Boolean valued models, we can define the notion of a *canonical name*:

Definition 5. *For each* $x \in \mathbf{V}$,

$$\hat{x} = \{\langle \hat{y}, 1\rangle : y \in x\}$$

Let ORD refer to the class of all ordinal numbers in $\mathbf{V}$. The main goal of this paper is to identify elements in $\mathbf{V}^{(\mathrm{PS}_3)}$ which behave almost similar to the classical ordinal numbers. It will be shown that there are more than one such elements in $\mathbf{V}^{(\mathrm{PS}_3)}$ corresponding to each $\alpha \in \mathrm{ORD}$ which will be named as α-like elements. But the non-classical behaviour of these elements will be discussed in §4.

For each $\alpha \in \mathrm{ORD}$ the α-like names in $\mathbf{V}^{(\mathrm{PS}_3)}$ are defined by transfinite recursion as follows.

Definition 6. *An element* $x \in \mathbf{V}^{(\mathrm{PS}_3)}$ *is called*

i) 0-like *if for every* $y \in \mathrm{dom}(x)$, *we have that* $x(y) = 0$; *and*
ii) α-like *if for each* $\beta \in \alpha$ *there exists* $y \in \mathrm{dom}(x)$ *which is* β*-like and* $x(y) \in \{1, 1/2\}$, *and for any* $z \in \mathrm{dom}(x)$ *if it is not* β*-like for any* $\beta \in \alpha$ *then* $x(z) = 0$.

Clearly, the canonical name $\hat{\alpha}$ is an α-like name for every $\alpha \in \mathrm{ORD}$.

3 Properties of α-like Elements

For each $\alpha \in \mathrm{ORD}$, there are many α-like names as the following results show.

Lemma 7. *For any* $x \in \mathbf{V}^{(\mathrm{PS}_3)}$ *and* $\alpha \in \mathrm{ORD}$, $[\![x = \hat{\alpha}]\!] = 1$ *if and only if* x *is* α*-like.*

Proof. The proof will be done by induction on the domain of $\hat{\alpha}$. We assume that we have shown the result for all elements in the domain of $\hat{\alpha}$. We know

$$[\![x = \hat{\alpha}]\!] = \bigwedge_{y \in \mathrm{dom}(x)} (x(y) \Rightarrow [\![y \in \hat{\alpha}]\!]) \wedge \bigwedge_{\hat{\beta} \in \mathrm{dom}(\hat{\alpha})} (1 \Rightarrow [\![\hat{\beta} \in x]\!])$$

[3] If $\mathbb{A}$ is a Boolean algebra or a Heyting algebra, then BQ_φ can be proved for all φ.

Hence $[\![x = \hat{\alpha}]\!] = 1$ if and only if both of the conjuncts are 1. The second conjunct is 1, i.e.,

$$\bigwedge_{\hat{\beta}\in\mathrm{dom}(\hat{\alpha})} (1 \Rightarrow [\![\hat{\beta} \in x]\!]) = 1;$$

if and only if for each $\hat{\beta} \in \mathrm{dom}(\hat{\alpha})$, $1 \Rightarrow [\![\hat{\beta} \in x]\!] = 1$ i.e.,

$$1 \Rightarrow \bigvee_{y\in\mathrm{dom}(x)} (x(y) \wedge [\![y = \hat{\beta}]\!]) = 1;$$

if and only if for each $\hat{\beta} \in \mathrm{dom}(\hat{\alpha})$ there exists $y \in \mathrm{dom}(x)$ such that $x(y) \in \{1, 1/2\}$ and $[\![y = \hat{\beta}]\!] = 1$; if and only if for each $\hat{\beta} \in \mathrm{dom}(\hat{\alpha})$ there exists $y \in \mathrm{dom}(x)$ such that y is β-like (by the induction hypothesis) and $x(y) \in \{1, 1/2\}$.

Again, since the first conjunct is 1, we have,

$$\bigwedge_{y\in\mathrm{dom}(x)} (x(y) \Rightarrow [\![y \in \hat{\alpha}]\!]) = 1;$$

if and only if for each $y \in \mathrm{dom}(x)$, $(x(y) \Rightarrow [\![y \in \hat{\alpha}]\!]) = 1$, i.e.,

$$(x(y) \Rightarrow \bigvee_{\hat{\beta}\in\mathrm{dom}(\hat{\alpha})} [\![y = \hat{\beta}]\!]) = 1;$$

if and only if for each $y \in \mathrm{dom}(x)$, if y is not β-like for any $\beta \in \alpha$ then by induction hypothesis it can be derived that $x(y) = 0$.

Hence combining the above results we get $[\![x = \hat{\alpha}]\!] = 1$ if and only if x is α-like and hence by the (meta-) induction the proof is done. □

Lemma 8. *For any $x \in \mathbf{V}^{(\mathrm{PS}_3)}$ and $\alpha \in \mathrm{ORD}$, $[\![x \in \hat{\alpha}]\!] = 1$ if and only if x is β-like for some $\beta \in \alpha$.*

Proof. Using Lemma 7, the following three statements are equivalent:

1. $[\![x \in \hat{\alpha}]\!] = 1$ if and only if $\bigvee_{\hat{u}\in\mathrm{dom}(\hat{\alpha})} [\![x = \hat{u}]\!] = 1$;
2. there exists $\hat{\beta} \in \mathrm{dom}(\hat{\alpha})$ such that $[\![x = \hat{\beta}]\!] = 1$; and
3. x is β-like for some $\beta \in \alpha$.

□

It is clear from the definition that for any $\alpha \in \mathrm{ORD}$, there are many α-like names in $\mathbf{V}^{(\mathrm{PS}_3)}$ in addition to $\hat{\alpha}$. Of course, we would desire that α-like names are equal and that for $\beta < \alpha$, β-like names are elements of α-like names (in the formal sense of $\mathbf{V}^{(\mathbb{A})}$):

Theorem 9. *Let $x \in \mathbf{V}^{(\mathrm{PS}_3)}$ be α-like for some $\alpha \in \mathrm{ORD}$. For any $y \in \mathbf{V}^{(\mathrm{PS}_3)}$, $[\![x = y]\!] = 1$ if and only if y is α-like.*

Proof. In [5], we proved that for any $x, y, z \in \mathbf{V}^{(\mathrm{PS}_3)}$,

$$[\![x = y]\!] \wedge [\![y = z]\!] \leq [\![x = z]\!].$$

Let x and y be two α-like elements in $\mathbf{V}^{(\mathrm{PS}_3)}$. So $[\![x = \hat{\alpha}]\!] \wedge [\![\hat{\alpha} = y]\!] \leq [\![x = y]\!]$. By Lemma 7 we have $[\![x = \hat{\alpha}]\!] = 1 = [\![y = \hat{\alpha}]\!]$, which implies $[\![x = y]\!] = 1$.

Conversely let $[\![x = y]\!] = 1$. By similar argument we can write, $[\![x = y]\!] \wedge [\![x = \hat{\alpha}]\!] \leq [\![y = \hat{\alpha}]\!]$ and hence $[\![y = \hat{\alpha}]\!] = 1$. Again by Lemma 7 it can be concluded that y is α-like. □

Theorem 10. *Let $x \in \mathbf{V}^{(\mathrm{PS}_3)}$ be α-like for some non-zero $\alpha \in$ ORD. For any $y \in \mathbf{V}^{(\mathrm{PS}_3)}$, $[\![y \in x]\!] \in \{1, 1/2\}$ if and only if y is β-like for some $\beta \in \alpha$.*

Proof. Let y be β-like for some $\beta \in \alpha$. Now

$$\begin{aligned}[\![y \in x]\!] &= \bigvee_{u \in \mathrm{dom}(x)} (x(u) \wedge [\![u = y]\!]) \\ &\geq x(v) \wedge [\![v = y]\!], \text{where } v \in \mathrm{dom}(x) \text{ is } \beta\text{-like and } x(v) \in \{1, 1/2\} \\ &\geq 1/2, \text{ by Theorem 9.}\end{aligned}$$

Conversely, let $[\![y \in x]\!] \in \{1, 1/2\}$, i.e.,

$$\bigvee_{u \in \mathrm{dom}(x)} (x(u) \wedge [\![u = y]\!]) \in \{1, 1/2\}.$$

Hence there exists some β-like $v \in \mathrm{dom}(x)$ such that $x(v) \in \{1, 1/2\}$ and $[\![v = y]\!] = 1$, where $\beta \in \alpha$. So by Theorem 9, it follows that y is also β-like. □

Let a binary class relation $\sim$ be defined on $\mathbf{V}^{(\mathrm{PS}_3)}$ by $x \sim y$ if and only if $\mathbf{V}^{(\mathrm{PS}_3)} \models x = y$, i.e., $[\![x = y]\!] = 1$. This relation is discussed in [5] where it is mentioned that $\sim$ is an *equivalence class relation.* Theorem 9 shows for each $\alpha \in$ ORD the collection of all α-like elements forms an equivalence class in $\mathbf{V}^{(\mathrm{PS}_3)}/\sim$. If x and y are two elements in the classes of α-like and β-like elements in $\mathbf{V}^{(\mathrm{PS}_3)}/\sim$ then $\alpha \in \beta$ is true in $\mathbf{V}$ implies $x \in y$ is valid in $\mathbf{V}^{(\mathrm{PS}_3)}$.

4 Ordinals in $\mathbf{V}^{(\mathrm{PS}_3)}$

We now rewrite the definitions of §2 in the language of set theory:

$$\mathrm{Trans}(x) = \forall y \forall z (z \in y \wedge y \in x \rightarrow z \in x)$$

$$\mathrm{LO}(x) = \forall y \forall z ((y \in x \wedge z \in x) \rightarrow (y \in z \vee y = z \vee z \in y))^4$$

$$\mathrm{WO}_{\in}(x) = \mathrm{LO}(x) \wedge \forall y (y \subseteq x \wedge \neg(y = \varnothing) \rightarrow \exists z (z \in y \wedge z \cap y = \varnothing))$$

$$\mathrm{ORD}(x) = \mathrm{Trans}(x) \wedge \mathrm{WO}_{\in}(x)$$

[4] LO(x) stands for the formula: x is a linear orderdered set with respect to $\in$.

where the following abbreviations are used in $\text{WO}_\in(x)$:

$$y \subseteq x := \forall t(t \in y \rightarrow t \in x),$$

$$\neg(y = \varnothing) := \exists z(z \in y),$$

$$(z \cap y = \varnothing) := \neg \, \exists w(w \in z \wedge w \in y).$$

Finally, we can connect the notion of α-like name to the set theoretic notion of ordinals:

Lemma 11. *Let $\alpha \in$ ORD and u be an α-like element in $\mathbf{V}^{(\text{PS}_3)}$. Then the following hold:*

i) $\mathbf{V}^{(\text{PS}_3)} \models \text{Trans}(u)$
ii) $\mathbf{V}^{(\text{PS}_3)} \models \text{LO}(u)$
iii) $\mathbf{V}^{(\text{PS}_3)} \models \text{WO}_\in(u)$

Proof. (*i*) We have to prove $[\![\forall y \forall z(z \in y \wedge y \in u \rightarrow z \in u)]\!] \in \{1, 1/2\}$. Since the truth table of $\Rightarrow$ in PS_3 does not contain $1/2$ it is sufficient to show $[\![\forall y \forall z(z \in y \wedge y \in u \rightarrow z \in u)]\!] = 1$.

Let us take any $z \in \mathbf{V}^{(\text{PS}_3)}$. Then,

$$\begin{aligned}[\![\forall y(y \in u \wedge z \in y \rightarrow z \in u)]\!] &= \bigwedge_{y \in \mathbf{V}^{(\text{PS}_3)}} ([\![y \in u]\!] \wedge [\![z \in y]\!] \Rightarrow [\![z \in u]\!]) \\ &= \bigwedge_{y \in \mathbf{V}^{(\text{PS}_3)}} ([\![y \in u]\!] \Rightarrow ([\![z \in y]\!] \Rightarrow [\![z \in u]\!])) \\ &= \bigwedge_{y \in \text{dom}(u)} (u(y) \Rightarrow ([\![z \in y]\!] \Rightarrow [\![z \in u]\!]))\end{aligned}$$

(since BQ_φ hold in $\mathbf{V}^{(\text{PS}_3)}$ for all negation-free formulas φ.)

For any $y \in \text{dom}(u)$ if $u(y) \neq 0$ then y is β-like for some non-zero $\beta \in \alpha$. Let for such an y, $[\![z \in y]\!] \in \{1, 1/2\}$. Therefore by Theorem 10, z is γ-like for some $\gamma \in \beta$. Clearly, $\gamma \in \alpha$. Therefore one more application of Theorem 10 provides $[\![z \in u]\!] \in \{1, 1/2\}$. Hence combining the above results we get

$$\bigwedge_{y \in \text{dom}(u)} (u(y) \Rightarrow ([\![z \in y]\!] \Rightarrow [\![z \in u]\!])) = 1$$

for any $z \in \mathbf{V}^{(\text{PS}_3)}$. This leads to the fact

$$[\![\forall y \forall z(y \in u \wedge z \in y \rightarrow z \in u)]\!] = 1, \text{ i.e., } \mathbf{V}^{(\text{PS}_3)} \models \text{Trans}(u).$$

(*ii*) Since for any $\alpha, \beta \in$ ORD exactly one of $\alpha \in \beta$, $\alpha = \beta$ and $\beta \in \alpha$ holds in $\mathbf{V}$, the proof can be derived easily by applying Theorems 9 and 10.

(*iii*) We already have $\mathbf{V}^{(\mathrm{PS}_3)} \models \mathrm{LO}(u)$ from (*ii*). So it is sufficient to prove that

$$[\![\forall y(y \subseteq u \wedge \neg(y = \varnothing) \to \exists z(z \in y \wedge z \cap y = \varnothing))]\!] = 1,^{5}$$

i.e., for any $y \in \mathbf{V}^{(\mathrm{PS}_3)}$ if $[\![y \subseteq u \wedge \neg(y = \varnothing)]\!] \in \{1, 1/2\}$ then $[\![\exists z(z \in y \wedge z \cap y = \varnothing)]\!] \in \{1, 1/2\}$. Now by definition and the fact that BQ_φ hold in $\mathbf{V}^{(\mathrm{PS}_3)}$ for all negation-free formulas φ,

$$[\![y \subseteq u]\!] = [\![\forall t(t \in y \to t \in u)]\!] = \bigwedge_{t \in \mathrm{dom}(y)} (y(t) \Rightarrow [\![t \in u]\!])$$

So, $[\![y \subseteq u]\!] \in \{1, 1/2\}$ if and only if for any $t \in \mathrm{dom}(y)$ if $y(t) \neq 0$ then $[\![t \in u]\!] \neq 0$, i.e., by Theorem 10 it can be concluded that t is β-like for some $\beta \in \alpha$. Again,

$$[\![\neg(y = \varnothing)]\!] = [\![\exists z(z \in y)]\!] = \bigvee_{z \in \mathbf{V}^{(\mathrm{PS}_3)}} \bigvee_{t \in \mathrm{dom}(y)} (y(t) \wedge [\![z = t]\!])$$

Therefore $[\![\neg(y = \varnothing)]\!] \in \{1, 1/2\}$ if and only if there exists $t \in \mathrm{dom}(y)$ such that $y(t) \in \{1, 1/2\}$.

Hence $[\![y \subseteq u \wedge \neg(y = \varnothing)]\!] \in \{1, 1/2\}$ if and only if there exists $t \in \mathrm{dom}(y)$ such that $y(t) \in \{1, 1/2\}$ and for each $t \in \mathrm{dom}(y)$ if $y(t) \in \{1, 1/2\}$ then t is β-like for some $\beta \in \alpha$.

Let us now find the value of $[\![\exists z(z \in y \wedge z \cap y = \varnothing)]\!]$ assuming $[\![y \subseteq u \wedge \neg(y = \varnothing)]\!] \in \{1, 1/2\}$. Let

$$\gamma = \min\{\beta \in \mathrm{ORD} \mid \text{there exists } t \in \mathrm{dom}(y) \text{ such that } y(t) \in \{1, 1/2\} \text{ and } t \text{ is } \beta\text{-like}\}.$$

By our assumption, $\gamma \geq 1$. There exists $t' \in \mathrm{dom}(y)$ such that $y(t') \in \{1, 1/2\}$ and t' is γ-like.

$$\begin{aligned}
&[\![\exists z(z \in y \wedge z \cap y = \varnothing)]\!] \\
&= [\![\exists z(z \in y \wedge \neg\exists w(w \in z \wedge w \in y))]\!] \\
&\geq [\![t' \in y \wedge \neg\exists w(w \in t' \wedge w \in y))]\!] \\
&= \bigvee_{t \in \mathrm{dom}(y)} (y(t) \wedge [\![t = t']\!]) \wedge (\bigvee_{w \in \mathrm{dom}(t')} (t'(w) \wedge [\![w \in y]\!]))^* \\
&\geq (y(t') \wedge [\![t' = t']\!]) \wedge [\bigvee_{w \in \mathrm{dom}(t')} (t'(w) \wedge \bigvee_{t \in \mathrm{dom}(y)} (y(t) \wedge [\![w = t]\!]))]^* \\
&\geq 1/2 \wedge [\bigvee_{w \in \mathrm{dom}(t')} (t'(w) \wedge \bigvee_{t \in \mathrm{dom}(y)} (y(t) \wedge [\![w = t]\!]))]^*.
\end{aligned}$$

---

[5] Since PS_3 satisfies the deductive principle: $((a \wedge b) \Rightarrow c) = (a \Rightarrow (b \Rightarrow c))$.

Claim 12. $[\bigvee_{w \in \mathrm{dom}(t')}(t'(w) \wedge \bigvee_{t \in \mathrm{dom}(y)}(y(t) \wedge [\![w = t]\!]))]^* = 1.$

Proof. It is sufficient to prove that

$$\bigvee_{w \in \mathrm{dom}(t')} (t'(w) \wedge \bigvee_{t \in \mathrm{dom}(y)} (y(t) \wedge [\![w = t]\!])) = 0.$$

Assume for some $w \in \mathrm{dom}(t')$, $t'(w) \in \{1, 1/2\}$. If possible let there exist $t \in \mathrm{dom}(y)$ such that both $y(t), [\![w = t]\!] \in \{1, 1/2\}$. By our assumption $y(t) \in \{1, 1/2\}$ implies t is β-like for some $\beta \in \alpha$. Since $[\![w = t]\!] \in \{1, 1/2\}$ by Theorem 9 we have w is β-like. Again since t' is γ like and $t'(w) \in \{1, 1/2\}$ therefore $\beta \in \gamma$ which contradicts the minimality of γ as $y(t) \in \{1, 1/2\}$. Hence the claim is proved. □

Therefore $[\![\exists z(z \in y \wedge z \cap y = \varnothing)]\!] \geq 1/2 \wedge 1 = 1/2$. This leads to the fact that for any $y \in \mathbf{V}^{(\mathrm{PS}_3)}$, if $[\![y \subseteq u \wedge \neg(y = \varnothing)]\!] \in \{1, 1/2\}$ then $[\![\exists z(z \in y \wedge z \cap y = \varnothing)]\!] \in \{1, 1/2\}$; i.e.,

$$[\![\forall y(y \subseteq u \wedge \neg(y = \varnothing) \to \exists z(z \in y \wedge z \cap y = \varnothing))]\!] = 1.$$

Hence we can conclude $\mathbf{V}^{(\mathrm{PS}_3)} \models \mathrm{WO}_\in(u)$. □

Combining (i) and (iii) of lemma 11 the following theorem can be derived.

Theorem 13. *Let* $\alpha \in \mathrm{ORD}$ *and* u *be an* α*-like element in* $\mathbf{V}^{(\mathrm{PS}_3)}$*. Then* $\mathbf{V}^{(\mathrm{PS}_3)} \models \mathrm{ORD}(u)$*.*

Theorem 13 shows any α-like element satisfies the classical definition of ordinal number. It is proved in [5] that the general Comprehension axiom scheme is not valid in $\mathbf{V}^{(\mathrm{PS}_3)}$. On the other hand it is a theorem of the paraconsistent set theory considered in [12]. As a consequence the collection of all ordinals becomes a set in that model. This fact leads us to the important question, whether the collection of elements which make the first order formula $\mathrm{ORD}(x)$ valid is a set in $\mathbf{V}^{(\mathrm{PS}_3)}$. The following theorem assures the answer is negative.

Theorem 14. *There is no set of all ordinals:*

$$\mathbf{V}^{(\mathrm{PS}_3)} \not\models \exists O\, \forall x(\mathrm{ORD}(x) \to x \in O).$$

Proof. Let $O \in \mathbf{V}^{(\mathrm{PS}_3)}$ be arbitrarily chosen. Then by definition, $\mathrm{dom}(O)$ is a set in $\mathbf{V}$. By Theorem 9, if $\alpha \neq \beta$ for any $\alpha, \beta \in \mathrm{ORD}$ then for any α-like u and β-like v, $\mathbf{V}^{(\mathrm{PS}_3)} \not\models u = v$. Hence u and v are not equal as a set in $\mathbf{V}$. We conclude that if for each $\alpha \in \mathrm{ORD}$ there exists an α-like u in $\mathrm{dom}(O)$ then $\mathrm{dom}(O)$ cannot be a set in $\mathbf{V}$ as the collection of all ordinals is not a set in $\mathbf{V}$. Hence there exists an $\alpha' \in \mathrm{ORD}$ such that there is no α'-like element in $\mathrm{dom}(O)$. Let u be an α'-like element. Then by Theorem 13, $[\![\mathrm{ORD}(u)]\!] \in \{1, 1/2\}$ but

$$[\![u \in O]\!] = \bigvee_{x \in \mathrm{dom}(O)} (O(x) \wedge [\![x = u]\!]) = 0.$$

Hence $[\![\forall x(\mathrm{ORD}(x) \to x \in O)]\!] = 0$. Since O is arbitrary we have

$$[\![\exists O\ \forall x(\mathrm{ORD}(x) \to x \in O)]\!] = 0.$$

So the theorem is proved. □

Non-Classical Behavior of α-like Elements

Let us consider an $\alpha \in \mathrm{ORD}$ and a formula $\varphi(x)$ with one free variable x. One may expect for any two α-like names u and v, they agree on whether the property φ holds of them.

But the ordinal-like elements behave non-classically: there exists a formula $\varphi(x)$ such that for any given non-zero $\alpha \in \mathrm{ORD}$ there exist two α-like elements u and v such that $\mathbf{V}^{(\mathrm{PS}_3)} \models \varphi(u)$ whereas $\mathbf{V}^{(\mathrm{PS}_3)} \not\models \varphi(v)$. For example, let $\varphi(x) := \neg\ \exists y(y \in x)$ and choose any non-zero $\alpha \in \mathrm{ORD}$. Fix any two α-like elements u and v as $ran(u) = \{1/2\}$ and $ran(v) = \{1\}$. Then clearly $[\![\varphi(u)]\!] = 1/2$, i.e., $\mathbf{V}^{(\mathrm{PS}_3)} \models \varphi(u)$. But it is easy to calculate that $[\![\varphi(v)]\!] = 0$, i.e., $\mathbf{V}^{(\mathrm{PS}_3)} \not\models \varphi(v)$.

In [5], it is stressed that Leibniz's law of the indiscernibility of identicals

$$\forall x \forall y(x = y \wedge \varphi(x) \to \varphi(y))$$

is not necessarily valid in $\mathbf{V}^{(\mathbb{A})}$ if $\mathbb{A}$ is a reasonable implication algebra. The above formula $\varphi(x) := \neg\ \exists y(y \in x)$ is one instance of that claim.

On the other hand it is also proved that for any instantiations of Leibniz's law with NFF-formulas φ is valid in general, and so α-like names agree on the validity of NFF-formulas.

5 Conclusion and Future Work

We have seen that the α-like names form equivalence classes in $\mathbf{V}^{(\mathrm{PS}_3)}/\sim$. However, due to the failure of Leibniz's law in $\mathbf{V}^{(\mathrm{PS}_3)}$, elements in the same $\sim$-equivalence class can instantiate different properties.

In future work, we plan to study the *natural numbers*, *rational numbers*, and *real numbers* in $\mathbf{V}^{(\mathbb{A})}$, together with their algebraic properties, as well as cardinal numbers in $\mathbf{V}^{(\mathbb{A})}$.

Acknowledgements. The author would like to thank Benedikt Löwe for the collaboration during his stay at the *Universiteit van Amsterdam* in June 2014 that was the basis of several ideas in this paper. The visit to Amsterdam was made possible by a travel grant from the *Indo-European Research Training Network in Logic (IERTNiL)*. He is also thankful to Mihir Kr. Chakraborty for providing important suggestions in some deeper issues and in shaping the final version of this paper. The author's research is partially funded by *CSIR*, Ph.D. fellowship, Govt. of India (09/028(0754)/2009-EMR-I).

References

1. Bell, J.L.: Set Theory: Boolean-Valued Models and Independence Proofs. Oxford University Press, Oxford (2005)
2. Carnielli, W.A., Marcos, J.: A Taxonomy of C-Systems. In: Carnielli, W.A., Coniglio, M.E., D'Ottaviano, I.M.L. (eds.) Paraconsistency: The Logical Way to the Inconsistent. Lecture Notes in Pure and Applied Mathematics, vol. 228, pp. 1–94. Marcel Dekker, New York (2002)
3. Coniglio, M.E., Da Cruz Silvestrini, L.H.: An Alternative Approach for Quasi-Truth. Logic Journal of the IGPL 22(2), 387–410 (2014)
4. Grayson, R.J.: Heyting-valued models for intuitionistic set theory. In: Fourman, M.P., Mulvey, C.J., Scott, D.S. (eds.) Applications of Sheaves, Proceedings of the Research Symposium on Applications of Sheaf Theory to Logic, Algebra and Analysis, University of Durham, Durham, July 9–21, 1977. Lecture Notes in Mathematics, vol. 753, pp. 402–414. Springer, Berlin (1979)
5. Löwe, B., Tarafder, S.: Generalised Algebra-Valued Models of Set Theory. Review of Symbolic Logic (accepted, 2014)
6. Ozawa, M.: Transfer Principle in Quantum Set Theory. Journal of Symbolic Logic 72(2), 625–648 (2007)
7. Ozawa, M.: Orthomodular-Valued Models for Quantum Set Theory. Preprint, ArXiv 0908.0367 (2009)
8. Takeuti, G., Titani, S.: Fuzzy Logic and Fuzzy Set Theory. Archive for Mathematical Logic 32(1), 1–32 (1992)
9. Tarafder, S., Chakraborty, M.K.: A Three-Valued Paraconsistent Logic Suitable for a Paraconsistent Set Theory (preprint, 2014)
10. Titani, S.: A Lattice-Valued Set Theory. Archive for Mathematical Logic 38(6), 395–421 (1999)
11. Titani, S., Kozawa, H.: Quantum Set Theory. International Journal of Theoretical Physics 42(11), 2575–2602 (2003)
12. Weber, Z.: Transfinite Numbers in Paraconsistent Set Theory. Review of Symbolic Logic 3(1), 71–92 (2010)

Extending Carnap's Continuum to Binary Relations

Alena Vencovská*

School of Mathematics, University of Manchester, Manchester M13 9PL, UK
alena.vencovska@manchester.ac.uk

Abstract. We investigate a binary generalization of Carnap's Continuum of Inductive Methods based on a version of Johnson's Sufficientness Postulate for polyadic atoms and determine the probability functions that satisfy it.

Introduction

The problem of drawing conclusions inductively has puzzled philosophers for centuries: how to use the available evidence to support a hypothesis to a certain degree, thus extending the classical deductive reasoning to allow us to reach less-than-certain but probable conclusions. A substantial contribution to this topic was made by logical positivism as represented by Rudolf Carnap and others during the earlier parts of the 20th century. They developed a formal framework in which rational assignments of probabilities to sentences could be studied, aiming to capture our reasoning about the world which was assumed to combine (just) elementary experiences and pure logic.

Carnap's ambitious program has been subsequently largely seen as a failure on the grounds that the framework cannot be made to correspond to the way we interpret the world and reason about it (see [1], [2]). Still, the advent of artificial reasoning agents justifies a re-examination of the purely formal and uninterpreted aspect of Carnap's proposal, a mode of reasoning which Carnap himself referred to as *Pure* Inductive Logic.[1]

While Carnap experimented with various formal frameworks to investigate this logic the most transparent seems to be when all statements are expressed in first order logic with a language involving countably many individuals and finitely many predicate or relation symbols, and principles are adopted for assigning belief values to these statements in a rational, logical fashion independently of any intended interpretation.

There are good arguments for identifying belief with subjective probability, and for identifying belief values based on some evidence with conditional probabilities. So Pure Inductive Logic works with probability functions. More formally,

* Supported by a UK Engineering and Physical Sciences Research Council (EPSRC) Research Grant R117181.

[1] For Carnap's approach, see for example [3], [4]. For more recent developments, see for example [5], [6]. [6] also contains an extensive bibliography of related works.

M. Banerjee and S.N. Krishna (eds.): ICLA 2015, LNCS 8923, pp. 207–217, 2015.

let L be a language with finitely many predicate or relation symbols (without equality) and with countably many constant symbols $a_1, a_2, a_3, \ldots$. We say that a function w which assigns real numbers between 0 and 1 to sentences of L is a probability function if for any sentences θ, ϕ and $\exists x\, \psi(x)$ the following conditions hold:

- If θ is logically valid then $w(\theta) = 1$.
- If θ and ϕ are mutually exclusive then $w(\theta \vee \phi) = w(\theta) + w(\phi)$.
- $w(\exists x\, \psi(x)) = \lim_{n\to\infty} w(\psi(a_1) \vee \psi(a_2) \vee \ldots \vee \psi(a_n))$.

The conditional probability of θ given ϕ is defined when $w(\phi) \neq 0$ as the ratio $\frac{w(\theta\wedge\phi)}{w(\phi)}$.

For any given language as above there are many probability functions, but some of them are better for inductive reasoning then others: for example, it is desirable that the conditional probability of a 'new' individual having some properties increases (or at least does not decrease) on the basis of another individual found having these properties. Similarly, it is desirable that to start with w does not code any unintended information about the individual constants, predicates or relations; any particular information about them should be supplied by the evidence on a case by case basis.

Accordingly, a number of arguably rational principles have been formulated that can be imposed on w. Not all of them are compatible and the question for Pure Inductive Logic is which combination(s) of these principles should be chosen, what are the probability functions satisfying them and what inferences they authorize.

Carnap and others in the 20th century studied the situation where there were only unary predicate symbols and no relation symbols of higher arities in the language. They independently identified a principle subsequently known as *Johnson's Sufficientness Postulate*, see [7], which yields a particular one-parameter family of probability functions c^L_λ for positive numbers λ (real or ∞): the *Carnap Continuum of Inductive Methods.* Their result is remarkable for its power and elegance, apparently reducing the choice of a rational probability function down to the choice of a single parameter λ, and it has long been seen as the corner stone of *unary* Pure Inductive Logic. The recent development of Pure Inductive Logic for languages containing relation symbols of higher arities therefore raises the major question of whether, and if so how, this continuum can, or should be, extended to these larger languages.

One possible answer to this is given in [8] where it is shown that for any positive λ the family of probability functions c^L_λ as we vary the unary language L has a natural continuation also to not necessarily unary L which preserves the key property of *Spectrum Exchangeability* (for details see [8]).

In this paper however we shall propose an alternative extension of Carnap Continuum (to binary languages) based on satisfying a natural generalization of the original Johnson's Sufficientness Postulate. In this sense then we would claim that it comes closer to capturing Johnson's and Carnap's original intuitions and insights.

Preliminaries

The most fundamental and generally accepted principle is that of *Constant Exchangeability, Ex,* which says that if $\theta(a_1, \ldots, a_m)$ is a sentence of L and $a'_1, \ldots, a'_m$ is any other choice of distinct constant symbols from amongst the $a_1, a_2, \ldots$ then $\theta(a_1, \ldots, a_m)$ and $\theta(a'_1, \ldots, a'_m)$ should have the same probability. We shall assume Ex. We will also need the *Principle of Regularity, Reg,* which requires any consistent quantifier free sentence of L to have non-zero probability.

For the purpose of this article we will restrict our attention to languages with finitely many unary predicate symbols $R_1, \ldots, R_p$, finitely many binary relation symbols $Q_1, \ldots, Q_q$ and no relation symbols of higher arities. Sentences $\Theta(a_1, \ldots, a_m)$ of the form

$$\bigwedge_{j=1}^{m} \bigwedge_{i=1}^{p} \pm R_i(a_j) \;\wedge \bigwedge_{\langle j,l\rangle \in \{1,\ldots,m\}^2} \bigwedge_{i=1}^{q} \pm Q_i(a_j, a_l) \tag{1}$$

where $\pm R_i(a_j)$ denotes one of $R_i(a_j)$, $\neg R_i(a_j)$ and similarly for $\pm Q_i(a_j, a_l)$, are called *state descriptions for* $a_1, \ldots, a_m$. Note that state descriptions for $a_1, \ldots, a_m$ are mutually exclusive and exhaustive (their disjunction is a tautology).

Any probability function w is uniquely determined by its values on state descriptions and in many situations it suffices to think of probability functions as functions defined on state descriptions and such that probabilities of state descriptions for $a_1, \ldots, a_m$ sum to 1, and probabilities of state descriptions for $a_1, \ldots, a_{m+1}$ which extend a given state description $\Theta(a_1, \ldots, a_m)$ sum to $w(\Theta(a_1, \ldots, a_m))$, see [5]. We shall use this in the present paper.

In the unary context, that is, when $q = 0$ and the language consists merely of the unary predicates $R_1, \ldots, R_p$, a state description for $a_1, \ldots, a_m$ is any sentence $\bigwedge_{j=1}^{m} \bigwedge_{i=1}^{p} \pm R_i(a_j)$. The formulae $\bigwedge_{i=1}^{p} \pm R_i(x)$ are called atoms and denoted $\alpha_1(x), \ldots, \alpha_{2^p}(x)$. Unary state descriptions are usually written as $\bigwedge_{j=1}^{m} \alpha_{h_j}(a_j)$, where $h_j \in \{1, \ldots, 2^p\}$.

Johnson's Sufficientness Postulate, JSP. $w\left(\alpha_i(a_{m+1}) \mid \bigwedge_{j=1}^{m} \alpha_{h_j}(a_j)\right)$ *depends only on* m *and* m_i*, where* m_i *is the number of times that* i *appears amongst the* h_j*.*

The classical result discussed in the Introduction tells us that as long as the language has at least two predicates ($p \geq 2$), the only probability functions satisfying Reg, Ex and JSP are the Carnap's c^L_λ functions ($0 < \lambda \leq \infty$) defined as follows[2]

$$c^L_\lambda\left(\alpha_i(a_{m+1}) \mid \bigwedge_{j=1}^{m} \alpha_{h_j}(a_j)\right) = \frac{m_i + \lambda 2^{-p}}{m + \lambda}$$

where m_i is as above.

---

[2] Note that the definition does yield the values of c^L_λ on all state descriptions and gives a probability function.

The Binary Case

When $q \neq 0$, a state description (1) cannot be expressed as a conjunction of sentences each involving only one constant as it is in the unary case with atoms, but it is possible to define *binary atoms* and proceed similarly[3].

For this purpose, we define $\beta_1(x), \dots, \beta_{2^{p+q}}(x)$ to be the formulae

$$\bigwedge_{i=1}^{p} \pm R_i(x) \quad \wedge \quad \bigwedge_{i=1}^{q} \pm Q_i(x,x)$$

and $\delta_1(x,y), \dots, \delta_{2^q}(x,y)$ to be the formulae $\bigwedge_{i=1}^{q} \pm Q_i(x,y)$. We define the binary atoms to be the formulae

$$\gamma_{[k,c,h,d]}(x,y) = \beta_k(x) \wedge \beta_c(y) \wedge \delta_h(x,y) \wedge \delta_d(y,x)$$

where $k, c \in \{1, \dots, 2^{p+q}\}$ and $h, d \in \{1, \dots, 2^q\}$.

The state description (1) can then be also expressed as

$$\bigwedge_{j=1}^{m} \beta_{v_j}(a_j) \wedge \bigwedge_{1 \leq j < l \leq m} \gamma_{h_{j,l}}(a_j, a_l)$$

where $v_j \in \{1, \dots, 2^{p+q}\}$ and $h_{j,l} \in \{\ [v_j, v_l, h, d] :\ h, d \in \{1, \dots, 2^q\}\ \}$. There is some redundancy in this expression but it is convenient for what follows.

We also need the concept of a partial state description which is a sentence $\Delta(a_1, \dots, a_m)$ of the form

$$\bigwedge_{j=1}^{m} \beta_{v_j}(a_j) \wedge \bigwedge_{\langle j,l \rangle \in A} \gamma_{h_{j,l}}(a_j, a_l) \tag{2}$$

where v_j and $h_{j,l}$ are as above and $A \subseteq \{\langle j,l \rangle :\ 1 \leq j < l \leq m\}$.

Clearly every state description is a partial state description. We define the signature of the (partial) state description (2) to be $\vec{m}\vec{n}$ such that

$$\vec{m} = \langle\, m_1, \dots, m_{2^{p+q}}\, \rangle,$$

where m_k is the number of $j \in \{1, \dots, m\}$ such that $v_j = k$ and

$$\vec{n} = \langle n_{[k,c,h,d]} : k, c \in \{1, \dots, 2^{p+q}\} \text{ and } h, d \in \{1, \dots, 2^q\} \rangle,$$

where $n_{[k,c,h,d]}$ is the number of $\langle j,l \rangle \in A$ such that

$$h_{j,l} = [k,c,h,d] \text{ or } h_{j,l} = [c,k,d,h]. \tag{3}$$

The atoms $\gamma_{[k,c,h,d]}$ and $\gamma_{[c,k,d,h]}$ are counted together because

$$\gamma_{[k,c,h,d]}(x,y) = \gamma_{[c,k,d,h]}(y,x) \tag{4}$$

[3] The concepts and results in this section come from [9].

and hence (2) implies $\gamma_{[k,c,h,d]}(a_j, a_l)$ just when it implies $\gamma_{[c,k,d,h]}(a_l, a_j)$. Since we wish the signature to record the numbers of (unordered) pairs of constants with certain behaviour and the decision to write (2) using *ordered* pairs $\langle a_j, a_l \rangle$ with $j < l$ is merely a matter of convention, $\gamma_{[k,c,h,d]}$ and $\gamma_{[c,k,d,h]}$ should play the same role. Note that $n_{[k,c,h,d]} = n_{[c,k,d,h]}$ and that the sum of the m_k is m. We remark that $\vec{n}$ uniquely determines $\vec{m}$.

Furthermore we define $n_{k,c}$ to be the number of $\langle j, l \rangle \in A$ such that (3) holds for some $h, d \in \{1, \ldots, 2^q\}$. We have

$$n_{k,c} = \sum_{h,d \in \{1,\ldots,2^q\}} n_{[k,c,h,d]} \quad (k \neq c), \qquad n_{k,k} = \sum_{\substack{h,d \in \{1,\ldots,2^q\} \\ h \leq d}} n_{[k,k,h,d]} \,. \tag{5}$$

It may seem that requiring a probability function to give state descriptions with the same signature equal probability is equivalent to Ex. However, this is not the case: the following principle is strictly stronger than Ex:

Binary Exchangeability, BEx. *For a state description $\Theta(a_1, \ldots, a_m)$ of L the probability $w(\Theta)$ depends only on the signature of Θ.*

Even so many probability functions do satisfy BEx, and there is a representation theorem for them similar to the de Finnetti representation theorem for probability function satisfying Ex, see [9]. We shall employ[4] the following result from [9]:

Theorem 1. *Let w be a probability function and assume that w satisfies BEx. Let $\Delta(a_1, \ldots, a_m)$ be a partial state description as in (2), $s, t, r, g \in \{1, \ldots, m\}$, $s < t$, $r < g$, $\langle r, g \rangle \notin A$ and γ a binary atom such that $\Delta \wedge \gamma(a_r, a_g) \wedge \gamma(a_s, a_t)$ is consistent. Then*

$$w(\gamma(a_r, a_g) \mid \Delta) \leq w(\gamma(a_s, a_t) \mid \Delta \wedge \gamma(a_r, a_g)) \,. \tag{6}$$

An Example. Let $p = q = 1$ so $L = \{R, Q\}$ where R is unary and Q is binary. We have

$$\begin{array}{ll}
\beta_1(x) = R(x) \wedge Q(x,x) & \delta_1(x,y) = Q(x,y) \\
\beta_2(x) = R(x) \wedge \neg Q(x,x) & \delta_2(x,y) = \neg Q(x,y) \\
\beta_3(x) = \neg R(x) \wedge Q(x,x) & \\
\beta_4(x) = \neg R(x) \wedge \neg Q(x,x) &
\end{array}$$

and the binary atom $\gamma_{[2,3,1,2]}(x,y)$ is the formula

$$R(x) \wedge \neg Q(x,x) \wedge \neg R(y) \wedge Q(y,y) \wedge Q(x,y) \wedge \neg Q(y,x).$$

[4] We remark that this theorem from a forthcoming paper is not essential for the result presented here (Theorem 2) in the sense that only a special case of it is needed in the proof and it could be added to the assumptions. Indeed the original Johnson's proof in [7] for the unary case introduces a corresponding postulate although it was subsequently shown to be unnecessary. To be precise, we could replace the usage of Theorem 1 by *assuming* that (6) from Theorem 1 holds when $\Delta(a_1, a_2, a_3) = \beta_k(a_1) \wedge \beta_c(a_2) \wedge \beta_k(a_3)$ (where $1 \leq k \leq c \leq 2^{p+q}$).

The sentence

$$\beta_1(a_1)\wedge\beta_1(a_2)\wedge\beta_1(a_3)\wedge\beta_4(a_4)\wedge \bigwedge_{1\leq j<l\leq 3} \gamma_{[1,1,1,1]}(a_j,a_l)\wedge\bigwedge_{j=1}^{3}\gamma_{[1,4,1,2]}(a_j,a_4) \quad (7)$$

is a state description for a_1,a_2,a_3,a_4 and the sentence

$$\beta_1(a_1)\wedge\beta_1(a_2)\wedge\beta_1(a_3)\wedge\beta_3(a_4)\wedge\gamma_{[1,1,1,1]}(a_1,a_2)\wedge\gamma_{[1,3,1,2]}(a_1,a_4) \quad (8)$$

is a partial state description for a_1,a_2,a_3,a_4.

Imagine a simple situation where individuals $a_1,a_2,\ldots$ could do just two things: think other individuals (also themselves) to be good cooks or not, and like to eat fish or not. If we interpret each a_j as a_j, $R(x)$ as 'x likes to eat fish' and $Q(x,y)$ as 'x thinks y is a good cook', then (7) says that a_1,a_2,a_3 all like to eat fish and think everybody including themselves to be good cooks and that a_4 does not like to eat fish and does not think anybody including him/herself to be a good cook; (8) says that a_1,a_2,a_3 all like to eat fish and each thinks him/herself to be a good cook, a_4 does not like to eat fish but thinks him/herself to be a good cook, a_1 and a_2 think each other to be good cooks and a_1 also thinks a_4 to be a good cook but a_4 does not think a_1 to be a good cook.

Now consider in light of this example what a binary variant of Johnson's Sufficientness Postulate might be. It appears reasonable that for a partial state description $\Theta(a_1,\ldots,a_m)$ as given by (2), the probability of an extension of it by some $\beta_k(a_{m+1})$ (how a new individual behaves in isolation) would depend only on the β_{v_j} (how other individuals behave in isolation) rather than the $\gamma_{h_{j,l}}$, and an extension of it by some $\gamma_{[k,c,h,d]}(a_s,a_t)$ for $1\leq s<t\leq m$, $\langle s,t\rangle\notin A$ (how a_s and a_t relate to each other given how each of them behaves in isolation) would depend only on those $\gamma_{h_{j,l}}$ where a_j and a_l behave in isolation just as a_s and a_t do.

Taking this a step further along the lines of the unary Johnson's Sufficientness Postulate and using the notation from page 210, this appears to lead to the requirement that the conditional probability of $\beta_k(a_{m+1})$ given (2) should depend only on m_k and m, and for $k=v_s$, $c=v_t$ the conditional probability of $\gamma_{[k,c,h,d]}(a_s,a_t)$ given (2) should depend only on $n_{[k,c,h,d]}$ and $n_{k,c}$.

However, there is a little catch when $k=c$ and $h\neq d$ for the following reason. Assuming again that $k=v_s$, $c=v_t$, for $k\neq c$ the conditional probability of $\gamma_{[k,c,h,d]}(a_s,a_t)$ given (2) equals the probability of increasing $n_{[k,c,h,d]}$ by 1 using a_s,a_t since only $\gamma_{[k,c,h,d]}(a_s,a_t)$, not $\gamma_{[c,k,d,h]}(a_s,a_t)$, is consistent with (2). Similarly when $k=c$ and $h=d$ because the two possibilities become the same. But when $h\neq d$, both $\gamma_{[k,k,h,d]}(a_s,a_t)$, $\gamma_{[k,k,d,h]}(a_s,a_t)$ can extend (2) and hence increasing $n_{[k,k,h,d]}$ by 1 using a_s,a_t can be done in two ways; the conditional probability of each should therefore arguably depend on $n_{[k,k,h,d]}$ and $n_{k,k}$ differently than when $k\neq c$ or $k=c, h=d$.

Binary Carnap Continuum

Motivated by the above, we will say that the atoms $\gamma_{[k,k,h,d]}$ where $h\neq d$ *double*, and we shall consider the following principle.

Binary Sufficientness Postulate, BSP. *For a partial state description $\Delta(a_1, \ldots, a_m)$ with signature $\vec{m}\vec{n}$ as on page 210:*

$$w(\beta_k(a_{m+1})|\Delta) \text{ depends only on } m_k \text{ and } m,$$

and for $1 \leq s < t \leq m$ with $\langle s,t\rangle \notin A$ and $v_s = k$, $v_t = c$,

$$w(\gamma_{[k,c,h,d]}(a_s,a_t)|\Delta) \text{ depends only on } n_{[k,c,h,d]} \text{ and } n_{k,c}, \text{ and on whether or not } \gamma_{[k,c,h,d]} \text{ doubles.}$$

Note that BSP implies BEx.

Theorem 2. *Let w be a probability function on L and assume $p+q \geq 2$, $q \geq 1$. Then w satisfies Ex, Reg and BSP just when there are $\mu, \lambda \in (0,\infty]$ such that for a partial state description $\Delta(a_1,\ldots,a_m)$ given by (2) with signature $\vec{m}\vec{n}$ we have*

$$w(\beta_k(a_{m+1})|\Delta) = \frac{m_k + \frac{\mu}{2^{p+q}}}{m+\mu}\,, \tag{9}$$

and for $s,t \in \{1,\ldots,m\}$, $s<t$, $\langle s,t\rangle \notin A$, $h,d \in \{1,\ldots,2^q\}$ and $k = v_s$, $c = v_t$ we have

$$w(\gamma_{[k,c,h,d]}(a_s,a_t)|\Delta) = \frac{n_{[k,c,h,d]} + \frac{\lambda}{2^{2q}}}{n_{k,c}+\lambda} \qquad [k \neq c \text{ or } (k=c \text{ and } h=d)]\,, \tag{10}$$

$$w(\gamma_{[k,k,h,d]}(a_s,a_t)|\Delta) = \frac{\frac{n_{[k,k,h,d]}}{2} + \frac{\lambda}{2^{2q}}}{n_{k,k}+\lambda} \qquad [k=c \;\text{ and } h \neq d]\,)\,. \tag{11}$$

Before proving the theorem, note that from (5) we can see that the above formulae uniquely define a probability function $C^L_{\lambda,\mu}$ satisfying Ex and Reg. Explicitly, for Δ as above,

$$\begin{aligned} C^L_{\lambda,\mu}(\Delta) \;&=\; \frac{\prod_k \prod_{j=0}^{m_k-1}(j+\frac{\mu}{2^{p+q}})}{\prod_{j=0}^{m-1}(j+\mu)} \prod_{k<c} \frac{\prod_{h,d}\prod_{j=0}^{n_{[k,c,h,d]}-1}(j+\frac{\lambda}{2^{2q}})}{\prod_{j=0}^{n_{k,c}-1}(j+\lambda)} \\ &\times \prod_k \frac{\prod_h \prod_{j=0}^{n_{[k,k,h,h]}-1}(j+\frac{\lambda}{2^{2q}})}{\prod_{j=0}^{n_{k,k}-1}(j+\lambda)} \prod_k \frac{\prod_{h<d}\prod_{j=0}^{n_{[k,k,h,d]}-1}(\frac{j}{2}+\frac{\lambda}{2^{2q}})}{\prod_{j=0}^{n_{k,k}-1}(j+\lambda)} \end{aligned}$$

where k,c range through $1,\ldots,2^{p+q}$ and h,d range through $1,\ldots,2^q$ and the empty product equals 1.

Proof of Theorem 2. Let w be a probability function satisfying Ex, Reg and BSP. Define w_1 on the unary language $L_1 = \{R_1,\ldots,R_p,P_1,\ldots,P_q\}$ by

$$w_1(\Theta) = w(\Theta^Q)$$

where Θ is a state description of L_1 and Θ^Q obtains from it by changing each $P_i(a_j)$ to $Q_i(a_j,a_j)$. w_1 extends to a probability function satisfying Ex, Reg and

JSP so by the classical unary result it is equal to c^L_μ for some $\mu \in (0,\infty]$. It follows that (9) holds for w.

Now fix k and c and assume $k \neq c$. Then none of the 2^{2q} of the $\gamma_{[k,c,h,d]}$ double. Let $g(r,n)$ stand for $w(\gamma_{[k,c,h,d]}(a_s,a_t)|\Delta(a_1,\dots,a_m))$ where $s,t \in \{1,\dots,m\}$ and $\Delta(a_1,\dots,a_m)$ is as above with $v_s = k$, $v_t = c$, $\langle s,t\rangle \notin A$, $n = n_{k,c}$ and $r = n_{[k,c,h,d]}$. Using Theorem 1 we can argue just as in the classical unary case (see [6, Chapter 17]) to obtain (10) for w with some $\lambda \in (0,\infty]$ for these particular k, c. Note that by BSP it also follows that the λ obtained for other pairs $k \neq c$ must be the same.

Adapting the same method we can also resolve the remaining case: Consider some fixed $k \in \{1,\dots,2^{p+q}\}$. Let $g(r,n)$ and $h(r,n)$ stand for $w(\gamma_{[k,k,h,d]}(a_s,a_t)|\Delta(a_1,\dots,a_m))$ $(h \neq d)$, $w(\gamma_{[k,k,h,h]}(a_s,a_t)|\Delta(a_1,\dots,a_m))$ respectively where $s,t \in \{1,\dots,m\}$ and $\Delta(a_1,\dots,a_m)$ is as above, with $v_s = v_t = k$, $s<t$, $\langle s,t\rangle \notin A$, $n = n_{k,k}$ and $r = n_{[k,k,h,d]}$ or $r = n_{[k,k,h,h]}$ respectively. Define $\chi = h(0,0)$ and $\xi = g(0,0)$ and let ν be the (unique) element in $(0,\infty]$ such that

$$h(1,1) = \frac{1+\chi\nu}{1+\nu}. \tag{12}$$

Such a unique ν exists since $0 < \chi = h(0,0) \leq h(1,1) < 1$ (by regularity and Theorem 1) and the function $x \mapsto \frac{1+\chi x}{1+x}$ is decreasing from 1 to χ as x runs from 0 to ∞. From

$$w\left(\bigvee_{h,d=1}^{2^q} \gamma_{[k,k,h,d]}(a_1,a_2) \mid (\beta_k(a_1)\wedge\beta_k(a_2))\right) = 1$$

we find that

$$2^q\chi + (2^{2q}-2^q)\xi = 1. \tag{13}$$

We shall first show that to conclude the proof it suffices to prove that for all $n \in \mathbb{N}$

$$h(r,n) = \frac{r+\chi\nu}{n+\nu} \quad \text{and} \quad g(r,n) = \frac{\frac{r}{2}+\xi\nu}{n+\nu} \qquad (r \in \{0,1,\dots,n\}). \tag{14}$$

This is because by BSP and the above result for $k \neq c$, $h(r,n) = \frac{r+\frac{\lambda}{2^{2q}}}{n+\lambda}$ for all $n \in \mathbb{N}$ and $r \in \{0,1,\dots,n\}$ so (14) forces $\nu = \lambda$ and $\chi = \frac{1}{2^{2q}}$ and hence by (13) also $\xi = \frac{1}{2^{2q}}$ so (9) and (10) follow.

Hence it remains to prove (14). For $n=0$ it follows by the definition of χ and ξ. Considering e.g.

$$w\left(\beta_k(a_1)\wedge\beta_k(a_2)\wedge\beta_k(a_3)\wedge\gamma_{[k,k,h,d]}(a_1,a_2)\wedge\gamma_{[k,k,h,h]}(a_1,a_3)\right)$$

we can see that $g(0,1)h(0,0) = h(0,1)g(0,0)$, that is,

$$\chi g(0,1) = \xi h(0,1). \tag{15}$$

From

$$w\left(\bigvee_{h,d}\gamma_{[k,k,h,d]}(a_1,a_3)\mid\beta_k(a_1)\wedge\beta_k(a_2)\wedge\beta_k(a_3)\wedge\gamma_{[k,k,h_0,d_0]}(a_1,a_2)\right)=1$$

with $h_0\neq d_0$ and $h_0=d_0$ respectively we obtain

$$2g(1,1)+2^q h(0,1)+(2^{2q}-2^q-2)g(0,1)=1 \tag{16}$$

and

$$h(1,1)+(2^q-1)h(0,1)+(2^{2q}-2^q)g(0,1)=1\,. \tag{17}$$

Using (12), (13) and (15) this yields

$$h(0,1)=\frac{\nu\chi}{1+\nu},\quad g(0,1)=\frac{\nu\xi}{1+\nu},\quad g(1,1)=\frac{\frac{1}{2}+\nu\xi}{1+\nu}$$

so (14) holds also for $n=1$.

Now assume (14) holds for n. For any $c,e,f\in\mathbb{N}$ with $c+e+f=n$ we find by considering e.g.

$$w\left(\gamma_{[k,k,h_1,d_1]}(a_2,a_3)\wedge\gamma_{[k,k,h_2,d_2]}(a_2,a_4)\mid\Delta\right)$$

where Δ is the conjunction of $\bigwedge_{j=1}^{n+1}\beta_k(a_j)$ and

$$\bigwedge_{j=2}^{c+1}\gamma_{[k,k,h_1,d_1]}(a_1,a_j)\wedge\bigwedge_{j=c+2}^{c+e+1}\gamma_{[k,k,h_2,d_2]}(a_1,a_j)\wedge\bigwedge_{j=c+e+2}^{n+1}\gamma_{[k,k,h_3,d_3]}(a_1,a_j)$$

by taking distinct pairs $\langle h_1,d_1\rangle$, $\langle h_2,d_2\rangle$, $\langle h_3,d_3\rangle$ with ($h_1=d_1$ and $h_2=d_2$) or ($h_1=d_1$ and $h_2\neq d_2$) respectively that

$$h(c,n+1)h(e,n)=h(e,n+1)h(c,n), \tag{18}$$

$$h(c,n+1)g(e,n)=g(e,n+1)h(c,n). \tag{19}$$

With c running from 1 to n and $e=0$ (18) yields

$$h(c,n+1)=\frac{h(c,n)}{h(0,n)}h(0,n+1)\qquad(c\in\{1\ldots n\}) \tag{20}$$

and hence by the inductive hypothesis we have

$$h(c,n+1)=\frac{c+\chi\nu}{\chi\nu}h(0,n+1)\qquad(c\in\{1\ldots n\})\,. \tag{21}$$

Considering $0\leq e,f\leq n+1$ with $e+f=n+1$ and taking distinct pairs $\langle h_1,d_1\rangle$, $\langle h_2,d_2\rangle$ with ($h_1=d_1$ and $h_2=d_2$) we find that

$$w\left(\bigvee_{h,d}\gamma_{[k,k,h,d]}\;\middle|\;\bigwedge_{j=1}^{n+2}\beta_k(a_j)\wedge\bigwedge_{j=2}^{e+1}\gamma_{[k,k,h_1,d_1]}(a_1,a_j)\wedge\bigwedge_{j=e+2}^{e+f+1}\gamma_{[k,k,h_2,d_2]}(a_1,a_j)\right)$$

equals 1 and hence

$$h(e,n+1)+h(f,n+1)+(2^q-2)h(0,n+1)+(2^{2q}-2^q)g(0,n+1)=1\,. \quad (22)$$

From this and (21) (using some $0<e,f<n+1$), from (19) with $c=e=0$ and from the inductive hypothesis we have

$$\left(\frac{e+\chi\nu}{\chi\nu}+\frac{f+\chi\nu}{\chi\nu}+(2^q-2)+(2^{2q}-2^q)\frac{\xi}{\chi}\right)h(0,n+1)=1$$

so

$$(\,(n+1+2\chi\nu)+(2^q-2)\chi\nu+(2^{2q}-2^q)\,\xi\nu)\,)\,h(0,n+1)=\chi\nu$$

and since $2^q\chi+(2^{2q}-2^q)\,\xi=1$, it follows that $h(0,n+1)=\frac{\chi\nu}{n+1+\nu}$. Hence by (21) and (19),

$$h(c,n+1)=\frac{c+\chi\nu}{n+1+\nu}\quad\text{and}\quad g(c,n+1)=\frac{\frac{c}{2}+\xi\nu}{n+1+\nu}\qquad(c\in\{0,1,\ldots,n\})\,.$$

From (22) (with $e=n+1$, $f=0$) we have

$$h(n+1,n+1)=\frac{n+1+\chi\nu}{n+1+\nu}\,.$$

Finally considering some $h_1\neq d_1$, from

$$w\left(\bigvee_{h,d}\gamma_{[k,k,h,d]}\;\middle|\;\bigwedge_{j=1}^{n+1}\beta_k(a_j)\wedge\bigwedge_{j=2}^{n+2}\gamma_{[k,k,h_1,d_1]}(a_1,a_j)\right)=1$$

we have

$$2g(n+1,n+1)+2^q\,h(0,n+1)+(2^{2q}-2^q-2)g(0,n+1)=1$$

which yields

$$g(n+1,n+1)=\frac{\frac{n+1}{2}+\xi\nu}{n+1+\nu}$$

completing the proof.

References

[1] Goodman, N.: A Query on Confirmation. Journal of Philosophy 43, 383–385 (1946)
[2] Quine, W.V.O.: Two Dogmas of Empiricism. The Philosophical Review 60, 20–43 (1951)
[3] Carnap, R.: A Basic System of Inductive Logic. In: Carnap, R., Jeffrey, R.C. (eds.) Studies in Inductive Logic and Probability, vol. I, pp. 33–165. University of California Press (1971)
[4] Carnap, R.: A Basic System of Inductive Logic, Part 2. In: Jeffrey, R.C. (ed.) Studies in Inductive Logic and Probability, vol. II, pp. 7–155. University of California Press (1980)

[5] Paris, J.B.: Pure Inductive Logic Workshop Notes for ISLA (2014), http://www.maths.manchester.ac.uk/~jeff/lecture-notes/ISLA.pdf
[6] Paris, J.B., Vencovská, A.: Pure Inductive Logic. To appear in the series Perspectives in Mathematical Logic. CUP (2015)
[7] Johnson, W.E.: Probability: The Deductive and Inductive Problems. Mind 41, 409–423 (1932)
[8] Landes, J., Paris, J.B., Vencovská, A.: A Characterization of the Language Invariant Families satisfying Spectrum Exchangeability in Polyadic Inductive Logic. Annals of Pure and Applied Logic 161, 800–811 (2010)
[9] Ronel, T., Vencovská, A.: The Principle of Binary Exchangeability (forthcoming)

Representing Imperfect Information of Procedures with Hyper Models

Yanjing Wang*

Department of Philosophy, Peking University, Beijing, China
y.wang@pku.edu.cn

Abstract. When reasoning about knowledge of procedures under imperfect information, the explicit representation of epistemic possibilities blows up the S5-like models of standard epistemic logic. To overcome this drawback, in this paper, we propose a new logical framework based on compact models without epistemic accessibility relations for reasoning about knowledge of procedures. Inspired by the 3-valued abstraction method in model checking, we introduce hyper models which encode the imperfect procedural information. We give a highly non-trivial 2-valued semantics of epistemic dynamic logic on such models while validating all the usual S5 axioms. Our approach is suitable for applications where procedural information is 'learned' incrementally, as demonstrated by various examples.

1 Introduction

Suppose there are four cities A, B, C, D which are connected by public transportation as the following leftmost map shows ($\xrightarrow{b}$ for bus and $\xrightarrow{t}$ for train):

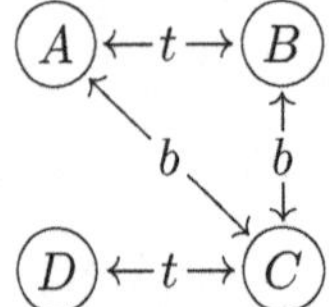
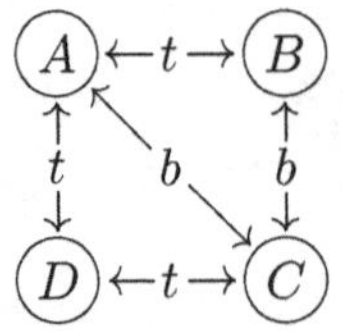
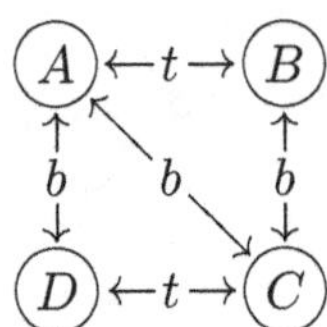
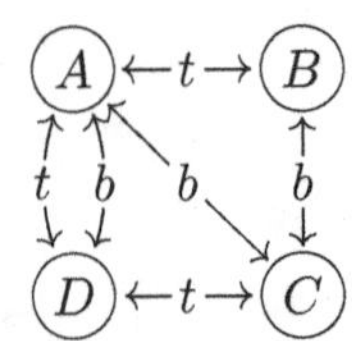

We may view the map as a Kripke model and use various modal logics such as Propositional Dynamic Logic (PDL) [12] to describe routes or more complicated trip plans from one city to others. Now suppose we are informed that A and D will also be connected next year, but it is not clear whether it will be a bus line or a train connection or even both. Then the new map can be any one of the three right-hand-side maps above. Although the new information is *imperfect*, we still can *know* that city D will become directly reachable from A since this is true in all the possible new maps, and it may be *possible* to reach C from A by train via D, since this is true in some possible maps.

As we have seen, imperfect information about the connectivity of the states introduces uncertainty. To encode such uncertainty, two-dimensional Kripke models are often used, which not only have labelled transitions but also epistemic accessibility

* The author thanks Floris Roelofsen and Ciyang Qing for their helpful comments on earlier versions of this paper. The research is partially supported by SSFC key project 12&ZD119.

M. Banerjee and S.N. Krishna (eds.): ICLA 2015, LNCS 8923, pp. 218–231, 2015.

relations between states, such as the epistemic temporal models in [10,4,11,15] and the models of imperfect information games in game theory (cf. e.g., [9]). In such epistemic logical frameworks, a proposition is known on a state if it is true on all the possible states linked with the current state by the epistemic relation.

However, explicitly representing possibilities as in the usual epistemic logical frameworks leads to the problem of state space explosion even in very simple case, if there is considerable ignorance. For example, if we have no idea how those four cities are connected by possibly four transportation methods (e.g., train, bus, flight, boat), then there are at least $(2^{4*3})^4 = 2^{48} > 10^{14}$ possible maps, which is in the order of the number of cells in a human body or the number of neuronal connections in a human brain. It is clear that in reality we do not go through all these possibilities in our mind to acquire knowledge. This leads to our first research question:

> *Can we have a compact alternative model of epistemic logic of procedures without explicitly representing all the possibilities?*

Moreover, it is important to incorporate new (imperfect) information about procedures into the current model which allows us to incrementally build up the model even from scratch. The new information may be given in a syntactic form which uses implicit quantifiers, such as '*there is* a bus going from A to *either* B or C, but I am not quite sure which due to the recent change of routes. On the other hand, from both B or C you should be able to reach D by *some* public transportation.' The imperfect information may also be given by a complicated procedure which is not just one-step, e.g., 'taking a bus then a train will get you home'. However, the semantic way of incorporating new information is usually done by essentially *eliminating* inconsistent possibilities according to the new information in epistemic frameworks such as dynamic epistemic logic [2], which again assumes that all the possibilities are represented in the model. Thus the following question is another challenge:

> *How to incorporate new procedural information semantically in an incremental fashion?*

The technical contribution of this paper is a new semantics-driven epistemic logical framework aiming at solving the above questions. Note that we may alternatively represent all the imperfect information syntactically but in the case of procedural information, the graphic models are more natural and compact for very complicated procedural information between states. Moreover, doing model checking is in general computationally more efficient than theorem proving. In the rest of this introduction, we will explain our main ideas.

Main Ideas

To handle the first question, we need a way to encode possibilities in a compact and implicit way. Our approach is inspired by the 3-valued abstraction technique in model checking (see [5] for an overview). To handle the problem of state space explosion in model checking, the abstraction technique makes a Kripke model smaller by abstracting away some information. Clearly, the smaller abstract model may not preserve the truth value of all the formulas in concern, but a suitable 3-valued semantics on abstractions can make sure the following:

- formulas that are true in the abstraction are also true in the original model
- formulas that are false in the abstraction are also false in the original model

In our point of view, an abstraction can be indeed viewed as a compact representation of potential concrete models which are consistent with the information represented in the abstract model. The information of transitions in the abstract model is typically encoded by two special kinds of transitions which are under- and over-approximations of the ones in the actual model [13]. This inspired us to use similar abstract transitions to encode imperfect procedural information. The final source of inspiration is from [3,1] where the approximations of the transitions are labelled by regular expressions and this helps us in dealing with arbitrarily complicated procedural information expressed by regular expressions.

Based on these ideas, we propose *hyper models* to encode the imperfect procedural information, and an epistemic PDL defined on such models, with the following features:

- Hyper models assemble possibilities in an implicit and compact way using abstract transitions (labelled by regular expressions) but not epistemic accessibility relations.
- They incorporate new information incrementally by *adding* new transitions.
- The semantics of our language is defined on the hyper models *directly*, and there is no need to unpack the hyper models into numerous possible Kripke models.
- The logic is still 2-valued and *all* the usual **S5** axioms are valid, but *not* the necessitation rule which may cause logical omniscience.

Of course, there is also a price to pay: not all the collections of Kripke models are representable by hyper models, to which we will come back in Section 4. In the rest of the paper, we first formalize imperfect procedural information in Section 2, and then introduce and discuss simple and full hyper models in Section 3. Finally we show that hyper models are indeed compact representations of Kripke models and point out future directions in Section 4.

2 Preliminaries

Kripke models are used to represent how states are connected by atomic actions:

Definition 1 (Kripke Model). *Given a set of basic propositional letters* $\mathbf{P}$*, and a set of atomic action symbols* Σ*, a Kripke model* $\mathcal{M}$ *over* $\mathbf{P}$ *and* Σ *is a tuple* $(S, \rightarrow, V)$ *where:*

- S *is a non-empty set of states.*
- $\rightarrow \subseteq S \times \Sigma \times S$ *is a binary relation over* S *labelled by action symbols from* Σ*.*
- $V: S \rightarrow 2^{\mathbf{P}}$ *assigns to each state a set of basic propositional letters.*

We write $s \xrightarrow{a} t$ *if* $(s, a, t) \in \rightarrow$*. Given* $w = a_1 a_2 \ldots a_n$ *we write* $s \xrightarrow{w} t$ *if there are* $s_0, s_1, \ldots, s_n$ *such that* $s_0 = s$ *and* $s_n = t$ *and* $s_k \xrightarrow{a_k} s_{k+1}$ *for all* $0 \leq k < n$*.*

Note that $\mathcal{M}$ may not be deterministic in the sense that for some $s \in S_{\mathcal{M}}$ and some $a \in \mathbf{\Sigma}$ there may be *more than* one t such that $s \xrightarrow{a} t$. Intuitively this means that doing a on state s may result in different states due to some external factors which are not modelled in $\mathcal{M}$.

The simplest procedure is a one-step atomic action $a \in \mathbf{\Sigma}$, based on which more complicated procedures are constructed as regular expressions:

$$\pi \quad ::= \quad a \mid \pi; \pi \mid \pi + \pi \mid \pi^*$$

where $a \in \mathbf{\Sigma}$. Intuitively, ; is the sequential composition, $+$ is the non-deterministic choice, and * is the iteration operation. Let $\Pi_{\mathbf{\Sigma}}$ denote the set of all regular expressions based on $\mathbf{\Sigma}$. The set of action sequences denoted by a regular expression π (notation: $\mathcal{L}(\pi)$) is defined as follows:

$$\begin{aligned}
&\mathcal{L}(a) = \{a\} \\
&\mathcal{L}(\pi; \pi') = \{wv \mid w \in \mathcal{L}(\pi) \text{ and } v \in \mathcal{L}(\pi')\} \\
&\mathcal{L}(\pi + \pi') = \mathcal{L}(\pi) \cup \mathcal{L}(\pi') \\
&\mathcal{L}(\pi^*) = \{\epsilon\} \cup \textstyle\bigcup_{n>0}(\mathcal{L}(\underbrace{\pi; \cdots ; \pi}_{n}))
\end{aligned}$$

where ϵ is the empty sequence. In the sequel, we abuse the notation by writing $w \in \pi$ for $w \in \mathcal{L}(\pi)$ and writing $\pi \subseteq \pi'$ for $\mathcal{L}(\pi) \subseteq \mathcal{L}(\pi')$.

We use the language of PDL to describe the procedures encoded in a Kripke model:

$$\phi \quad ::= \quad \top \mid p \mid \neg\phi \mid \phi \wedge \phi \mid \langle\pi\rangle\phi$$

where $\langle\pi\rangle\phi$ formulas are interpreted on pointed Kripke model $\mathcal{M}, s$ as follows (cf. e.g., [7]):

$$\boxed{\mathcal{M}, s \Vdash \langle\pi\rangle\phi \iff \text{there exists a } t \text{ such that } s \xrightarrow{w} t \text{ for some } w \in \pi \text{ and } t \Vdash \phi}$$

We say $s_0, s_1, \ldots, s_n$ is an *execution* of π if there exist $w = a_1, a_2, \ldots, a_n \in \pi$ such that $s_k \xrightarrow{a_{k+1}} s_{k+1}$ for $0 \leq k < n$. In another word, $\langle\pi\rangle\phi$ is true at s iff there is an execution of π from s to a ϕ-*state*, i.e., a state where ϕ holds.

Now, we formalize a piece of imperfect procedure information as a Hoare-like triple $\langle\phi, X, \psi\rangle$ where ϕ and ψ are PDL formulas and X is one of $\pi^\exists$ or $\pi^\forall$ where $\pi \in \Pi_{\mathbf{\Sigma}}$. ϕ and ψ denote the *precondition* (initial states) and the *postcondition* (goal states) of the procedure respectively, and π denotes the procedure quantified by $\exists$ or $\forall$. Intuitively, the quantifiers work as follows:

- $\langle\phi,\ \pi^\exists,\ \psi\rangle$ says that if ϕ holds then there *exists* an execution of π which can make ψ true, e.g., "One of these two buses will get you to the university from home."
- $\langle\phi,\ \pi^\forall,\ \psi\rangle$ says that if ϕ holds then *all* the executions of π will make sure ψ, e.g., "All the buses departing here will get you to the university."

The *correctness* of a piece of information $\langle\phi, X, \psi\rangle$ is defined below, given a Kripke model $\mathcal{M}$:

- $\langle\phi,\ \pi^\exists,\ \psi\rangle$ is correct iff ($\forall t \Vdash \phi, \exists w \in \pi \exists t' : t \xrightarrow{w} t'$ and $t' \Vdash \psi$) iff $\mathcal{M} \Vdash \phi \to \langle\pi\rangle\psi$
- $\langle\phi,\ \pi^\forall,\ \psi\rangle$ is correct iff ($\forall t \Vdash \phi, \forall w \in \pi \forall t'$: if $t \xrightarrow{w} t'$ then $t' \Vdash \psi$) iff $\mathcal{M} \Vdash \phi \to [\pi]\psi$

3 Hyper Models

Here is an example to motivate our definition of hyper models:

You have no idea how four cities A, B, C, D are connected (by train or bus). Now suppose that someone tells you that there is a bus going to either B or C from A, there is a train connection from C to D, and all the buses departing from B are going to C. Now what do you know about the route from A to D?

Again, let b denote the bus connections and let t denote the train connections. Let p_x be the basic proposition denoting the location of town x for $x \in \{A, B, C, D\}$. The imperfect procedural information can be formalized as:

$$\langle p_A, b^{\exists}, p_B \vee p_C \rangle, \quad \langle p_C, t^{\exists}, p_D \rangle, \quad \langle p_B, b^{\forall}, p_C \rangle.$$

The simple-minded learning process is to add those information as special transitions in the map, as illustrated below (note that the $b^{\exists}$ transition is from A to $\{B, C\}$):

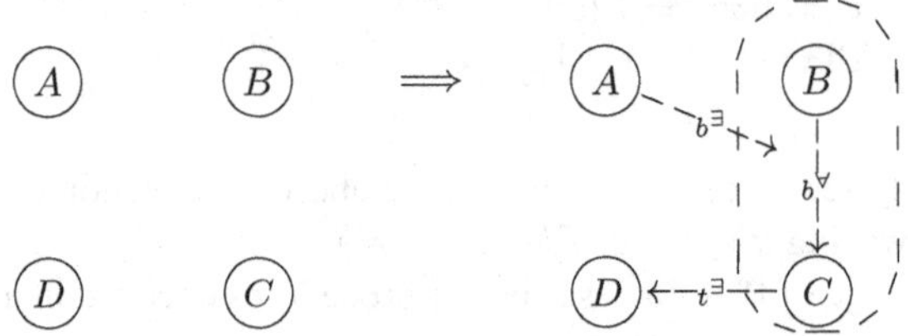

Given that the information is truthful, the real situation is still not yet determined, for example, the following are three of the possible actual situations consistent with the information available:

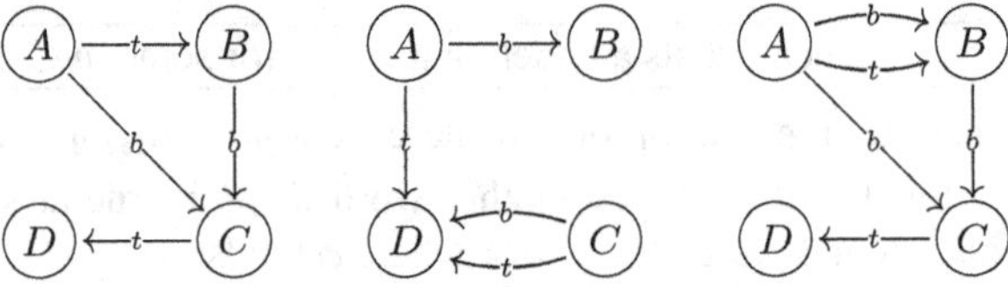

However, the agent should *know* the following, which may help him to go to D from A:

There is a bus from A to either B or C, and if it reaches C then D can be reached by a train, otherwise take any bus (if available) from B to get C first in order to reach D finally.

In the rest of this section, we will introduce hyper models formally, and a semantics for *epistemic PDL (EPDL)* based on them to reason about knowledge of procedures.

3.1 Models with Simple Procedural Information

This subsection is a technical warm-up for the next one. We only consider simple procedures ($a^{\forall}$ or $a^{\exists}$) based on singleton sets of initial states. To represent such information, we introduce the *simple hyper models* based on Kripke models with extra transitions labelled by $a^{\forall}$ or $a^{\exists}$ from a *single* state to a *set* of states:

Definition 2 (Simple Hyper model). *A simple hyper model is a tuple* $(S, \rightarrow, \rightarrow_\exists, \rightarrow_\forall, V)$ *where:*

- $(S, \rightarrow, V)$ *is a Kripke model w.r.t.* $\mathbf{\Sigma}$.
- $\rightarrow_\exists \subseteq S \times \mathbf{\Sigma} \times 2^S$ *is a labelled binary relation from a state to a set of states.*
- $\rightarrow_\forall \subseteq S \times \mathbf{\Sigma} \times 2^S$ *is a labelled binary relation from a state to a set of states.*
- *for all* $s \in S, T \subseteq S$*:* $s \xrightarrow{a}_\exists T$ *implies that there exists* $t \in T$ *such that* $s \xrightarrow{a} t$.
- *for all* $s \in S, T \subseteq S$*:* $s \xrightarrow{a}_\forall T$ *implies that for all* $t \in S$*:* $s \xrightarrow{a} t$ *implies* $t \in T$.

In (simple) hyper models, $\rightarrow$ represents the *actual* transitions between the states, and $\rightarrow_\exists$ and $\rightarrow_\forall$ represent the available imperfect procedural information to an agent. The last two conditions are crucial to guarantee the correctness of procedural information in the model. Note that this model is from the modeller's point of view, and the agent's knowledge only depends on $\rightarrow_\forall$ and $\rightarrow_\exists$ but not $\rightarrow$, as it will become clear in the semantics of the logic. Here we include the actual transitions in order to validate whether our logic, to be defined later, is a proper epistemic logic, e.g., whether everything the agent knows is actually true. When representing the agent's procedural information only, we can simply leave out the actual transitions given $\rightarrow_\forall$ and $\rightarrow_\exists$ are reliable. Note that $s \xrightarrow{a}_\forall \emptyset$ denotes 'negative' information: there is no a-transition from s. On the other hand, it is impossible to have $s \xrightarrow{a}_\exists \emptyset$ due to the first correctness condition.

Note that, the transitions $\rightarrow_\exists$ and $\rightarrow_\forall$ are *not* defined by $\rightarrow$. As an example, recall the model we mentioned at the beginning of this section (now with the actual transitions):

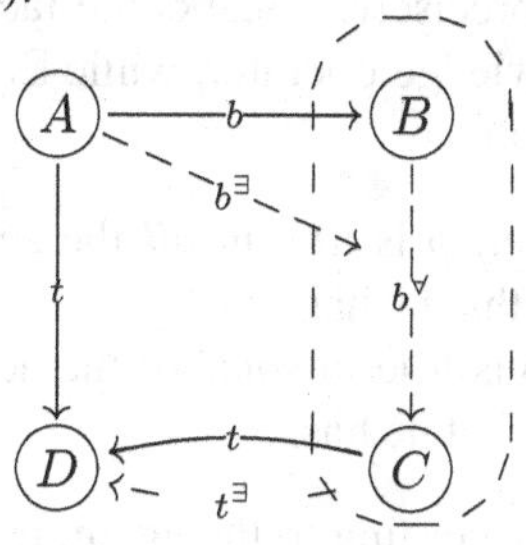

where:

- $S = \{A, B, C, D\}$,
- $\rightarrow = \{(A, b, B), (A, t, D), (C, t, D)\}$,
- $\rightarrow_\exists = \{(A, b, \{B, C\}), (C, t, \{D\})\}$,
- $\rightarrow_\forall = \{(B, b, \{C\})\}$,
- for all $s, v \in \{A, B, C, D\}$, $p_s \in V(v)$ iff $s = v$.

It is easy to verify that the last two correctness conditions are satisfied, e.g., for $A \xrightarrow{b}_\exists \{B, C\}$ we have $A \xrightarrow{b} B$. On the other hand, although $A \xrightarrow{t} D$, there is no $\xrightarrow{t}_\exists$ nor $\xrightarrow{t}_\forall$ from A to D.

Remark 1. Some readers may wonder whether further conditions on $\rightarrow_\exists$ and $\rightarrow_\forall$ should apply, to which we will come back in Section 4. For now, let us keep everything simple to understand the merit of the framework.

A fragment of EPDL is used to talk about the knowledge of a single agent on simple hyper models:

Definition 3 (Epistemic Action Language EAL). *Given a countable set of propositional variables* $\mathbf{P}$*, a finite sets of atomic actions* $\mathbf{\Sigma}$*, the formulas of EAL are given by:*

$$\phi \ ::= \ \top \mid p \mid \neg\phi \mid (\phi \wedge \phi) \mid K\phi \mid \langle a \rangle \phi$$

where $p \in \mathbf{P}$ *and* $a \in \mathbf{\Sigma}$.

As usual, we define $\bot$, $\phi \vee \psi$, $\phi \to \psi$, $\hat{K}\phi$ and $[a]\phi$ as the abbreviations of $\neg\top$, $\neg(\neg\phi \wedge \neg\psi)$, $\neg\phi \vee \psi$, $\neg K \neg\phi$ and $\neg\langle a\rangle\neg\phi$ respectively.

Definition 4 (Semantics). *The semantics of EAL on a simple hyper model $\mathcal{M} = (S, \to, \overset{\cdot}{\to}_\exists \to_\forall, V)$ is given by the following satisfaction relation w.r.t. a mode $x \in \{0, \Box, \Diamond\}$:*

$$
\begin{array}{rl}
\mathcal{M}, s \vDash \phi \Leftrightarrow & \mathcal{M}, s \vDash_0 \phi \\
\mathcal{M}, s \vDash_x p \Leftrightarrow & p \in V(s) \\
\mathcal{M}, s \vDash_x \phi \wedge \psi \Leftrightarrow & \mathcal{M}, s \vDash_x \phi \textit{ and } \mathcal{M}, s \vDash_x \psi \\
\mathcal{M}, s \vDash_x K\phi \Leftrightarrow & \mathcal{M}, s \vDash_\Box \phi \\
\mathcal{M}, s \vDash_x \neg\phi \Leftrightarrow & \begin{cases} \mathcal{M}, s \nvDash_0 \phi \textit{ IF } x = 0 \\ \mathcal{M}, s \nvDash_\Diamond \phi \textit{ IF } x = \Box \\ \mathcal{M}, s \nvDash_\Box \phi \textit{ IF } x = \Diamond \end{cases} \\
\mathcal{M}, s \vDash_x \langle a\rangle\phi \Leftrightarrow & \begin{cases} \exists t \in S : s \overset{a}{\to} t \textit{ and } \mathcal{M}, t \vDash_0 \phi & \textit{IF } x = 0 \\ \exists T \subseteq S : s \overset{a}{\to}_\exists T \textit{ and } \forall t \in T : \mathcal{M}, t \vDash_\Box \phi & \textit{IF } x = \Box \\ \forall T \subseteq S : s \overset{a}{\to}_\forall T \textit{ implies } \exists t \in T : \mathcal{M}, t \vDash_\Diamond \phi & \textit{IF } x = \Diamond \end{cases}
\end{array}
$$

We say that ϕ is valid in $\mathcal{M}$ ($\mathcal{M} \vDash \phi$) if for any $s \in S_\mathcal{M}$: $\mathcal{M}, s \vDash \phi$. ϕ is valid if for any $\mathcal{M}$: $\mathcal{M} \vDash \phi$.

Clearly, this clumsy-looking semantics needs a good explanation. First of all, $\vDash_\Box$ and $\vDash_\Diamond$ are used as auxiliary semantics in order to define $\vDash$ ($\vDash_0$). 0, $\Box$ and $\Diamond$ can be viewed as *contexts* in evaluating the formulas. More precisely, 0 marks the factual mode: evaluating formulas outside the scope of any knowledge operator, while $\Box$ and $\Diamond$ denote the *knowledge modes* with the following intentions:

- $\vDash_\Box \phi$: the agent thinks that ϕ is *necessarily* true, i.e., ϕ is true in *all* the actual situations consistent with the procedural information that he has.
- $\vDash_\Diamond \phi$: the agent thinks that ϕ is *possibly* true, i.e., ϕ is true in *some* of the actual situations consistent with the procedural information that he has.

The alternations of $\Diamond$ and $\Box$ are triggered by negations: according to the agent, if ϕ is necessarily true then it is not possible to be not true, and if it is possible to be not true then it is not necessarily true. The clause for $K\phi$ says that the agent knows ϕ iff he thinks ϕ is necessarily true. Careful readers may wonder about the fact that $\mathcal{M}, s \vDash_0 p \iff \mathcal{M}, s \vDash_\Box p \iff \mathcal{M}, s \vDash_\Diamond p$, which means that a basic proposition is true iff the agent thinks that it is necessarily true iff the agent thinks that it is possibly true. This is because we assume the agent does not have any uncertainty about the basic propositions on the states. There is only uncertainty about the transitions ("*The agent knows all the cities but does not know how they are connected*").[1]

Note that the above semantics coincides with the standard possible world semantics on formulas without the K-operator. When evaluating epistemic formulas, things get more complicated. Note that we have a non-standard semantics for negation, thus it is worth working out the semantics for abbreviations, e.g., $\mathcal{M}, s \vDash_x \phi \to \psi$ may not be

[1] Actually we can incorporate uncertainty about the basic propositions by adding over- and under-approximations of the truth values of them. We leave it to future work.

equivalent to $\mathcal{M}, s \vDash_x \phi$ implies $\mathcal{M}, s \vDash_x \psi$ depending on x. We summarize the results as follows (the readers are strongly encouraged to work out these by themselves):

$$\mathcal{M}, s \vDash_x \phi \vee \psi \Leftrightarrow \mathcal{M}, s \vDash_x \phi \text{ or } \mathcal{M}, s \vDash_x \psi$$

$$\mathcal{M}, s \vDash_x \phi \to \psi \Leftrightarrow \begin{cases} \mathcal{M}, s \vDash_0 \phi \text{ implies } \mathcal{M}, s \vDash_0 \psi & \text{IF } x = 0 \\ \mathcal{M}, s \vDash_\Diamond \phi \text{ implies } \mathcal{M}, s \vDash_\Box \psi & \text{IF } x = \Box \\ \mathcal{M}, s \vDash_\Box \phi \text{ implies } \mathcal{M}, s \vDash_\Diamond \psi & \text{IF } x = \Diamond \end{cases}$$

$$\mathcal{M}, s \vDash_x \hat{K}\psi \Leftrightarrow \mathcal{M}, s \vDash_\Diamond \psi$$

$$\mathcal{M}, s \vDash_x [a]\phi \Leftrightarrow \begin{cases} \forall t : s \xrightarrow{a} t \text{ implies } \mathcal{M}, t \vDash_0 \phi & \text{IF } x = 0 \\ \exists T \subseteq S : s \xrightarrow{a}_\forall T \text{ and } \forall t \in T : \mathcal{M}, t \vDash_\Box \phi & \text{IF } x = \Box \\ \forall T \subseteq S : s \xrightarrow{a}_\exists T \text{ implies } \exists t \in T : \mathcal{M}, t \vDash_\Diamond \phi & \text{IF } x = \Diamond \end{cases}$$

Now we see clearly that the agent knows ϕ iff ϕ is necessarily true to him, and ϕ is considered possible by the agent iff ϕ is possibly true to him. Let us also unravel the cases for $K\langle a\rangle\phi$ and $K[a]\phi$ to see the merit of the semantics more clearly:

$$\begin{array}{lcl} \mathcal{M}, s \vDash K\langle a\rangle\phi & \iff & \exists T \subseteq S : s \xrightarrow{a}_\exists T \text{ and } \forall t \in T : \mathcal{M}, t \vDash_\Box \phi \\ \mathcal{M}, s \vDash K[a]\phi & \iff & \exists T \subseteq S : s \xrightarrow{a}_\forall T \text{ and } \forall t \in T : \mathcal{M}, t \vDash_\Box \phi \\ \mathcal{M}, s \vDash \hat{K}\langle a\rangle\phi & \iff & \forall T \subseteq S : s \xrightarrow{a}_\forall T \text{ implies } \exists t \in T : \mathcal{M}, t \vDash_\Diamond \phi \\ \mathcal{M}, s \vDash \hat{K}[a]\phi & \iff & \forall T \subseteq S : s \xrightarrow{a}_\exists T \text{ implies } \exists t \in T : \mathcal{M}, t \vDash_\Diamond \phi \end{array}$$

The best way to understand the semantics is by looking at examples. Recall the model we mentioned earlier, we can verify the formulas on the right-hand side:

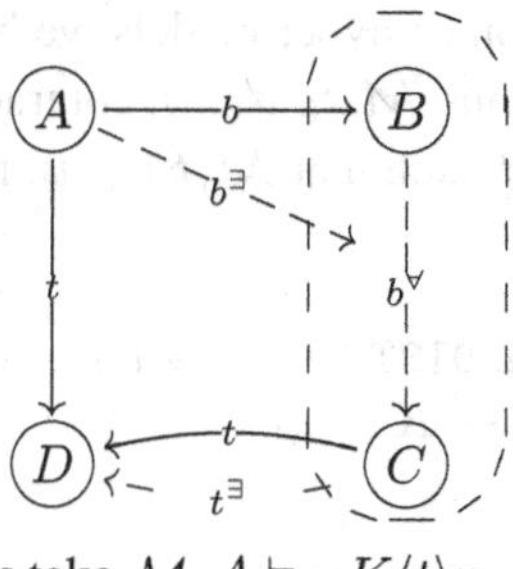

$$\begin{array}{l} \mathcal{M} \vDash p_A \to \langle b\rangle(p_B \vee p_C) \\ \mathcal{M} \vDash p_B \to [b]p_C \\ \mathcal{M} \vDash p_C \to \langle t\rangle p_D \\ \mathcal{M}, A \vDash \langle t\rangle p_D \wedge \neg K\langle t\rangle p_D \wedge \neg K\neg\langle t\rangle p_D \\ \mathcal{M}, B \vDash [b]\neg p_D \wedge K[b]\neg p_D \\ \mathcal{M}, C \vDash \langle t\rangle p_D \wedge K\langle t\rangle p_D \\ \mathcal{M}, A \vDash K\langle b\rangle((p_C \to \langle t\rangle p_D) \wedge (p_B \to [b]p_C)) \end{array}$$

Let us take $\mathcal{M}, A \vDash \neg K\langle t\rangle p_D \wedge \neg K\neg\langle t\rangle p_D$ as an example:

$\mathcal{M}, A \vDash \neg K\langle t\rangle p_D \wedge \neg K\neg\langle t\rangle p_D$
$\iff \mathcal{M}, A \vDash_0 \neg K\langle t\rangle p_D \wedge \neg K\neg\langle t\rangle p_D$
$\iff \mathcal{M}, A \vDash_0 \neg K\langle t\rangle p_D$ and $\mathcal{M}, A \vDash_0 \neg K\neg\langle t\rangle p_D$
$\iff \mathcal{M}, A \nvDash_0 K\langle t\rangle p_D$ and $\mathcal{M}, A \nvDash_0 K\neg\langle t\rangle p_D$
$\iff \mathcal{M}, A \nvDash_\Box \langle t\rangle p_D$ and $\mathcal{M}, A \nvDash_\Box \neg\langle t\rangle p_D$
$\iff$ (not $(\exists T \subseteq S : A \xrightarrow{t}_\exists T$ and $\forall v \in T : \mathcal{M}, v \vDash_\Box p_D)$) and $\mathcal{M}, A \vDash_\Diamond \langle t\rangle p_D$
$\iff$ (it is not the case that $A \xrightarrow{t}_\exists \{D\}$) and ($\forall T \subseteq S : A \xrightarrow{t}_\forall T$ implies $\exists v \in T : \mathcal{M}, v \vDash_\Diamond p_D$)

Since there are no $\xrightarrow{t}_\exists$ nor $\xrightarrow{t}_\forall$ transitions from A, $\mathcal{M}, A \vDash \neg K\langle t\rangle p_D \wedge \neg K\neg\langle t\rangle p_D$.

In the above model $\mathcal{M}$, it seems that $\mathcal{M} \vDash K\phi \to \phi$, namely the knowledge is true. This is not accidental. We will show that the usual **S5** axioms are all valid. To prove it, we first show that if ϕ is necessarily true then it is true, and if it is true then it is possibly true.

Lemma 1. *For all the pointed simple hyper model $\mathcal{M}, s$, any ϕ the following two hold:*

1. $\mathcal{M}, s \vDash_\Box \phi$ *implies* $\mathcal{M}, s \vDash_0 \phi$
2. $\mathcal{M}, s \vDash_0 \phi$ *implies* $\mathcal{M}, s \vDash_\Diamond \phi$

Therefore $\mathcal{M}, s \vDash_\Box \phi$ *implies* $\mathcal{M}, s \vDash_\Diamond \phi$.

Proof. We prove the two claims simultaneously by induction on the structure of ϕ. $\phi = p$ or $\phi = \phi_1 \wedge \phi_2$: trivial. For $\phi = \neg\psi$: Suppose $\mathcal{M}, s \vDash_\Box \neg\psi$ then according to the semantics, $\mathcal{M}, s \nvDash_\Diamond \psi$ thus by IH (2nd claim), we have $\mathcal{M}, s \nvDash_0 \psi$ namely $\mathcal{M}, s \vDash_0 \neg\psi$. Similar for the second claim. For $\phi = K\psi$: Suppose $\mathcal{M}, s \vDash_\Box K\psi$ then $\mathcal{M}, s \vDash_\Box \psi$ thus according to the semantics $\mathcal{M}, s \vDash_0 K\psi$. Similarly, suppose $\mathcal{M}, s \vDash_0 K\psi$ then $\mathcal{M}, s \vDash_\Box \psi$, thus $\mathcal{M}, s \vDash_\Diamond K\psi$ according to the semantics.

For $\phi = \langle a \rangle\phi$: Suppose $\mathcal{M}, s \vDash_\Box \langle a \rangle\psi$ then there exists $T \subseteq S$ $s \xrightarrow{a}_\exists T$ and $T \vDash_\Box \psi$. According to the definition of hyper models, there is a $t \in T$ such that $s \xrightarrow{a} t$ and $\mathcal{M}, t \vDash_\Box \psi$. By IH we have there is a t such that $s \xrightarrow{a} t$ and $\mathcal{M}, t \vDash_0 \psi$ thus $\mathcal{M}, s \vDash_0 \langle a \rangle\psi$. Now for the second claim, suppose $\mathcal{M}, s \vDash_0 \langle a \rangle\psi$ then there is a t_0 such that $s \xrightarrow{a} t_0$ and $\mathcal{M}, t_0 \vDash_0 \psi$. By IH, $\mathcal{M}, t_0 \vDash_\Diamond \psi$. In order to show that $\mathcal{M}, s \vDash_\Diamond \langle a \rangle\psi$, we prove that for all $T \subseteq S : s \xrightarrow{a}_\forall T$ implies that there is a $t \in T$ such that $\mathcal{M}, t \vDash_\Diamond \psi$. Suppose not, then there exists a T_0 such that $s \xrightarrow{a}_\forall T_0$ and for all $t \in T_0 : \mathcal{M}, t \nvDash_\Diamond \psi$. Since $s \xrightarrow{a}_\forall T_0$, by the definition of hyper models we have for all $t : s \xrightarrow{a} t$ implies $t \in T_0$. Since $s \xrightarrow{a} t_0$, $t_0 \in T_0$ thus $\mathcal{M}, t_0 \nvDash_\Diamond \psi$, contradiction. Therefore for all $T \subseteq S : s \xrightarrow{a}_\forall T$ implies there is a t such that $\mathcal{M}, t \vDash_\Diamond \psi$, namely, $\mathcal{M}, s \vDash_\Diamond \langle a \rangle\psi$.

Theorem 1. *The following* **S5** *axiom schemas are valid:* $\texttt{DIST} : K(\phi \to \psi) \to (K\phi \to K\psi)$, $\texttt{T} : K\phi \to \phi$, $\texttt{4} : K\phi \to KK\phi$, $\texttt{5} : \neg K\phi \to K\neg K\phi$

Proof. For DIST:

$$\begin{aligned}
& \mathcal{M}, s \vDash K(\phi \to \psi) \to (K\phi \to K\psi) \\
\iff & \mathcal{M}, s \vDash_0 K(\phi \to \psi) \text{ implies } \mathcal{M}, s \vDash_0 (K\phi \to K\psi) \\
\iff & \mathcal{M}, s \vDash_\Box \phi \to \psi \text{ implies} \\
& (\mathcal{M}, s \vDash_\Box \phi \text{ implies } \mathcal{M}, s \vDash_\Box \psi) \\
\iff & (\mathcal{M}, s \vDash_\Diamond \phi \text{ implies } \mathcal{M}, s \vDash_\Box \psi) \text{ implies} \\
& (\mathcal{M}, s \vDash_\Box \phi \text{ implies } \mathcal{M}, s \vDash_\Box \psi)
\end{aligned}$$

Now suppose ($\mathcal{M}, s \vDash_\Diamond \phi$ implies $\mathcal{M}, s \vDash_\Box \psi$) and $\mathcal{M}, s \vDash_\Box \phi$. Since $\mathcal{M}, s \vDash_\Box \phi$ then by Theorem 1, we have $\mathcal{M}, s \vDash_\Diamond \phi$. Thus $\mathcal{M}, s \vDash_\Box \psi$.

For T:

$$\begin{aligned}
& \mathcal{M}, s \vDash K\phi \to \phi \\
\iff & \mathcal{M}, s \vDash_0 K\phi \text{ implies } \mathcal{M}, s \vDash_0 \phi \\
\iff & \mathcal{M}, s \vDash_\Box \phi \text{ implies } \mathcal{M}, s \vDash_0 \phi
\end{aligned}$$

From Theorem 1, it is clear that $\mathcal{M}, s \vDash K\phi \to \phi$.

For 4:

$$\begin{aligned} & \mathcal{M}, s \vDash K\phi \rightarrow KK\phi \\ \iff & \mathcal{M}, s \vDash_0 K\phi \text{ implies } \mathcal{M}, s \vDash_0 KK\phi \\ \iff & \mathcal{M}, s \vDash_\Box \phi \text{ implies } \mathcal{M}, s \vDash_\Box \phi \end{aligned}$$

For 5:

$$\begin{aligned} & \mathcal{M}, s \vDash \neg K\phi \rightarrow K\neg K\phi \\ \iff & \mathcal{M}, s \nvDash_0 K\phi \text{ implies } \mathcal{M}, s \vDash_0 K\neg K\phi \\ \iff & \mathcal{M}, s \nvDash_\Box \phi \text{ implies } \mathcal{M}, s \vDash_\Box \neg K\phi \\ \iff & \mathcal{M}, s \nvDash_\Box \phi \text{ implies } \mathcal{M}, s \nvDash_\Diamond K\phi \\ \iff & \mathcal{M}, s \nvDash_\Box \phi \text{ implies } \mathcal{M}, s \nvDash_\Box \phi \end{aligned}$$

Based on Lemma 1, we also have the following results:[2]

Proposition 1. *For any pointed hyper model $\mathcal{M}, s$ and any EAL formula ϕ the following hold:*

1. *$\mathcal{M}, s \nvDash_\Box \phi \wedge \neg\phi$, i.e., $\vDash_\Box$ is consistent.*
2. *$\mathcal{M}, s \vDash_\Diamond \phi \vee \neg\phi$, i.e., $\vDash_\Diamond$ is complete.*
3. *$\vDash$ is consistent and complete.*

Proof. Lemma 1 says that for any $\mathcal{M}, s$ any formula EAL ϕ: $\mathcal{M}, s \vDash_\Box \phi$ implies $\mathcal{M}, s \vDash_\Diamond \phi$.

For (1): Suppose $\mathcal{M}, s \vDash_\Box \neg\phi$ then $\mathcal{M}, s \nvDash_\Diamond \phi$ therefore $\mathcal{M}, s \nvDash_\Box \phi$. Thus $\mathcal{M}, s \nvDash_\Box \phi \wedge \neg\phi$.

For (2): $(\mathcal{M}, s \vDash_\Diamond \phi \text{ or } \mathcal{M}, s \vDash_\Diamond \neg\phi) \iff (\mathcal{M}, s \vDash_\Diamond \phi \text{ or } \mathcal{M}, s \nvDash_\Box \phi) \iff (\mathcal{M}, s \vDash_\Box \phi \text{ implies } \mathcal{M}, s \vDash_\Diamond \phi)$. (3) is trivial by definition.

One way to interpret the above results is that the knowledge is consistent, and at least one of ϕ and $\neg\phi$ is considered possible by the agent. On the other hand, $\vDash_\Box$ is not complete and $\vDash_\Diamond$ is not consistent, which can be demonstrated by the following simple example in which the agent has no information ($\xrightarrow{a}_\exists$ and $\xrightarrow{a}_\forall$ are empty):

$$s \xrightarrow{\;a\;} t$$

According to the semantics, $\mathcal{M}, s \vDash_\Diamond \langle a\rangle\top \wedge \neg\langle a\rangle\top$, and equivalently we have $\mathcal{M}, s \nvDash_\Box \neg\langle a\rangle\top$ and $\mathcal{M}, s \nvDash_\Box \langle a\rangle\top$. The "inconsistency" of $\vDash_\Diamond$ does not cause the inconsistency of $\vDash$ due to the semantics of the negation. Clearly, $\langle a\rangle\top \vee \neg\langle a\rangle\top$ is valid but $K(\langle a\rangle\top \vee \neg\langle a\rangle\top)$ is not valid in the above model, thus:

Proposition 2. *The rule of necessitation ($\vDash \phi$ infers $\vDash K\phi$) is not valid.*

On the other hand, our K operator is more *constructive* than the standard epistemic operator, demonstrated by the fact that K operator actually distributes over both $\vee$ and $\wedge$. In our setting, $K(\phi \vee \psi)$ should be read as 'the agent knows whether ϕ *or* ψ'. Correspondingly, $\hat{K}(\phi \wedge \psi)$ should be read as ' the agent considers both ϕ and ψ possible'.

Proposition 3. *The following are valid:*

$$\mathtt{DIST}\wedge : K(\phi \wedge \psi) \leftrightarrow K\phi \wedge K\psi \quad \mathtt{DIST}\vee : K(\phi \vee \psi) \leftrightarrow K\phi \vee K\psi$$

Since there is no uncertainty of basic propositions in the hyper models, the following holds:

[2] The use of the words *consistent* and *complete* are due to the convention in the abstraction literature c.f. e.g., [3].

Proposition 4. $\mathtt{INV}: (p \to Kp) \wedge (\neg p \to K\neg p)$ *is valid.*

To completely axiomatize the logic, it is also important to have axioms controlling the interactions between K and $[a]$. Here we observe that the axiom of *perfect recall* (PR) is valid: $K[a]\phi \to [a]K\phi$ while the converse (*no learning*) is invalid.[3]

Proposition 5. $\mathtt{PR}: K[a]\phi \to [a]K\phi$ *is valid.*

We leave it for future work whether $\mathtt{PR}, \mathtt{INV}, \mathtt{DIST}\wedge, \mathtt{DIST}\vee$, $\mathtt{DIST}$, $\mathtt{T}, 5, 4$ on top of a propositional calculus are enough to completely axiomatize EAL over simple hyper frames.

3.2 Models with Arbitrary Procedural Information

In this subsection, we consider arbitrary procedural information. Intuitively, the correct information $\langle\phi, \pi^\forall, \psi\rangle$ ($\langle\phi, \pi^\exists, \psi\rangle$) can be incorporated by adding to the model $\mathcal{M}$ a transition labelled by $\pi^\forall$ ($\pi^\exists$) from $\{\mathcal{M}, s \mid s \vDash \phi\}$ to $\{\mathcal{M}, t \mid t \vDash \psi\}$ and this leads to the definition of unrestricted hyper models (recall that Π_Σ is the set of regular expressions based on Σ):

Definition 5 (Hyper Model). *An hyper model is a tuple* $(S, \to, \to_\exists, \to_\forall, V)$ *where:*

- $(S, \to, V)$ *is a Kripke model*
- $\to_\exists \subseteq 2^S \times \Pi_\Sigma \times 2^S$ *is a labelled binary relation from a set of states to a set of states.*
- $\to_\forall \subseteq 2^S \times \Pi_\Sigma \times 2^S$ *is a labelled binary relation from a set of states to a set of states.*
- *for all* $T, T' \subseteq S$*:* $T \xrightarrow{\pi}_\exists T'$ *implies that for all* $t \in T$ *there exists* $w \in \pi$ *and* $t' \in T'$ *such that* $t \xrightarrow{w} t'$.
- *for all* $T, T' \subseteq S$*:* $T \xrightarrow{\pi}_\forall T$ *implies that for all* $t \in T$ *all* $w \in \pi$*:* $t \xrightarrow{w} t'$ *implies* $t' \in T'$.

Again, the last two conditions guarantee that the information incorporated in the models are correct. The class of simple hyper models can be viewed as a subclass of hyper models, where the transitions are all in the shapes of $\{s\} \xrightarrow{a}_\exists T$ and $\{s\} \xrightarrow{a}_\forall T$.

Now we can consider the full EPDL language.

Definition 6 (Epistemic PDL). *Given a countable set of* propositional variables $\mathbf{P}$, *a finite set of* action symbols Σ, *the formulas of EPDL language are defined by:*

$$\begin{array}{rcl} \phi & ::= & \top \mid p \mid \neg\phi \mid (\phi \wedge \phi) \mid K\phi \mid \langle\pi\rangle\phi \\ \pi & ::= & a \mid \pi;\pi \mid \pi + \pi \mid \pi^* \end{array}$$

where $a \in \Sigma$ *and* $p \in \mathbf{P}$.

[3] Versions of these axioms appear in temporal epistemic logic and dynamic epistemic logic (cf. [4,14]).

Definition 7 (Semantics). *The semantics of EPDL on hyper models $\mathcal{M} = (S, \to, \to_\exists, \to_\forall, V)$ is defined similarly as the semantics of EAL on hyper models, with the following clauses replacing the clauses for $\langle a \rangle \phi$ formulas in the case of EAL:*

$\mathcal{M}, s \vDash_0 \langle \pi \rangle \phi \Leftrightarrow \exists t \in S : s \xrightarrow{w} t$ *for some* $w \in \pi$ *such that* $\mathcal{M}, t \vDash_0 \phi$

$\mathcal{M}, s \vDash_\Box \langle \pi \rangle \phi \Leftrightarrow \exists T_0, \dots, T_k \subseteq S, \exists \pi_1, \dots, \pi_k \in \Pi_\Sigma : (s \in T_0, T_0 \xrightarrow{\pi_1}_\exists \cdots \xrightarrow{\pi_k}_\exists T_k, (\pi_1; \cdots; \pi_k) \subseteq \pi$ *and* $\forall t \in T_k : \mathcal{M}, t \vDash_\Box \phi)$

$\mathcal{M}, s \vDash_\Diamond \langle \pi \rangle \phi \Leftrightarrow \forall T_0, \dots, T_k \subseteq S, \forall \pi_1, \dots, \pi_k \in \Pi_\Sigma : ((s \in T_0, T_0 \xrightarrow{\pi_1}_\forall \cdots \xrightarrow{\pi_k}_\forall T_k, \pi \subseteq (\pi_1; \cdots; \pi_k))$ *implies* $(\exists t \in T_k : \mathcal{M}, t \vDash_\Diamond \phi))$

It is not hard to see that on simple hyper models the above semantics coincides with the semantics of EAL on $\langle a \rangle \phi$ formulas. This justifies our abuse of $\vDash$ for both EPDL and EAL. The semantics says that the agent knows that $\langle \pi \rangle \phi$ at s if there is a 'refinement' of π $((\pi_1; \cdots; \pi_k) \subseteq \pi)$ such that each π_i step can be realized by a $\xrightarrow{\pi_i}_\exists$ transition, and in the end it will certainly reach a ϕ-state. Note that it is not necessary that $\pi_1; \cdots; \pi_k = \pi$, since we just need to guarantee there exists an execution of π.

We now prove the analogue of Lemma 1.

Lemma 2. *For all the pointed hyper model $\mathcal{M}, s$, any EPDL formula ϕ the following two hold: (1) $\mathcal{M}, s \vDash_\Box \phi$ implies $\mathcal{M}, s \vDash_0 \phi$; (2) $\mathcal{M}, s \vDash_0 \phi$ implies $\mathcal{M}, s \vDash_\Diamond \phi$. Therefore $\mathcal{M}, s \vDash_\Box \phi$ implies $\mathcal{M}, s \vDash_\Diamond \phi$.*

Proof. We only need to show the case of $\langle \pi \rangle \psi$. Suppose $\mathcal{M}, s \vDash_\Box \langle \pi \rangle \psi$ then:
$\exists T_0, \dots, T_k \subseteq S, \exists \pi_1, \dots, \pi_k \in \Pi : s \in T_0, T_0 \xrightarrow{\pi_1}_\exists \cdots \xrightarrow{\pi_k}_\exists T_k, \pi_1; \cdots; \pi_k \subseteq \pi$ and $\forall t \in T_k : \mathcal{M}, t \vDash_\Box \psi$.
Let $t_0 = s$. By the definition of hyper model, there exist $t_i \in T_i$ and $w_i \in \pi_i$ for $1 \leq i \leq k$ such that $t_{i-1} \xrightarrow{w_i} t_i$. It is clear that $w_1 \cdots w_k \in \pi_1; \cdots; \pi_k$. Since $\pi_1; \cdots; \pi_k \subseteq \pi$, $w_1 \cdots w_k \in \pi$. Thus by IH, $\mathcal{M}, s \vDash_0 \langle \pi \rangle \psi$.

Now for the second claim, suppose $\mathcal{M}, s \vDash_0 \langle \pi \rangle \psi$ then there is a t° such that $s \xrightarrow{w} t^\circ$ for a $w \in \pi$ and $\mathcal{M}, t^\circ \vDash_0 \psi$. By IH, $\mathcal{M}, t^\circ \vDash_\Diamond \psi$. Now suppose towards contradiction that $\mathcal{M}, s \nvDash_\Diamond \langle \pi \rangle \psi$ then according to the semantics we have:
$\exists T_0, \dots, T_k \subseteq S, \exists \pi_1, \dots, \pi_k \in \Pi : s \in T_0, T_0 \xrightarrow{\pi_1}_\forall \cdots \xrightarrow{\pi_k}_\forall T_k, \pi \subseteq \pi_1; \cdots; \pi_k$ and $\forall t \in T_k : \mathcal{M}, t \nvDash_\Diamond \psi$
Obviously, if we can show that $t^\circ \in T_k$ then a contradiction is derived. In the following we prove that $t^\circ \in T_k$. Since $\pi \subseteq \pi_1; \cdots; \pi_k$ and $w \in \pi$, $w \in \pi_1; \cdots; \pi_k$. Therefore there exist $w_i \in \pi_i$ for $1 \leq i \leq k$ such that $w = w_1; \cdots; w_k$ (w_i can be an empty string). According to the definition of hyper model, if $s \xrightarrow{w_1 \cdots w_i} t$ then $t \in T_i$ for all $1 \leq i \leq k$. In particular if $s \xrightarrow{w_1 \cdots w_k} t$ then $t \in T_k$. Now it is clear that $t^\circ \in T_k$.

Based on this lemma and the proof of Theorem 1, the following theorem holds immediately.

Theorem 2. `DIST`, `T`, `4`, *and* `5` *are valid for EPDL on hyper models.*

It is easy to verify that the EPDL analogies of Proposition 1 and Proposition 2 also hold.

4 Discussion and Future work

So far, we have only laid out the basics of an alternative semantics for EPDL based on hyper models where epistemic relations are replaced by two approximations of the actual transitions. In this section, we discuss some subtle issues about the semantics and point out further directions.

First of all, we justify that hyper models are indeed compact representations of a collection of Kripke models. On the one hand, each hyper model $\mathcal{M}$ can be unfolded into a set (call it $Unf(\mathcal{M})$) of Kripke models over the same set of states on which the imperfect information given by $\to_\exists$ and $\to_\forall$ transitions in the hyper model is correct, i.e., satisfying the last two conditions in the definition of hyper models. Based on Lemma 2, we can easily show that the knowledge in any hyper model $\mathcal{M}$ are truthful to $Unf(\mathcal{M})$, and $\mathcal{M}$ encodes all the possibilities in $Unf(\mathcal{M})$:

Proposition 6. *For every PDL formula ϕ and every s in any hyper model $\mathcal{M}$:*

- *if $\mathcal{M}, s \vDash K\phi$ then $\mathcal{N}, s \vDash \phi$ for every $\mathcal{N} \in Unf(\mathcal{M})$*
- *if $\mathcal{N}, s \vDash \phi$ for some $\mathcal{N} \in Unf(\mathcal{M})$ then $\mathcal{M}, s \vDash \hat{K}\phi$*

Note that even in very simple cases, $|Unf(\mathcal{M})|$ may be *exponential* in the size of the hyper model and Σ. For example, let $S = \{s, t\}$, $V(p) = t$, and let π_Σ be the 'sum' of all actions in Σ, then the hyper model with $\{s\} \xrightarrow{\pi_\Sigma}_\forall \{t\}$ as the only transition has $2^{|\Sigma|}$ epistemically possible Kripke models to realize all the $\phi_\Delta = \bigwedge_{a\in\Delta}\langle a\rangle p \wedge \bigwedge_{b\notin\Delta}\neg\langle b\rangle p$ formulas at s for each $\Delta \subseteq \Sigma$. If the hyper model does not provide any procedural information (i.e., deleting the only transition) then $|Unf(\mathcal{M})| = 2^{|S|\cdot|\Sigma|\cdot|S|}$.

On the other hand, we may ask: is every set of concrete models (over a given set of states S) representable by a hyper model over S? Unfortunately, the answer is negative. For example, take the set of two Kripke models over $S = \{s, t\}$: $s \xrightarrow{a} t$ and $s \xrightarrow{b} t$. Let $\phi = (\langle a\rangle\top \wedge \neg\langle b\rangle\top) \vee (\langle b\rangle\top \wedge \neg\langle a\rangle\top)$. It is clear that $K\phi$ is true w.r.t. this set of models and state s. However, over S, no matter how the $\to_\forall$ and $\to_\exists$ transitions are chosen, we cannot make sure $K\phi$ holds on s since it would imply the knowledge of one of the disjunctions according to Proposition 3. The problem lies in the fact that we treat disjunction in the scope of K as "knowing whether" due to the semantics for Boolean connectives. We cannot really specify certain kinds of *inter-dependency* between the transitions. We may have a hyper model over s with the following transition $\{s\} \xrightarrow{(a+b)^\exists} \{t\}$, but we cannot make sure $\xrightarrow{a}$ and $\xrightarrow{b}$ are mutually exclusive between s and t. One potential solution is to make the labels of the transitions more expressive to specify conditional information, but we suspect a 'satisfactory' solution will in turn introduce another kind of 'state-explosion' on the transitions in hyper models. We need to find the balance between expressiveness and complexity.

Moreover, although the logic validates the T axiom, which makes sure everything we know is truthful, it is not very clear how much information we can 'know' by the semantics of EPDL on the hyper models. It seems that by requiring more conditions on the hyper model we may get more from the hyper models. Let us illustrate this in the simple hyper models. Below lists some intuitive closure properties that we may impose and their corresponding logical properties:

($s \xrightarrow{a}_\forall T_1$ and $s \xrightarrow{a}_\forall T_2$) implies $s \xrightarrow{a}_\forall T_1 \cap T_2$	$K[a]\phi \wedge K[a]\psi \to K[a](\phi \wedge \psi)$
($s \xrightarrow{a}_\forall T_1$ and $s \xrightarrow{a}_\exists T_2$) implies $s \xrightarrow{a}_\exists T_1 \cap T_2$	$K[a]\phi \wedge K\langle a\rangle\psi \to K\langle a\rangle(\phi \wedge \psi)$
for each s each a, there exists $T \subseteq S$ such that $s \xrightarrow{a}_\forall T$	$K[a]\top$

In fact, we only need one and only one $\xrightarrow{a}_{\forall}$ outgoing arrow from each state for each $a \in \Sigma$ since all the $\xrightarrow{a}_{\forall}$ targeting sets can be intersected together, and the 'default' transition would be $s \xrightarrow{a}_{\forall} S$ where S is the set of all possible states. Note that, in general, the above requirements may help to bring back the missing reasoning power within the scope of K, due to the lack of necessitation rule for K in the logic.

To conclude, our epistemic framework is more compact and constructive compared to the standard possible-world approach of epistemic logic, in the sense that the hyper model resembles a collection of Kripke models and we can incrementally extend the model even from scratch by adding new imprecise information. At the same time, we pay the price that we are not able to represent all the collections of Kripke models since certain dependency of transitions is not encoded in the hyper models. To use the logic, we may make use of the 3-valued model checking algorithms (e.g., [6]), we leave out the exact complexity analysis to future work. Finally, it is also a natural next step to go probabilistic, as probabilities can be seen as another form of abstraction of qualitative information, as remarked in [8].

References

1. Chen, T., van de Pol, J., Wang, Y.: PDL over accelerated labeled transition systems. In: Proceedings of TASE 2009, pp. 193–200. IEEE Computer Society Press, Los Alamitos (2008)
2. van Ditmarsch, H., van der Hoek, W., Kooi, B.: Dynamic Epistemic Logic (Synthese Library), 1st edn. Springer (2007)
3. Espada, M., van de Pol, J.: Accelerated modal abstractions of labelled transition systems. In: Johnson, M., Vene, V. (eds.) AMAST 2006. LNCS, vol. 4019, pp. 338–352. Springer, Heidelberg (2006)
4. Fagin, R., Halpern, J., Moses, Y., Vardi, M.: Reasoning about knowledge. MIT Press (1995)
5. Grumberg, O.: 2-valued and 3-valued abstraction-refinement in model checking. In: Logics and Languages for Reliability and Security, pp. 105–128 (2010)
6. Grumberg, O., Lange, M., Leucker, M., Shoham, S.: When not losing is better than winning: Abstraction and refinement for the full mu-calculus. Information and Computation 205(8), 1130–1148 (2007)
7. Harel, D., Kozen, D., Tiuryn, J.: Dynamic Logic. The MIT Press (2000)
8. Huth, M.: Abstraction and probabilities for hybrid logics. ENTCS 112, 61–76 (2005)
9. Kuhn, H.W.: Extensive games and the problem of information. In: Kuhn, H.W., Tucker, A.W. (eds.) Contributions to the Theory of Games, pp. 196–216. Princeton University Press (1953)
10. Moore, R.C.: A formal theory of knowledge and action. Tech. rep., DTIC Document (1984)
11. Parikh, R., Ramanujam, R.: Distributed processes and the logic of knowledge. In: Proceedings of Conference on Logic of Programs, pp. 256–268. Springer, London (1985)
12. Pratt, V.R.: Semantical considerations on floyd-hoare logic. Tech. rep., Cambridge, MA, USA (1976)
13. Shoham, S., Grumberg, O.: 3-valued abstraction: More precision at less cost. Information and Computation 206(11), 1313–1333 (2008)
14. Wang, Y., Cao, Q.: On axiomatizations of public announcement logic. Synthese 190, 103–134 (2013)
15. Wang, Y., Li, Y.: Not all those who wander are lost: Dynamic epistemic reasoning in navigation. In: Advances in Modal Logic, pp. 559–580 (2012)

Author Index